# Basic and Clinical Perspectives in Vision Research

## A Celebration of the Career of Hisako Ikeda

# Basic and Clinical Perspectives in Vision Research

## A Celebration of the Career of Hisako Ikeda

Edited by

Jon G. Robbins
*King's College*
*University of London*
*London, England*

Mustafa B. A. Djamgoz
*Imperial College of Science, Technology, and Medicine*
*University of London*
*London, England*

and

Anthony Taylor
*Sherrington School of Physiology*
*St. Thomas' Hospital*
*London, England*

Plenum Press • New York and London

Library of Congress Cataloging-in-Publication Data

On file



*Cover illustration:* calligraphy by Judith F. Dowling of the Japanese kanji for retina, Mōmaku. Mō, the upper character, means "net"; maku, the lower character, means "membrane."

Proceedings of an international symposium on Basic and Clinical Perspectives in Vision Research: A Celebration of the Career of Hisako Ikeda, held August 9–10, 1993, in London, England

ISBN 0-306-45202-2


A Division of Plenum Publishing Corporation
233 Spring Street, New York, N. Y. 10013



10 9 8 7 6 5 4 3 2 1



Printed in the United States of America

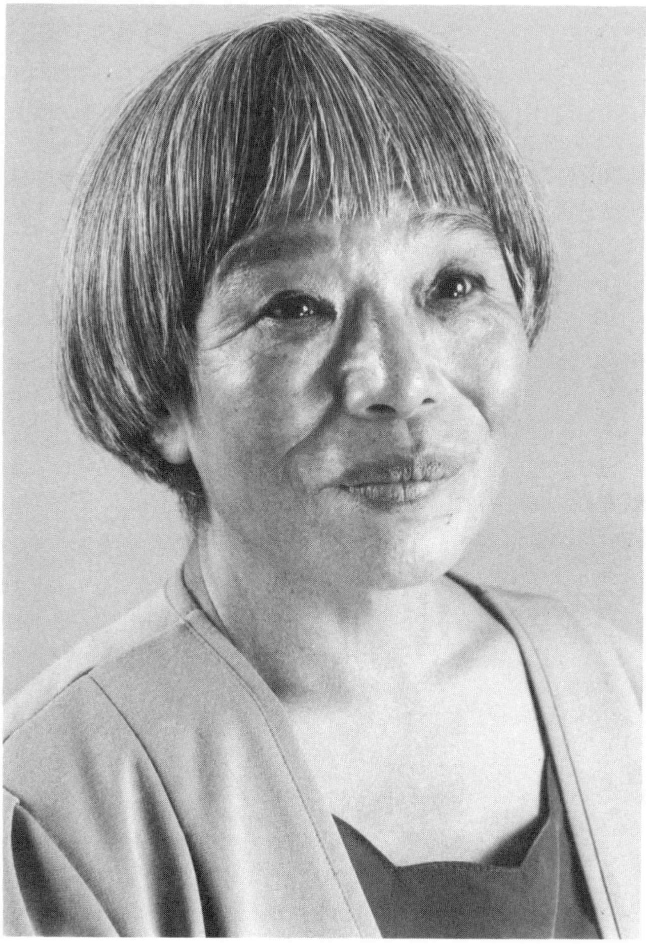

Hisako Ikeda

*We wish you a long and happy retirement*

This book arose from a meeting held in August of 1993 at St. Thomas' Hospital, London to mark the retirement of Professor Hisako Ikeda B.A., M.A., Ph.D., D.Sc. After forty years in visual science she has contributed significantly to our knowledge of the visual system not only through her wealth of publications but through her interactions with scientists, clinicians, students, and patients. We have all been under her "spell"—Anthony Taylor was head of the Department, Mustafa Djamgoz was a colleague, and Jon Robbins was a Ph.D. student. We and others have benefited from, on the one side, her appetite for hard work and on the other, her great loyalty as a friend. We hope this book goes some way toward acknowledging her achievements and indicating the great esteem in which she is held by the scientific and clinical community.

Jon Robbins, Mustafa Djamgoz, and Tony Taylor, 1995

# FOREWORD

I have been asked to write a brief foreword to this volume honoring Hisako Ikeda, providing a review of the accomplishments in our field over the past four decades, when Hisako was an active participant. This I am delighted to do. It has been a most exciting time in vision research and Hisako has been right in the middle of much of the excitement, publishing on a wide variety of topics and providing much new data and many new insights.

Hisako's research career can be divided by decades into four quite distinct areas of inquiry. In the 1950s, as a student in Japan, her research interests were psychophysical in nature, and she was concerned with visual illusions, figural aftereffects, and motion detection. In the 1960s, after her move to London, she began electrophysiological studies. Much of her work in the 1960s was concerned with the electroretinogram (ERG), its components, and the use of this electrical response for evaluating spectral sensitivities of the eye and retinal degenerations. This work represented the beginning of her electrodiagnostic clinical work, which continued until her retirement.

In the 1970s, Hisako began recording single cells in the visual system using microelectrodes, first from the ganglion cells of the retina, then from cortical and lateral geniculate neurons. Of particular interest to Hisako and her colleagues was the distinction between sustained and transient neurons along the visual pathways. These electrophysiological studies led her next into studies of amblyopia and the effects of visual deprivation on visual system neurons.

In the 1980s Hisako altered research directions once again and began studying retinal pharmacological mechanisms, focusing on the excitatory and inhibitory amino acid neurotransmitters and on the neuromodulator, dopamine. She also became interested in demyelination of the optic nerve caused by virus infection and how axonal transport is affected by demyelinations. Her studies on retinal pharmacology and the optic nerve continued well into the 1990s.

How does Hisako's work relate to the major advances in our field over these 40 years? In short, very well indeed! She began her research career at about the time of Kuffler's and Barlow's classic studies showing the center-surround nature of the receptive fields of ganglion cells. This set the stage for the exploration of receptive field properties of neurons at many levels of the visual system, from ganglion cells to cortical neurons, and to the notion of parallel processing pathways within the visual system. In the retina, sustained X (cat) or P (monkey) ganglion cells can be distinguished from transient Y (cat) or M (monkey) ganglion cells, as Hisako and others have shown, and it is thought that these two classes of cells serve different visual functions. The sustained cells, with smaller receptive fields, are believed to be involved in mediating form and color vision, whereas the transient responding cells are thought to be responsible for signaling movement. A segregation of pathways mediating various aspects of vision is now believed to extend throughout the visual system, and this realization has profoundly influenced our thinking of the nature of perception. Explanations of how we perceive motion and why we see visual illusions as we do, topics studied by Hisako early in her career, can now be suggested.

The seminal studies by Hubel and Wiesel in the 1960s on the receptive fields of cortical neurons led to the finding that profound changes can occur in the cortex of young animals as a result of visual deprivation. Severe loss of acuity (amblyopia) can be induced following surprisingly short periods of form deprivation. Although the mechanisms underlying amblyopia are still not fully understood, intriguing and detailed studies by Hisako and many others have provided much information on the nature of changes that occur along the visual pathway following visual deprivation, and have resulted in the important conclusion that the cortex is surprisingly plastic.

In the mid 1960s, substantial advances in our understanding of retinal mechanisms were beginning to be made. Electron microscopy allowed investigators to observe the functional contacts (synapses) made between retinal neurons and the first views of synaptic pathways in the retina were revealed. Intracellular recordings from the various retinal neurons were made, providing insights with regard to how the receptive fields of ganglion cells are formed. Bipolar cells were shown to have a center-surround receptive field organization, much like ganglion cells, with the center response resulting from direct receptor–bipolar cell interactions and the surround response resulting from inhibitory horizontal cell activity. Many amacrine cells were found to respond vigorously to moving stimuli but only transiently to stationary ones, suggesting that amacrine cells and inner plexiform layer mechanisms are responsible for movement detection by ganglion cells.

These studies set the stage for exploration of the pharmacology of the retina in the 1970s and 1980s by many investigators including Hisako. The retina, like the rest of the brain, was found to contain a surprisingly large number of neuroactive substances, perhaps as many as 30 or more. However, relatively few substances, mainly amino acids, appear to be involved in mediating fast-excitatory and -inhibitory synaptic interactions in the retina. L-glutamate appears to be the transmitter released by photoreceptor, bipolar, and ganglion cells, whereas γ-aminobutyric acid and glycine are the main inhibitory neurotransmitters released in the retina. The majority of substances released at synaptic sites in the retina are thought to act as neuromodulators. Their primary function is not to initiate activity but to modify it, and dopamine represents the best studied representative of this group of substances. So, for example, dopamine has been shown to modify both electrical and chemical synaptic transmission in the retinas of many species, and in this volume Hankins and Ikeda show that dopamine decreases the electrical coupling between horizontal cells in the rat!

Over the span of her career, wonderful progress has been made in understanding and using the ERG and other electrodiagnostic tests in the clinic, and again Hisako has contributed significantly to this work. Building on the pioneering work of Granit in the late 1930s, Noell and Brown provided important new analyses of the ERG in the 1950s and 1960s. The value of the ERG and other visual system potentials for diagnosing and analyzing retinal degenerations became clear in the 1960s, and by the 1970s we understood how many of the components of the ERG are generated. The 1980s saw the elucidation of how the absorption of light by a visual pigment molecule, such as rhodopsin, leads to excitation of photoreceptor cells, and molecular biological studies revealed that, in certain forms of retinitis pigmentosa, retinal degeneration is caused by a gene defect that causes a single amino acid substitution in the rhodopsin molecule. Classifying, analyzing, and evaluating various types of retinal degenerations electrophysiologically, a field in which Hisako has long toiled, is now rightfully coming into its own.

I end this foreword on a personal note. I have long admired Hisako's range of research techniques and interests and her enormous enthusiasm and energy. Whenever we have met, the conversations have been lively and insightful. I particularly wish to acknowledge her wonderful hospitality during my sabbatical year in Cambridge in 1978–79. She invited me as a special guest to the Physiological Society meeting at St. Thomas's that year, and I will always treasure memories of the society's dinner that evening in an excursion boat that travelled up and down the Thames. Later in the spring, our family visited Hisako and her husband John at their lovely cottage in Kent, and I remember very well the spirited croquet games on the back lawn and the wonderful excursions to nearby sights. Hisako, you have made our field and our lives much richer and we are very grateful. Happy retirement!

John E. Dowling
Cambridge, Massachusetts

*The editors would like to thank all the contributors and the participants who made the meeting and this book possible. We particularly would like to express our gratitude to Judith Dowling who allowed us to use her calligraphy (for the word "retina" in Japanese) on the cover of this book.*

# CONTENTS

# SECTION I

## Ocular Structure, Function and Disease

1

# A STUDY OF PUPIL RESPONSE COMPONENTS IN HUMAN VISION

John L. Barbur

Applied Vision Research Centre,
Department of Optometry and Visual Science, City University
Northampton Square, London EC1V 7DD, United Kingdom

## INTRODUCTION

### Background

The accepted classical view of the pupil response to light is that the ambient light level determines largely the steady-state size of the pupil (Lowenstein et al, 1964) and that rapid increments in light flux on the retina cause a brisk constriction of the pupil, that is often described as the dynamic pupil light reflex response (Stark & Shermann, 1957; Alexandridis, 1985; Lowenfeld, 1993). The afferent pathways involved in the control of the pupil in man have been associated with subcortical projections and this is consistent with clinical observations which suggest that the pupils continue to respond normally to sudden changes in room illumination even when the patients are corticaliy blind (Brindley et al, 1969). The response of the pupil often represents an important measurement in neurological and ophthalmologic examinations, but the poor understanding of the visual pathways involved and the kind of stimulus characteristics that cause pupillary responses limit the usefulness and potential use of such tests. Animal studies have contributed significantly to our understanding of the afferent neural pathways that subserve the Pupil Light Reflex response (PLR) in man and helped to reinforce the accepted classical view. Lesion studies combined with measurements of pupil responses before and after surgery in monkey have demonstrated the direct involvement of the pretectal olivary nuclei and the Edinger-Westphal nuclei in both the steady state control of pupil size and the dynamic PLR response (Pierson & Carpenter, 1974). In rat, the control of the pupil response is somewhat less complicated and functionally more consistent with the regulation of retinal illuminance (Clarke & Ikeda, 1985a,b). The pupil diameter follows an almost linear relationship when plotted as a function of the logarithm of retinal illuminance over a wide range. The existence of neurones in the pretectal olivary nuclei of the rat that respond only to increments in retinal illuminance is also consistent with these findings. Some neurones were found to respond almost linearly (on logarithmic axes), over some six log units change in retinal illuminance. Fewer neurones located in the posterior pretectal nucleus were also found to respond in a similar fashion to decrements in retinal illuminance (Clarke & Ikeda, 1985a). In

*Basic and Clinical Perspectives in Vision Research*, Edited by Jon Robbins et al.
Plenum Press, New York, 1995

man, the response of the pupil is more complicated the regulation of the afferent innervation may not be restricted entirely to subcortical projections with visual signals also coming indirectly via the visual cortex (Alexandridis, 1985; Loewenfeld, 1993). The linear relationship between the PLR response amplitude and the log of retinal illuminance is somewhat more limited and is complicated by factors such as the size, background adaptation and the eccentricity of the stimulus (Alexandridis, 1985; Lowenstein et al, 1964). Another complication is the absence of PLR responses in patients with post-geniculate lesions when small test stimuli are employed (Harms, 1951; Cibis et al, 1975; Barbur et al, 1988; Kardon, 1992) and the clear demonstration of pupil responses to stimulus attributes that are normally associated with cortical projections such as stimulus structure (Slooter & van Noren, 1980; Ukai, 1985; Barbur & Forsyth, 1986; Barbur, 1988), colour (Kohn & Clynes, 1969; Saini & Cohen, 1979; Young & Alpern, 1980; Barbur, 1991c) and movement (Barbur et al, 1992a). New findings in relation to visual pathways that drive the PLR response in man have been obtained recently from investigations in normal subjects and in patients with damaged primary visual cortex. The results suggest that the PLR is driven by at least two separate components, a subcortical component that determines largely the size of the pupil and a second, cortical component, that contributes significantly to the dynamic PLR response and is absent in patients with post-geniculate lesions. In addition, new findings on pupil responses to stimulus structure, colour and movement in normal subjects and in patients with damaged cortical projections have also emerged and will be presented here. The results show that both Pupil Grating Responses (PGR) and Pupil Colour Responses (PCR) are absent when the stimuli are presented in the periphery of the visual field and are restricted to regions that are cortically blind. The same visual stimuli do however elicit systematic pupil responses when imaged in the corresponding regions of the sighted visual field. Pupil Motion Responses (PMR), on the other hand, can be elicited even in the absence of a direct geniculostriate projection, but such responses are of significantly reduced amplitude.

## DESCRIPTION OF EXPERIMENTAL METHODS

The continuous recording of pupil diameter was carried out on the P_SCAN system which was designed for simultaneous measurements of pupil size and the two-dimensional movements of the eyes (Barbur et al, 1987). The measurement of pupil diameter and its centre co-ordinates relies on numerical methods that are statistically equivalent to fitting the best circle to the pupil and this results in a measurement precision of 0.01 mm (Barbur, 1991c). Although the random pupil noise fluctuations can be of much larger amplitude, the precision of a single measurement must be adequate to resolve the small signal that one wishes to enhance by means of averaging (Lynn, 1992). The visual stimuli were generated on a 20" high resolution, colour display (Hewlett Packard Model D1187A) which was calibrated for both luminance and chromaticity co-ordinates using a Gamma Scientific Model DR-2 telespectroradiometer. The calibration data makes possible the generation of any specified luminance and chromaticity (within the limits imposed by the phosphors of the display) using standard colorimetric transformations (Wyszecki & Stiles, 1982).

### Measurement of the Pupil Light Reflex

The stimulus for triggering a PLR response was a disc of radius 4 deg and was presented onto a uniform background field of luminance 24 cd m $^{-2}$. The stimulus was presented either in the foveal region or in the periphery when the centre of the test flash was 6 deg above the horizontal meridian and 11 deg away from fixation. The parameter that was varied systematically was the luminance of the test flash.

A novel stimulus that isolates different components of the PLR response was also developed. The technique employed is based on spatiotemporal background perturbation methods that have been used successfully to reduce selectively the sensitivity of neural mechanisms involved in the detection of moving stimuli (Barbur & Ruddock, 1980; Barbur et al, 1981). The test pattern designed for investigation of PLR responses consists of an area of $9^o$ x $9^o$ divided into smaller square elements of size $0.7^o$ x $0.7^o$. In order to achieve spatially localised, contrast masking, we employed spatial modulation of the luminance of each check in the array and varied systematically the amplitude of Random Luminance Modulation (RLM). The spatial modulation was dynamic (i.e., the luminance of each check changing randomly in time), but the space-averaged luminance of the whole stimulus field remained constant and equal to that of the surrounding background. In order to trigger a PLR response, the $9^o$ x $9^o$ square stimulus was presented as an incremental flash against a uniform background field of luminance 24 cd $m^{-2}$. The luminance contrast of the square test pattern was constant (i.e., $\delta L/L_b = 0.3$) and the random luminance modulation of each individual check continued throughout the stimulus flash. The effect of RLM modulation is to mask the detection of spatial patterns defined by luminance contrast and hence this has been labelled *pattern* masking. In order to generate *light* flux masking, the mean luminance of the whole test pattern was also modulated independently, in the same sequence, causing random temporal changes in the mean light flux level of the test stimulus (i.e., the $9^o$ x $9^o$ square). The use of large amplitude local pattern and light flux masking in the stimulus abolishes any form of PLR response. The RLM values employed in this study were in the range 0% to 60%.

## Measurement of Pupil Colour Responses

Our principal concern when measuring pupil colour responses was to ensure that the sudden generation of a suprathreshold coloured stimulus in the centre of a uniform achromatic background field does not trigger a PLR response. This situation may arise when the coloured stimulus is also detected by achromatic contrast mechanisms or when rod receptor signals are involved and contribute to the PLR pathway. This is more likely to be the case when the background luminance is low, and when the stimulus is presented in the periphery of the visual field (Barbur et al, 1994a). In visual psychophysics, the detection of achromatic contrast signals when coloured stimuli are involved is often made ineffective by various masking techniques (Hardy et al, 1954; Reffin et al, 1991). To study the use of chromatic signals in human vision without having to worry about the detection of luminance contrast components, we developed a spatiotemporal, luminance masking technique (Barbur et al, 1992b; 1993) that makes use of isoluminant stimuli embedded in a pattern of random luminance modulation. The principle of this method is based largely on the spatial (Barbur & Ruddock, 1980) and the temporal (Barbur et al, 1981) background perturbation techniques developed to investigate the spatiotemporal properties of achromatic vision. Using this technique it is possible to mask the detection of any achromatic signals triggered by the onset of the coloured stimulus and this has been demonstrated successfully in normal trichromats and in dichromats (Barbur et al, 1994b). The stimulus designed for this test is somewhat similar to that developed to isolate the light flux component of the PLR response and consists of an array of rectangular checks generated at the centre of a uniform background field. Random, spatiotemporal luminance modulation of each check in the array was generated throughout the stimulus presentation time. In addition to the ongoing achromatic luminance modulation, a chromatic displacement step was introduced in a predetermined pattern of checks that caused the formation of vertical bars for a fixed duration, following an initial period of achromatic spatiotemporal noise. The luminance of each check changes randomly every 0.05 s and takes one of 64 possible values, equally spaced, both above and below that of the uniform background field (usually set to 34 cd $m^{-2}$ and x, y - chromaticity co-ordinates of 0.305 and 0.323). The coloured test pattern consists of vertical bars generated by a shift of chromaticity towards the spectrum locus, away from background chromaticity, in one of twelve equally spaced directions. The amplitude of chromatic displacement employed was about ten times larger than the average chromatic threshold measured psychophysically. The PCR amplitude for each direction of chromatic displacement investigated was obtained by averaging 32 traces.

## Measurement of Pupil Motion Responses

Dynamically generated random dot patterns similar to those described by Baker and Braddick (1982) were used to generate coherent movement so as to avoid the detection of sudden changes in retinal illuminance at motion onset. Detailed description of the stimulus configuration employed has been given elsewhere (Barbur et al, 1992a; Sahraie, 1993). The display background for PMR studies subtended a visual angle of $17^{o}$ x $13^{o}$ and had a luminance of 13 cd $m^{-2}$. The single dots had a luminance of 76 cd m-2 and each dot subtended a visual angle of $0.22^{o}$ x $0.26^{o}$. The coherent movement interval was always preceded and followed by random motion of the same step size and inter-frame-interval. Each PMR trace was obtained by averaging 64 single measurements.

# RESULTS

## Pupil Light Reflex

The classic PLR response was investigated in several hemianopes, with the stimulus presented separately both in the normal and the cortically blind areas of the visual field. The results obtained showed the same characteristics in all subjects and were similar to those shown in Fig. 1. The PLR response amplitude was defined as the difference in pupil diameter just before the onset of constriction and the point of maximum constriction (see Fig. 1). In every hemianope tested, the PLR amplitude was either reduced with the stimulus presented in the blind hemifield (by comparison to the sighted hemifield) or was completely absent for small stimuli. When the change in light flux level on the retina caused by the stimulus was increased by increasing the size and/or the luminance of the test flash, the response amplitudes with the stimulus in the blind hemifield increased and in the limit the pupil constrictions generated were similar to those obtained by stimulating comparable regions of the retina in the sighted hemifield. Fig. 2a shows pupil response traces obtained with the more complex stimulus consisting of a large area of small checks formed by independent, random luminance modulation of each check. The inset shows the RLM amplitude employed and the rectangular pulse along the abscissa shows the time of stimulus presentation. For an RLM value of 0 %, there is no modulation and the test flash is presented on a uniform background field. The change in pupil diameter caused by the test flash is plotted as a function of RLM amplitude in Fig. 2b. For a test flash of luminance contrast, $\delta L/L_b = 0.3$, the results show that the decrease in pupil diameter asymptotes to a steady value as the RLM amplitude is increased. In order to assess the variance of repeated measurements and to compare this to the variance of pupil noise fluctuations computed from the pupil responses that are averaged for each stimulus (Barbur, 1991c), we measured PLR responses, with and without RLM masking, in the same subject at discrete intervals over a period of several hours. The mean difference in response amplitude computed from the six sets of measurements together with the corresponding pupil noise standard deviation (i.e., $\pm 2\sigma$) are shown in Fig. 2d. The results

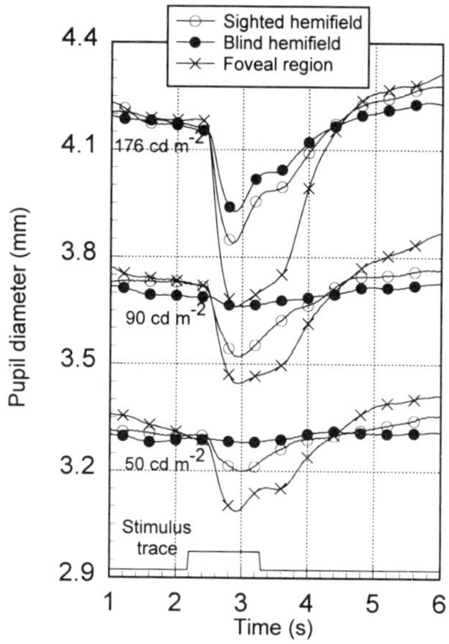


**Figure 1.** Comparison of PLR responses measured foveally and in the sighted and blind hemi-fields of a subject with hemianopia caused by unilateral degeneration of his primary visual cortex (Barbur et al, 1980; 1993). The test stimulus was a disc of radius 4 deg and was presented onto a uniform background field of luminance 24 cd $m^{-2}$. The centre of the test flash was 6 deg above the horizontal meridian and 11 deg away from fixation. Similar data were also obtained as a function of stimulus size when the luminance of the test flash remained unchanged. The results show that although the responses elicited from the blind hemi-field are always of reduced amplitude by comparison with the sighted field, the observed difference is reduced when the stimulus causes a significant change in total light flux on the retina. This is usually the case when the stimulus is of large size and high luminance. Similar tests were also carried out in three other subjects with hemianopia caused by post-geniculate lesions. The results obtained showed consistent reductions in PLR response amplitude when comparing corresponding regions in the sighted and blind hemifields. These results and the findings of Fig. 2 can be explained by postulating the existence of two afferent components for the PLR response with different spatial properties (Cole et al, 1995).

suggest that the response variability observed over several hours can be accounted for completely in terms of the variance associated with random noise fluctuations in the pupil. The observed response saturation with increasing RLM amplitude (as shown in Fig. 2b) suggests that two pathways may be involved in driving the PLR response. One pathway that signals small field, local luminance increments, and the other that sums up increments in light flux over much larger areas of the visual field and is ineffective when small stimuli of low luminance are involved. The much reduced PLR amplitude observed in patients with damaged post-geniculate pathways is consistent with the absence or decreased sensitivity of one or both of these pathways. If this hypothesis is correct, then one would expect a PLR response in the blind hemifield of reduced amplitude by comparison with the sighted hemifield. The RLM modulation with the stimulus in the blind hemifield should either have no effect on the observed responses or should eliminate completely the residual PLR response. To test this hypothesis and its predictions, PLR responses were measured in the sighted and the blind hemifields of a hemianope using the same stimulus conditions, with and without RLM masking. The results presented in Fig. 3 show that the RLM modulation has practically no effect on PLR response amplitude and this suggests that the PLR component that is normally masked by small field, local contrast modulation is absent in the hemianope.

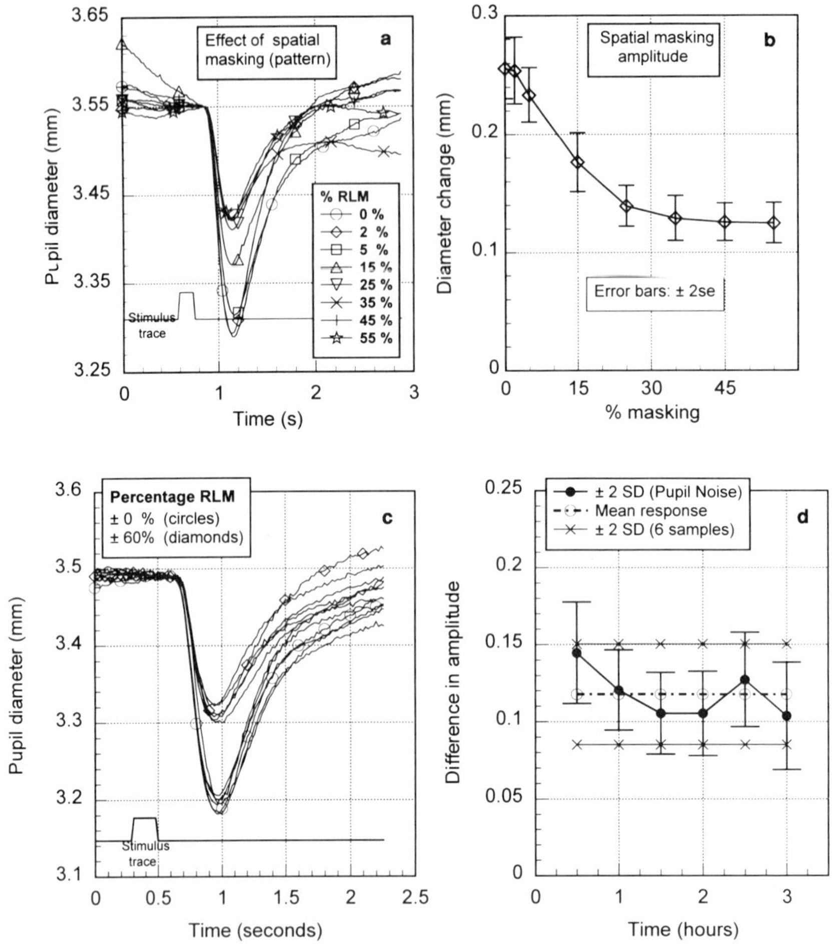


**Figure 2.** Pupil responses elicited with a 9° x 9° achromatic square of contrast 0.3 (i.e., $\delta L/L_b$), presented against a uniform background field of luminance 24 cd $m^{-2}$. The square stimulus is made of small checks (i.e., 0.7° x 0.7°) that are defined by dynamic, local luminance contrast modulation. The space-averaged luminance of the stimulus remains unchanged when spatial modulation is employed, but the amplitude of random luminance contrast modulation of each check can be varied systematically. Each pupillogram represents the average of 16 measurements, and shows how the response amplitude is affected by the level of Random Luminance Modulation (RLM), as indicated in section a. The pupil constriction, $\delta d$, was extracted from each average trace and is plotted against RLM amplitude in section b. The results show that the PLR amplitude decreases asymptotically to a steady level with increasing RLM masking, but the pupil response cannot be reduced below this level when only pattern RLM masking is employed. An analysis of the pupil noise fluctuations in repeated measurements of PLR response is shown in sections c & d. Six sets of repeated measurements were obtained with and without spatiotemporal modulation. Each set employed only two interleaved stimuli (i.e., RLM values of 0% & 60 %). The luminance contrast of the test flash in these repeated tests was 0.4 and 32 pupil response traces were averaged for each stimulus condition. The results made possible an examination of the nature of the statistical fluctuations involved. We wish to establish whether the observed differences in sequential tests in the same subject can be accounted for in terms of inherent pupil noise fluctuations or whether the changes observed when the same test is repeated at a different time need to be attributed to other factors. The pupil response amplitude, $\delta d_i$, was calculated from each trace so as to eliminate the fluctuations in steady-state pupil size during the test (Barbur, 1992c). The 32 estimates of $\delta d_i$ (i = 1 to 32) obtained for each stimulus condition were then used to compute the variance of pupil noise fluctuations (i.e., $\sigma_N^2$). The variance of the difference in response amplitudes measured with and without spatiotemporal luminance masking was then calculated and this provides a measure of Standard Deviation (SD) for the cortical

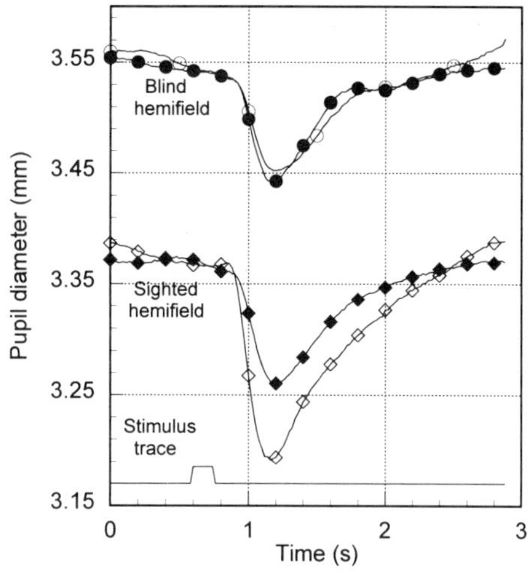


**Figure 3.** Comparison of PLR responses with and without spatial pattern masking (i.e., solid & open symbols, respectively) measured at an eccentricity of 10° in the sighted (diamonds) and the blind hemifield (circles) of a hemianope (adapted from Barbur et al, 1995). The test pattern was a 9° x 9° square divided into small checks of about 0.7° x 0.7°. The RLM amplitude, defined as the range of allowable Random Luminance Modulation of each check luminance wrt that of the uniform background field, was ± 45%. The luminance contrast of the large test pattern was 0.3 (i.e., $\delta L/L_b$) and was presented against a uniform background field of luminance 24 cd $m^{-2}$. The spatial modulation is dynamic, the luminance of each check changing randomly in time, but the space averaged luminance of the stimulus remains constant and equal to that of the surrounding background. The results show that the spatial pattern masking has no effect on the pupil response measured in the blind hemifield. This result is consistent with the absence of a central component, but the pattern masking reduces significantly the PLR response in the sighted field (solid diamonds).

## Pupil Colour Responses

Previous studies have demonstrated that the RLM technique isolates the use of pure chromatic signals even when the coloured test target is well above threshold for detection. This is defined as the difference in the chromaticity of the test stimulus with respect to the adjacent achromatic background field required for threshold detection (Barbur et al, 1994b). Suprathreshold coloured stimuli were therefore employed. Each chromatic displacement was always towards a selected point on the spectrum locus and results in the generation of coloured vertical bars. Fig. 4a shows data obtained for different directions of chromatic displacement in the CIE (x, y) - chromaticity diagram. Each trace represents the average of

---

component of the PLR response amplitude (i.e., SD = $\sqrt{(\sigma^1{}_2 + \sigma^2{}_2)}/32$. The mean amplitudes and the corresponding pupil noise fluctuations (i.e., ± 2SD range) are shown in section d for each set of measurements (solid circles). The dash / dotted line and crosses show the mean of the six sets of measurements and the range determined by the corresponding standard deviation (i.e., ± 2SD range). The results show that the pupil noise variance remains constant over several hours and that no systematic changes are involved. The errors on each measured response, estimated with respect to the mean response (i.e., circles in section b wrt the dash / dotted line) is well within the range determined statistically on the basis of inherent pupil noise fluctuations. The crosses show the ± 2SD range based on the standard deviation of the six estimates. The results suggest that the measured pupil noise variance can be used to predict the statistical significance level of any change in the cortical component of the PLR response amplitude when repeated measurements are involved (Sections a & b have been adapted from Barbur et al, 1995).

32 pupil responses for the same direction of chromatic displacement. The PCR amplitude was then extracted from each trace by computing the difference between the diameter of the pupil just before constriction and the point of maximum constriction. This response amplitude is plotted in Fig. 4b along the direction of the corresponding chromatic displacement in the chromaticity diagram (solid circles), together with the thresholds for detection of coloured bars measured psychophysically. Although the PCR data correlate well with the reciprocal of the chromatic thresholds measured psychophysically, the constriction of the pupil reflects the processing of suprathreshold chromatic signals. The fixed chromatic displacement that was used to trigger PCR responses in this experiment (i.e., 0.06) is about 10 times larger than the average threshold chromatic displacement measured psychophysically. Pupil colour responses were also obtained with the stimulus presented separately both in the sighted and the blind hemifields of a hemianope. Twelve, equally spaced directions of chromatic displacement were used and each trace represents the average of 32 measurements for each direction of chromatic displacement. The results obtained in the sighted hemifield are similar to those measured in the foveal region, but of reduced amplitude (Fig. 4c). No significant pupil colour responses could be elicited when the small coloured stimulus employed in this study was presented in the hemianopes's blind hemifield.

## Pupil Grating Responses

PGR data were obtained using sinusoidal gratings of space-averaged luminance that was either equal to or lower than the surrounding uniform background field. The lower luminance gratings consisted only of dark bars with no increment in luminance anywhere on the retina. The onset of a lower luminance grating is accompanied by a net decrease in space-averaged luminance whilst the restoration of the uniform background field at grating offset causes a net increase in luminance and this in turn generates a PLR response. Typical PGR traces are shown in Fig. 5a. The experiments were repeated as a function of grating spatial frequency and the corresponding pupil response amplitudes measured at grating onset are shown in Fig. 5b. The results show that the onset of both equal or lower luminance gratings causes a constriction of the pupil and that the response amplitude elicited by lower luminance gratings is about half that elicited by equal luminance gratings and that this effect is independent of spatial frequency. Fig. 5c shows a comparison of PGR traces measured in response to gratings presented foveally and in the blind and the sighted hemifields of a hemianope. The foveal traces show the largest response amplitudes, but significant responses can also be elicited when the same gratings are presented in the subject's sighted hemifield. The same gratings presented to the blind hemifield failed to elicit a significant response.

## Pupil Motion Response

PMR traces were measured to the onset of coherent movement generated in a dynamic random dot pattern. The pupil response amplitudes triggered by the onset of coherent motion were generally smaller than the maximum constrictions elicited with gratings or coloured stimuli. The stimulus field was restricted to two rectangular patches positioned symmetrically on each side of fixation so as to minimise unwanted eye movements. Fig. 6a shows averaged PMRs for stimulus speeds in the range 1.3 to 9.5 deg $s^{-1}$. Other speeds were not investigated because of the limitations imposed by existing display hardware. The coherent movement was preceded and followed by random motion of the dots. No significant pupil responses were elicited by the offset of coherent motion. When the random dot stimuli were presented to the more peripheral regions of the visual field, the PMR traces elicited were of reduced amplitude. Fig. 6b shows a comparison of PMRs elicited with the stimulus presented in the blind and the sighted hemifield of the same hemianope that has participated in the

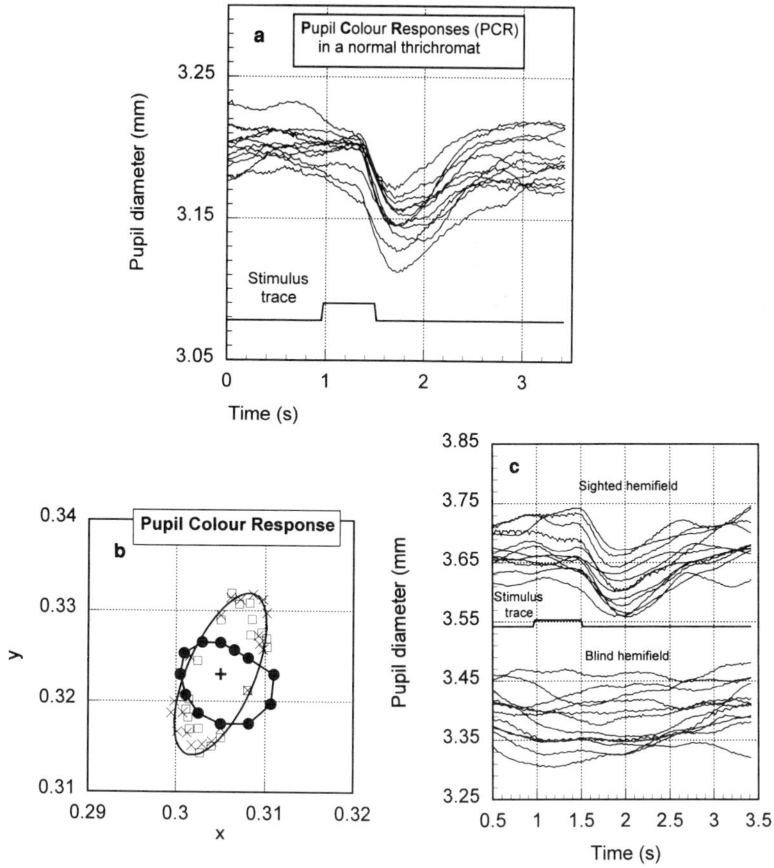


**Figure 4.** Pupil responses to chromatic signals in a normal trichromat (a, b) and in the blind and sighted hemifields of a hemianope (section c). Section a shows averaged response traces for a range of stimulus colours (i.e., the direction of chromatic displacement, away from background field chromaticity in the x, y diagram). An RLM amplitude of ± 25% was used to mask the detection of any achromatic luminance contrast components (Barbur et al, 1994b). The RLM modulation of the checks continued throughout each trace, but the coloured pattern was only presented for 0.55 s as indicated by the rectangular pulse along the abscissa. The Pupil Colour Response (PCR) amplitude, $\delta d$, was computed from each trace and represents the difference between the pupil diameter before constriction starts and the point of maximum constriction. Section b shows a comparison of chromatic threshold discrimination measurements (crosses) and the corresponding pupil colour response amplitudes extracted from the traces shown in section a (circles). The smooth curve represents the best fit ellipse to the chromatic discrimination thresholds (crosses) measured with ± 25% RLM. This RLM amplitude ensures that the subject cannot use achromatic contrast signals to detect the test pattern (Barbur et al, 1992;1994b). The remaining data plotted as open symbols were obtained with lower RLM amplitudes. The distance between the centre cross (i.e., the chromaticity co-ordinates of the surrounding achromatic background field, ($x_b$,$y_b$: 0.305, 0.323)) and each of the solid circles plotted is proportional to the actual constriction of the pupil for each direction of chromatic modulation investigated. The PCR data mirror the chromatic threshold sensitivity, but reflect the processing of suprathreshold chromatic stimuli (i.e., a fixed chromatic displacement of 0.06 units in the CIE x, y - diagram, ~ about 6 times larger than the major semi-axis of the ellipse). Section c shows a comparison of pupillary responses to chromatic stimuli when presented in the sighted and the blind hemifields of a hemianopic subject. Twelve, equally spaced directions of chromatic displacement were used and each trace represents the average of 32 measurements for the same direction of chromatic displacement. The results obtained in the sighted hemifield are similar to those measured in the foveal region, but of reduced amplitude.

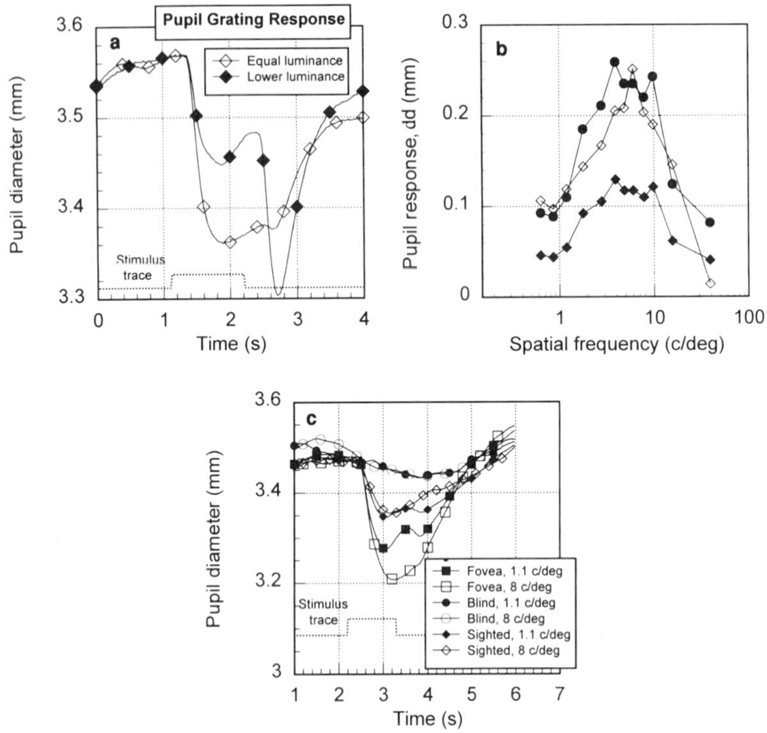


**Figure 5.** Pupil responses to achromatic, sinusoidal gratings. The gratings consist of either bright and dark bars of space-averaged luminance equal to that of the uniform background field (i.e., equal luminance condition), or only of dark bars of lower space-averaged luminance. The gratings subtended a visual angle of $6^{\circ} \times 6^{\circ}$ and were presented in the centre of a uniform background field of luminance, $L_b$, equal to 34 cd $m^{-2}$. Each trace represents the average of 32 measurements. Single grating responses for gratings of spatial frequency 4.9 cycles / deg are shown in section a. Section b shows the pupil constriction amplitude, δd, elicited by bright and dark bars (i.e., sinusoidal gratings of equal space averaged luminance) and also by dark bars only (i.e., gratings consisting only of local decrements in luminance) as a function of spatial frequency. The data represent the averaged results for three subjects. The onset of a lower luminance grating causes an overall reduction in light flux level over the stimulus. In spite of this reduction in luminance, the pupil always constricts initially, but the amplitude of constriction is about half that elicited with gratings of equal luminance. This effect can be seen by doubling the response amplitudes elicited with lower luminance gratings in section b. Section c shows PGR data measured in a hemianope. The gratings were of equal space-averaged luminance and were presented either in the foveal region or at an eccentricity of 10° in the blind and the sighted hemifields. The results show PGR responses of reduced, but significant amplitude in the sighted hemifield. No significant responses could be elicited in the blind hemifield.

PCR and PGR studies. The subject's left visual cortex was damaged as a result of accidental head injury when 8 years old. MRI scans show complete damage to V1 with the exception of a region at the pole that corresponds to the subject's macular sparing (Barbur et al, 1993). The same brain imaging study has also revealed the presence of significant activation of V5 and associated areas when the subject's task was to discriminate the direction of a moving stimulus presented to his blind hemifield. Activation of V5 and other pre-striate areas under such conditions demonstrates that some signals of retinal origin can reach pre-striate areas in man, even in the absence of V1. The results of our PMR studies are consistent with these earlier findings in that a small, but statistically significant PMR amplitude was elicited in response to the onset of coherent movement in the blind hemifield (Fig. 6b).

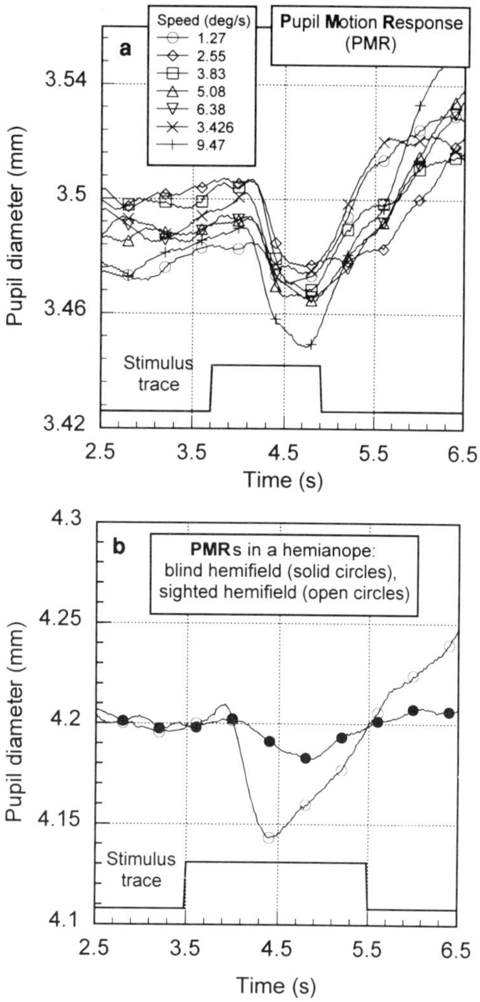


**Figure 6.** Pupil responses to the onset of coherent movement, generated in a dynamic random dot pattern, under stimulus conditions that do not cause a PLR response. The display background subtended a visual angle of 16.8° x 12.7° and was of luminance 13 cd $m^{-2}$. Ninety, discrete elements, each subtending a visual angle of 0.2° x 0.24° were used to generate each stimulus. The luminance of each dot was 75 cd $m^{-2}$ and the stimulus consisted of two rectangular areas of 7° x 5.2° angular subtense that were positioned symmetrically on either side of fixation (section a). The onset of coherent movement was preceded by dynamic noise, when each dot changed randomly in position at a predetermined rate. Coherent motion was then generated by discrete displacement of each dot, in the same direction for a period of time as indicated by the rectangular pulse along the abscissa. The same spatial step increment in dot position was used during the dynamic noise and the coherent motion stage of the stimulus. The offset of coherent motion was then followed by random dynamic noise. Each trace represents the average of 64 measurements (adapted from Barbur et al, 1992a). Section b shows similar data measured in a hemianope with the test stimulus presented either in the sighted or in the blind hemifield. In these experiments the stimulus occupied a single rectangular area of 8.5° x 9.3° that was centred 10° away from the vertical meridian. The Pupil Motion Response (PMR) amplitude is somewhat reduced in the visual periphery in normal subjects by comparison with the foveal region as was also the case for the hemianope. The PMR amplitude following stimulation of the blind hemifield is very small, but the response is statistically significant when 64 traces are averaged.

# DISCUSSION

The results presented in this manuscript suggest that in addition to increments in retinal illuminance, the processing of a number of other stimulus attributes such as spatial structure, colour and movement can also trigger systematic pupillary responses. The results obtained from studies of PLR response components in subjects with hemianopia caused by post-geniculate lesions suggest that in addition to the subcortical pathway, the PLR response is also driven by a cortical component. The extent to which the pupil grating, colour and motion responses are mediated independently from the cortical component of the PLR response is difficult to assess on the basis of our current knowledge. A simplifying assumption is that pupil light reflex responses can be separated from pupil responses to stimulus attributes that normally involve the activity of the pre-striate areas of the visual cortex (i.e. non-PLR responses). The hypothesis that was suggested to explain the existence of non-PLR components (Barbur & Forsyth, 1986) implies that such responses arise not as a result of direct activation of an afferent pupillary pathway, but simply as a result of a transient disturbance of the steady cortical inhibitory pathway that regulates the activity of subcortical nuclei involved in the control of the pupil. A novel stimulus defined by its spatial structure, colour or motion components generates a large change in the steady state level of neural activity in the cortex, when presented suddenly to the eye. This in turn causes a transient perturbation of the supranuclear inhibitory input to the parasympathetic pupillomotor pathways (Alexandridis, 1985). The transient reduction of an inhibitory input results in a corresponding increase in parasympathetic innervation of the sphincter muscle of the iris and hence the observed constriction of the pupil. The amplitude of constriction is likely to be proportional to the level of cortical activity generated and hence the dependence of non-PLR response amplitudes on stimulus size, spatial frequency, eccentricity, chromatic saturation and contrast. In addition to the principal projection from the primary visual cortex to several pre-striate areas, it is now well established that some pre-striate areas can also receive small projections from some neurones in the interlaminar layers of the LGN or indirectly via subcortical nuclei (see Stoerig & Cowey, 1993 for review). It is therefore possible, at least in principle, to elicit non-PLR responses associated with sudden changes in neural activity in pre-striate areas even in the absence of direct projections to these areas from V1. The non-geniculostriate projections to pre-striate areas are however small and therefore the level of neural activity generated in these areas in the absence of V1 may be below the threshold required to generate pupillary responses. The optimisation of stimulus parameters so as to generate maximum innervation of pre-striate areas via subcortical projections may therefore be needed if non- PLR responses are to be elicited in the absence of V1. Some small but significant pupil motion responses have been demonstrated in the absence of V1 (Fig. 6b). This PMR response has not been entirely unexpected given the significant activation of V5 and other pre-striate areas in response to moving stimuli presented to the subject's blind hemifield (Barbur et al, 1993). Non-PLR responses are therefore likely to reflect sudden changes of activity in pre-striate areas of the visual cortex either as a result of direct projections to these areas via the primary visual cortex or via subcortical pathways. Current experimental findings support this hypothesis and help to rule out alternative explanations. The following are some of the alternatives that have been ruled out on the basis of available experimental findings.

The pupil response to chromatic stimuli of the kind employed in these studies reflects changes in the contribution that rod signals make to the PLR response. This cannot be the case for the majority of the stimulus conditions employed since no PCRs can be elicited in patients with cerebral achromatopsia (Barbur et al, 1992c) or in dichromats when the direction of chromatic modulation is restricted to the corresponding colour confusion lines (Barbur et al, 1993). Since these subjects have normal rod vision, pupil colour responses cannot be explained in terms of rod-mediated PLR signals.

The pupil response to sinusoidal gratings reflects PLR responses driven by local increments in light flux on the retina. If this hypothesis was true, then it would be difficult to explain why the onset of a lower luminance grating that causes no increase in light flux level at any point on the retina, also causes the pupil to constrict. The amplitude of this constriction varies systematically with grating spatial frequency. No PGRs can be elicited in hemianopes for either equal and lower luminance gratings, an observation that is also inconsistent with a subcortical PLR response component driven by local luminance increments.

The non-PLR responses reflect transient increases in accommodation triggered by the onset of the stimulus. It is therefore assumed that increments in accommodation cause the observed constrictions of the pupil via the near-reflex response. Simultaneous measurements of accommodation rule out this possibility since the mean and standard deviation of accommodation fluctuations remain statistically invariant when sampled just before stimulus presentation, at the time of maximum pupil constriction and several seconds after stimulus offset (Keenleyside, 1990). It is also difficult to explain why such stimuli always cause a constriction even when different viewing distances are employed.

The pupil constriction is caused by innervation from chromatic mechanisms feeding into a PLR pathway before the lateral geniculate nucleus. This is a more subtle explanation that was put forward to explain certain findings in relation to pupil colour responses. It is based on elegant experiments which have demonstrated that sudden exchanges of scotopically balanced coloured stimuli (Saini & Cohen, 1978; Young & Alpern, 1980), or the step removal of one primary colour from a colour mixture (Kohn & Clynes, 1969) cause a constriction of the pupil irrespective of the direction of the chromatic exchange. These studies have been followed by several other investigations which suggest that both spectrally-opponent, parvocellular ($P_\beta$ type neurones) and broad-band, magnocellular ($P_\alpha$ type neurones) are involved in mediating different aspects of the pupil response (Young & Kennish, 1993; Kimura & Young 1995). The implicit assumption in all these studies is that chromatic signals are extracted and contribute to the PLR response before the level of the lateral geniculate body. This is unlikely to be the case since subcortical projections and particularly that to the pretectum consist of broad-band, $P_\alpha$ type neurones (Perry & Cowey, 1984). Results of pupillary studies in patients with damaged pre-striate areas also support this alternative view. If cortical processing of chromatic signals was not involved in the generation of pupil colour responses then it would be very difficult to explain why PCRs are absent in patients with achromatopsia caused by bilateral lesions to ventro-medial, extra-striate visual cortex (Barbur et al, 1992c).

## Clinical Applications

The development of human vision is affected significantly by maturation of several factors that include more peripheral optical or receptor immaturities (Jacobs and Blakemore, 1988) as well as the maturation of the neural processes associated with the cortical and subcortical pathways (Fielder et al., 1985; Braddick et al., 1986; Dubowitz et al., 1986). The discovery of non-PLR responses and the ability to isolate the light flux component of the PLR response are of great fundamental interest in studies of visual development and may also have useful clinical applications. Several studies have suggested that subcortical neural pathways are likely to be involved in early visual function (Atkinson, 1984; 1992) when the cortical component of the PLR response may be absent. The contribution which subcortical pathways make to visual performance is thought to be greater in the first months of life, but maturation may also affect the functions mediated by subcortical pathways (Atkinson 1984; Braddick et al, 1992). Results of current studies in normal subjects and in patients with damaged central visual pathways suggest that non-PLR pupil responses can only be observed when pre-striate areas become involved in the processing of the attributes of the visual stimulus which they represent. The PCR has never been studied in infants although it can be used reliably to diagnose dichromacy in adults. This test involves the measurement of small constrictions of the pupil when chromatic patterns that are restricted in direction of chromatic displacement to the colour confusion lines of each class of dichromat are employed. The new spatiotemporal, background masking techniques make possible the use of saturated colours without having to set isoluminance and this is a significant advantage in infant vision studies. The measurement and averaging of pupil responses under conditions of restricted movements of the head must, however, be made easier to carry out, if such tests are to become of greater clinical use. The PGR, on the other hand, has already been investigated in infants and shown to follow closely the development of visual acuity, but these tests remain difficult to carry out clinically (Cocker et al, 1991; 1994). Improved experimental techniques that will make it possible to measure non-PLR responses in infants may therefore be of great fundamental and clinical interest. The existing findings suggest that two components of the

PLR response could be isolated and studied separately. The two components may have different time-courses of visual maturation. We do not know how the central and the subcortical components of the PLR response are affected in infants with delayed visual maturation. Such data will be useful in supporting or disproving hypothesis concerning the role of subcortical projections in the maturation of the cortical route (Fielder & Nicholson, 1990).

The ability to isolate cortical and subcortical components in the PLR response and to study their different spatiotemporal properties will undoubtedly help to understand and interpret certain clinical observations in relation to abnormal pupil responses. In the Marcus Gunn pupil, for example, the dynamic PLR response to flashed stimuli can be relatively unaffected when each eye is stimulated in isolation, but the steady-state size of the pupils is larger when the stimulus is viewed monocularly with the affected eye (Gunn, 1902; Levatin, 1959; Levatin et al, 1973). Abnormal pupil responses have been reported in cases of recovering optic nerve disorders, even when colour vision and visual acuity (measured clinically) were found to be normal (Taylor, 1990). In such subjects it is of great interest to establish which of the two PLR components is abnormal. Pupil responses observed in patients with scotomas caused by damaged central visual pathways or the pattern of response associated with the Marcus Gunn pupil can also be explained simply by the selective loss of one of the two components of the PLR response. In the Marcus Gunn pupil, fibres associated with the large-field, light flux component that contributes most to the control of pupil size is abnormal, but the remaining component that mediates mostly the dynamic pupil light reflex response is relatively unaffected. If this hypothesis is correct then the type of pupil response abnormality will depend largely on the properties of the visual stimulus employed and the extent to each of the two PLR pathways have been damaged. A better understanding of the spatial and temporal properties of the two components of the PLR response will undoubtedly help to explain the varying patterns of abnormal response observed clinically. These preliminary results are promising, but the significance and potential use of these findings in clinical neuro-ophthalmology remains to be explored.

## ACKNOWLEDGMENTS

I would like to thank my colleagues P Forsyth, A Goodbody, A Harlow, V Cole, A Sahraie and C Williams for useful discussions and help the experimental work reported here. I would also like to thank my subjects for their patience with many hours of experimental observations. The investigations reported here were supported by a research grant awarded by the Wellcome Trust.

## REFERENCES

Alexandridis, E. (1985) The Pupil, Springer-Verlag, New-York.

Atkinson, J. (1984) Human visual development over the first 6 months of life. A review and a hypothesis., *Human Neurobiol.* 3:61-74.

Atkinson, J. (1992) Early visual development: differential functioning of parvocellular and magnocellular pathways, *Eye* 6:129-135.

Baker C L & Braddick O J (1982) The basis of area and dot number effects in random dot motion perception, *Vision Res.* 22:1253-1260.

Barbur, J. L., Birch, J., Harlow, A. J. and Plant, G. (1992c) The pupil colour response (evidence for involvement of central neural mechanisms), *Perception* 21:74.

Barbur, J. L., Ruddock, K. H. and Waterfield, V. A. (1980) Human visual responses in the absence of the geniculo -calcarine projection, *Brain* 03:905-928.

Barbur, J. L., Thomson, W. D. and Forsyth, P. M. (1987) A new system for the simultaneous measurement of pupil size and two-dimensional eye-movements., *Clin. Vis. Sci.* 2:131-142.

Barbur, J. L., Birch, J. and Harlow, J. A. (1993) Colour vision testing using spatiotemporal luminance masking. In Colour vision deficiencies XI (Ed. B. Drum) 417-426. Kluwer Academic Publishers, Dordrecht.

Barbur, J. L., Harlow, A .J., Sahraie, A.,Stoerig, P. and Weiskrantz, L. (1994a). Responses to chromatic stimuli in the absence of V1: pupillometric and psychophysical studies. In: Noninvasive Assessment of the Visual System Technical Digest. Optical Society of America, Washington, D.C. 2:312-315.

Barbur, J. L., Watson, J. D. G., Frackowiak, R. S. J. and Zeki S. (1993). Conscious visual perception without V1, *Brain* 16:1293-1302.

Barbur, J, L,, Holliday, I. K. and Ruddock, K. H. (1981). The spatial and temporal organisation of motion perception units in human vision, *Acta Psychologica* 48:35-47.

Barbur, J. L., Birch, J. and Harlow, J. A. (1992b) Threshold and suprathreshold responses to chromatic stimuli using psychophysical and pupillometric methods. In NAVS Technical Digest OSA, Washington, D.C. 1:51-54.

Barbur J. L. (1991a) Pupillary responses to grating stimuli, *J. Psychophysiol.* 5:259-263.

Barbur, J. L. (1991b) Pupillary responses to stimulus structure and colour: possible mechanisms Noninvasive Assessment of the Visual System Technical Digest, Optical Society of America, Washington, D.C. 1:68-71.

Barbur, J. L., Cole, V., Harlow, J. and Sahraie, A.(1995) Pupil Colour and light reflex Responses in the periphery of the visual field, *Invest. Ophthalmol. Vis. Sci.* 36:660.

Barbur, J. L. (1991c). Methods for studying the behaviour of the pupil, *J. Psychophysiol.* 5:223-239.

Barbur, J. L. and Forsyth, P. M. (1986). Can the pupil response be used as a measure of the visual input associated with the geniculo-striate pathway, *Clin. Vis. Sci.* 1:107-111.

Barbur, J. L. (1988). Spatial frequency-based methods and their use in clinical psychophysics. Clin. Vis. Sci.2:225-233.

Barbur, J. L., Forsyth, P. M. and Thomson, D. W. (1988). A new system for the simultaneous measurement of pupil size and two dimensional eye-movements, *Clin. Vis. Sci.*2:31-142.

Barbur, J. L. and Ruddock, K. H. (1980). Spatial characteristics of movement detecting mechanisms in human vision. I. Chromatic stimuli, *Biological Cybernetics* 37: 77-92.

Barbur, J. L., Harlow, A. J. and Plant, G. (1994b). Insights into the different exploits of colour in the visual cortex, *Proc. R. Soc. Lond. B,* 258:327-334.

Barbur, J. L., Harlow, J. A. and Sahraie, A. (1992a) Pupillary responses to stimulus, structure, colour and movement, *Ophthal. Physiol. Optics* 12:137-141.

Barbur, J. L., Keenleyside, M. and Thomson, D. W. (1988). Investigation of central visual processing by means of pupillometry, In Seeing Colour and Contour', Eds. Kulikowski, J. J., Dickinson, C. M. and Murray, I. J. Proceedings of The Third International Symposium of the Northern Eye Institute pp 431-451.

Braddick, O., Atkinson J., Hood, B., Harkness, W. and Vargha-Khadem, F. (1992). Possible blindsight in infants lacking one cerebral hemisphere, *Nature* 360:461-463.

Braddick, O. J., Wattam-Bell, J. and Atkinson, J. (1986). Orientation-specific cortical responses develop in early infancy, *Nature* 320:617-619.

Brindley, G. S., Gautier-Smith ,P. C., and Lewin, W. (1969). Cortical blindness and the functions of the non-geniculate fibres of the optic tracts, *J. Neurol. Neurosurg. Psychiatr.* 32:259-264.

Cibis, G. W., Campos, E. C. and Aulhorn, E. (1975). Pupillary hemiakinesia in suprageniculate lesions, *Arch. Ophthalmol.* 12: 1322-1327.

Clarke, R. J. and Ikeda, H. (1985a). Luminance and darkness detectors in the olivary and posterior pretectal nuclei and their relationship to the pupillary light reflex in the rat. I. Studies using steady luminance levels, *Exp. Brain Res.* 57:224-232.

Clarke, R. J. and Ikeda, H. (1985b). Luminance detectors in the olivary pretectal nucleus and their relationship to the pupillary light reflex in the rat. II. Studies using sinusoidal light., *Exp. Brain Res.* 59:83-90.

Cocker, K. D. and Moseley, M. J. (1991). The pupil grating response and visual acuity. In: JOSA Technical Digest on Non-invasive Assessment of the Visual System, Optical Society of America, Washinton DC. 1:113-116.

Cocker, K. D., Moseley, M. J., Bissenden, J. G. and Fielder, A. R. (1994) Visual acuity and pupillary tresponses to spatial structure in infants, *Invest. Ophthalmol. Vis. Sci.* 35:2620-2625.

Cole, V., Barbur, J. L. and Harlow, H. J. (1995). Pupil light reflex components isolated using luminance contrast and light flux masking techniques, *Invest. Ophthalmol. Vis. Sci.* 36:940.

Cowey, A. and Stoerig, P. (1991). The neurobiology of blindsight, *Trends in Neurosci.*14:140-145.

Dubowitz, L. M. S., Mushin, J., de Vreis, L. and Arden, G. B. (1986) Visual function in the newborn infant: is it cortically mediated?, *Lancet* May 17:1139-1141.

Fielder, A. R., Russell-Eggitt, I. R., Dodd, K. L. and Mellor, D. H. (1985) Delayed visual maturation, *Trans. Ophthalmol. Soc. UK,* 104:653-661.

Fielder, A. R. and Nicholson, J. (1990). Delayed visual maturation—ophthalmic and nadiodevelopmental aspects, *Dev. Med. Child Neurol.*32:872-881.

Gunn, R. M. (1902). Functional or hysterical amblyopia, *Ophthalmol. Rev.* 21: 271-280.

Hardy, L. H., Rand, C. and Rittler, M. C. (1954). H-R-R Polychromatic Plates, *J. Opt. Soc. Am.* 44:509-523.

Harms, H. (1951) Hemianopische Pupillenstarre., *Klin Monatsbl Augenheilkd* 118: 133-147

Jacobs, D. S .and Blakemore, C. (1988). Factors limiting the postnatal development of visual acuity in the monkey, *Vision Res.*28: 947-958.

Kardon, R. H.(1992). Pupil Perimetry. Editorial review, Curr. Opin. Ophthalmol. 3:565-570.

Keenleyside, M. S. (1990). Pupillometry and assessment of visual function. PhD thesis, University of Oxford.

Kimura, E. and Young, R. S. L.(1995). Nature of the pupillary responses evoked by Chromatic Flashes on a white background, *Vision Res.* 35: 897-906.

Kohn, M. and Clynes, M. (1969). Colour dynamics of the pupil., *Ann.New York Acad. Sci.* 156:931-950.

Levatin, P., Prasloski, P. F. and Collen, M. F. (1973). The swinging flashlight test in multiphasic screening for eye disease, *Can. J. Ophthalmol.* 8:356-360.

Levatin, P. (1959). Pupillary escape in disease of the retina and optic nerve, *Arch. Ophthalmol.* 62:768-779.

Loewenfeld, I. E. (1993). The Pupil: Anatomy, physiology, and clinical applications. Vol 1, Iowa State University Press.

Lowenstein, O., Kawabata, H. and Loewenfeld, I. E. (1964). The pupil as indicator of retinal activity, *Am. J. Ophthalmol.* 57: 569-596.

Lynn, P. A. (1992). Digital signals, processors and noise. Macmillan New Electronics Series, Macmillan Press Ltd.

Perry, V, H, and Cowey, A. (1984). Retinal ganglion cells that project to the superior colliculus and pretectum in the macaque monkey, *Neurosci.* 12:1125-1137.

Pierson, R. J. and Carpenter, M. B. (1974). Anatomical analysis of pupillary reflex pathways in the rhesus mankey. J. Comp. Neurol. 158:121-144.

Reffin, J. P., Astell, S. and Mollon, J. D.(1991). Trials of a computer-controlled colour vision test that preserves the advantages of pseudoisochromatic plates. In Colour vision deficiencies X, pp 69-76. Eds. B. Drum, J. D. Moreland and A. Serra, Kluwer Academic Publishers, Dordrecht, Netherlands.

Sahraie, A. (1993) Some aspects of the pupil response in relation to stimulus movement and colour. PhD thesis, City University.

Saini, V. D. and Cohen, G. H. (1979). Using colour substitution pupil response to expose chromatic mechanisms, *J. Optical Soc. Am.* 26:1029-1035

Slooter, J. H. and van Noren, D. (1980). Visual acuity measured with pupil responses to checkerboard stimuli, *Invest. Ophthalmol. Vis.Sci* 19:105-108.

Stark, L. and Shermann, P. M. (1957). A servoanalytic study of consensual pupil reflex to ligh, *J. Neurophysiol.* 20:17-26.

Stoerig, P. and Cowey, A. (1993) Blindsight: neurons and behaviour, *Prog. Brain Res.* 95:445-459.

Taylor, D. S. I. (1990) Pediatric Ophthalmology, Blackwell. Oxford pp 561.

Ukai, K. (1985). Spatial pattern as a stimulus to the pupillary system, *J. Opt. Soc. Am.* 2:1094-1100.

Wyszecki, G. and Stiles, W. S. (1982). Colour Science, Concepts and Methods, Quantitative Data and Formulae, 2nd edition, John Wiley, New-York.

Young, R. S. L. and Alpern, M. (1980). Pupil responses to foveal exchange of monochromatic lights, *J. Opt. Soc. Am.* 70:697-706.

Young, R. S. L. and Kennish, J. (1993). Transient and sustained components of the pupil response evoked by achromatic spatial patterns, *Vision Res.* 33:2239-2252.

# FACTORS REGULATING THE BLOOD FLOW IN THE OPTIC NERVE HEAD

Sohan Singh Hayreh

Department of Ophthalmology
University of Iowa
Iowa City, Iowa

Ischemic disorders of the optic nerve head (ONH) constitute an important cause of blindness or seriously impaired vision. Acute ONH ischemia results in anterior ischemic optic neuropathy (AION) - a seriously blinding disease. There is overwhelming evidence that glaucomatous optic neuropathy is a chronic ischemic disorder of the ONH. Glaucomatous optic neuropathy is seen in primary open angle glaucoma and low (or normal) tension glaucoma. Both AION and glaucomatous optic neuropathy are common disorders, seen more commonly in middle aged and older than in young persons. For a proper understanding of the pathogenesis and management of these important and highly prevalent ischemic disorders of the ONH, it is imperative to know its blood supply and various factors which regulate the blood flow in the ONH.

## BLOOD SUPPLY OF THE OPTIC NERVE HEAD

The following very brief account of the blood supply of the ONH is essentially based on my anatomical (Hayreh, 1958; 1963a;1964; Singh & Dass, 1960a; 1960b), histological (Hayreh, 1958; 1963a; 1963b; 1964; Singh & Dass, 1960a; 1960b), experimental (Hayreh & Perkins, 1968; 1971 Hayreh, 1969; 1970; 1974a; 1974b; 1974c; 1975; 1990; Hayreh et al 1970; Hayreh & Baines, 1972a; 1972b; 1973; Hayreh & Weingeist, 1980; Hayreh & Chopdar, 1982; Kishi et al, 1985; Hayreh et al, 1986a; 1986b) and clinical (Hayreh & Perkins, 1968; Hayreh, 1969; 1974d; 1974e; 1975; 1981; 1985; 1987 1990; Hayreh & Walker, 1967) studies in health and disease. A detailed account is given elsewhere (Hayreh 1978; 1989).

For purposes of description of blood supply, the anterior part of the optic nerve can be divided into the following 4 parts from front to back, with the central retinal vessels lying in their center (Hayreh, 1978) (Fig. 1):

### The Surface Nerve Fiber Layer

This is the most superficial layer and is mostly supplied by the retinal arterioles (Fig.1-A). In some eyes, however, its temporal part may be supplied by the posterior ciliary artery (PCA) circulation from the prelaminar region. When a cilio-retinal artery is present, that sector then is usually supplied by the PCA circulation.

## The Pre-laminar Region

This is sandwiched between the surface nerve fiber layer in front and the lamina cribrosa behind, and is situated at level of the choroid. It is supplied by the PCA circulation via the centripetal branches from the peripapillary choroid, with no contribution by the peripapillary choriocapillaris or the central retinal artery (Fig. 1).

## The Lamina Cribrosa

This is situated at level of the sclera. It is supplied by centripetal branches arising directly from the short PCAs (or from the "Circle of Zinn and Haller", when that incomplete Circle exists), as well as by some branches arising from the recurrent pial branches from the peripapillary choroid (Fig. 1). The central retinal artery gives no branches in this region. Based on a recent post-mortem cast study, Ujiie and Takahashi (1987) described that lamina cribrosa is supplied by three arterial systems:

i. peripapillary choroid,
ii. intrascleral arterial network (the so-called "Circle of Zinn and Haller"),
iii. branches from the short PCAs running along the optic nerve.

## The Retrolaminar Region

This part lies immediately posterior to the lamina cribrosa and is supplied by two vascular systems (Fig. 1):

*Peripheral Centripetal Vascular System.* This is the primary source of blood supply to this part. It is composed of pial vascular plexus surrounding this part of the optic nerve. It is supplied mainly by the recurrent pial branches arising from the peripapillary choroid and from the "Circle of Zinn and Haller", when present, and may also be supplied by pial branches from the central retinal artery and other orbital arteries.

*Axial Centrifugal Vascular System.* This is an inconsistent and minor source of blood supply to this part. It consists of intraneural branches of the central retinal artery which, when present, are highly variable (Sihgh & Dass, 1960b).

Olver et al.(1990) described the "Circle of Zinn and Haller" and its role in the supply of the retrolaminar region from their cast study. They saw it as an elliptical microscopic arteriolar intrascleral anastomosis, formed by branches of the medial and lateral "para-optic short PCAs". They found that the ellipse was divided into superior and inferior parts by the entry points of these branches into the eye, and that the two parts supplied the corresponding parts of the retrolaminar optic nerve by their pial branches. They found morphological variations in the Circle in terms of form, position and branches between subjects and even between the two eyes of the same subject. They postulated that this supplies an anatomical basis for the altitudinal visual field defects commonly seen in AION. This is certainly a very interesting finding; however, no other author using the same technique has so far described such a pattern. Also, in my post-mortem injection studies, in the few specimens where I could outline the "Circle of Zinn and Haller" by dissection of the sclera, I did not see such an arrangement in any specimen (see Fig. 18 in Hayreh, 1963b). The other problem is that one cannot deduce the *in vivo* functional blood flow pattern from the anatomical pattern seen on

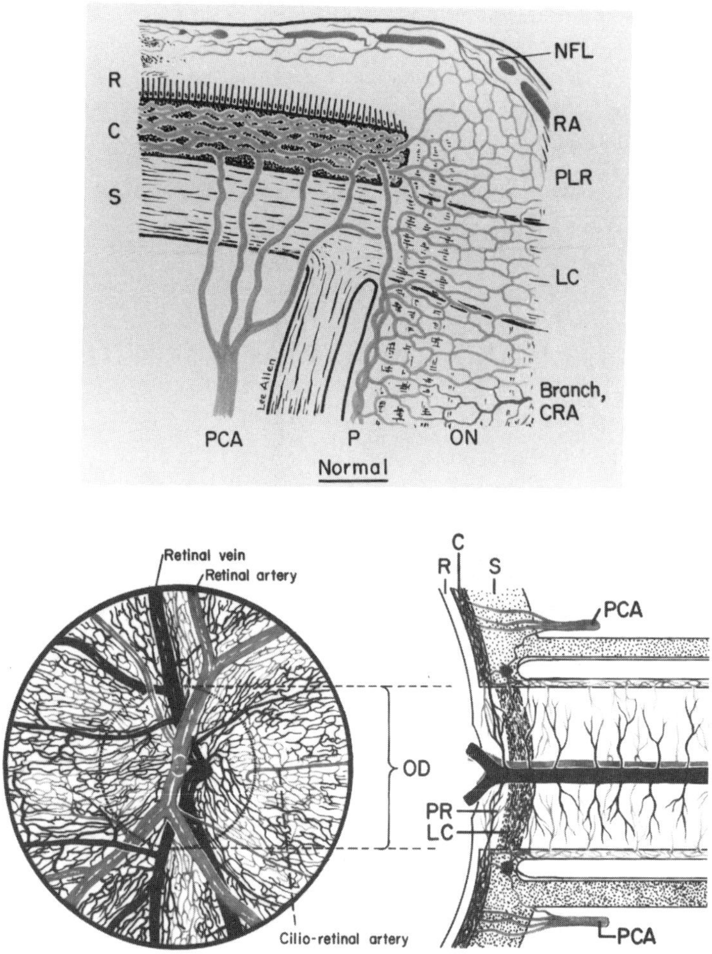


**Figure 1.** (A,B) Schematic representations of blood supply of the optic nerve head and retrolaminar optic nerve. C= Choroid, CRA= Central Retinal Artery, CRV= Central Retinal Vein, LC= Lamina cribrosa, NFL= surface nerve fiber layer of the disc, OD= Optic disc, ON= Optic Nerve, P= Pia, PCA= Posterior Ciliary Artery, PLR/PR= Prelaminar Region, R= Retina, RA= Retinal Arteriole, S= Sclera. (Reproduced from Hayreh, 1978).

post-mortem cast studies. This has been amply demonstrated by our *in vivo* studies (Hayreh, 1969; 1970; 1974a; 1974b; 1972c; 1974d; 1974e 1975a; 1975b; 1981; 1985; 1989; 1990; Hayreh et al, 1969b; 1970; Hayreh & Baines, 1972a; 1972b; Hayreh & Chopdar, 1982) on the choroidal vascular bed; all previous post-mortem injection studies on the choroidal vascular bed showed it to be a freely anastomosing vascular bed, but our *in vivo* studies revealed the strictly segmental nature of the choroidal vascular bed, including the peripapillary choroid. The same holds good for other ocular vascular beds. I have stressed the importance of this serious discrepancy between post-mortem anatomical patterns and the *in vivo* functional blood flow pattern (Hayreh, 1970; 1974a; 1974b; 1974c; 1975; 1990; 1974d; 1974e 1985; 1989). Olver et al (1990) also rightly stressed that vascular casting does not directly tell us about functional blood flow. Moreover, the marked variation in the anatomical pattern of the "Circle of Zinn and Haller", seen by me and also reported by Olver et al (1990), indicates that the superior and inferior elliptical pattern described by Olver et al (1990) is not a universal finding.

From this description it is obvious that PCAs, via the peripapillary choroid or the short PCAs, are the main source of blood supply to the ONH, with the peripapillary choroid playing the most important role in the blood supply. Our *in vivo* studies in health and disease have shown that the PCAs and peripapillary choroid have a segmental blood flow (Hayreh, 1969; 1970; 1974d; 1974e; 1975a; 1975b; 1981; 1985; 1990; Hayreh & Baines, 1972a; Hayreh & Chopdar, 1982). There is further evidence that the *in vivo* blood supply to the anterior part of the optic nerve is strictly segmental in nature.

## FACTORS CONTROLLING THE BLOOD FLOW IN THE OPTIC NERVE HEAD

Nutrition of the ONH depends upon its blood flow. The blood flow in the ONH can be calculated by using the Poiseuille's equation (for the flow of fluids through stiff tubes). This equation is as follows:

Flow = Perfusion pressure/Resistance to flow

In the eye and ONH: *Perfusion pressure* = Mean blood pressure (BP) minus intraocular pressure, and *Mean blood pressure* = Diastolic BP + 1/3 (Systolic minus diastolic BP).

From this equation, it is evident that blood flow in the ONH depends upon the (a) systemic BP, (b) intraocular pressure, and (c) vascular resistance. In addition to these, the presence or absence of autoregulation of blood flow in the ONH plays an important role.

## PATHOPHYSIOLOGY OF THE VARIOUS VASCULAR FACTORS CONTROLLING THE BLOOD FLOW IN THE ONH

### Efficiency of the Autoregulation in the Optic Nerve Head

Among the ocular tissues, it is well-established that the retinal circulation normally has efficient autoregulation(Porsa, 1941; Dollery et al, 1963; Russell, 1963; Alm & Bill, 1972; 1973; Bill, 1974). The presence of efficient autoregulation in the normal ONH has been reported recently by a large number of experimental and clinical studies (Ernest, 1974; 1976; 1977;1979;Geijer & Bill, 1979; Sossi & Anderson, 1983; Weinstein et al, 1983; Robert & Maurer, 1983; Bill, 1985; Pillunat et al, 1985a; 1985b; 1986a; 1986b; 1987; Sperber & Bill, 1985; Bill & Sperber, 1987 Robert et al, 1988; 1989; Novack et al, 1990) although it was denied in earlier studies (Alm & Bill, 1973). The choroid has no autoregulation. The goal of autoregulation in a tissue is to maintain a relatively constant blood flow during changes in its perfusion pressure. Autoregulation operates only over a critical range of perfusion pressure so that with a rise or fall of perfusion pressure beyond the critical range, the autoregulation becomes ineffective and results in breakdown of autoregulation. Thus, autoregulation does not protect a tissue all the time but only within a certain critical range of perfusion pressure. Autoregulation is considered a feature of the terminal arterioles although the exact mechanism and site are still not fully understood. To regulate the blood flow with the rise or fall of BP beyond normal levels, the arterioles normally constrict or dilate respectively. It is possible that capillaries may also exercise autoregulation via vasoconstrictors and vasodilators released by the capillary endothelial cell, such as endothelin and nitric oxide.

The blood flow autoregulation in the ONH may break down under a number of circumstances, as is evident from the following:

*When Perfusion Pressure Goes Below or above a Critical Range.* The level of perfusion pressure below which this autoregulation in the ONH breaks down has been reported variably in different experimental studies, and was found to vary from 15 to 30 mm Hg in young healthy animals (Ernest, 1977; Geijer & Bill, 1979; Sossi & Anderson, 1983; Bill, 1985; Sperber & Bill, 1985; Novack et al, 1990). Breakdown of autoregulation also occurs in accelerated or malignant arterial hypertension (Hayreh et al, 1986a).

*Effect of Arterial Hypertension on Blood Flow Autoregulation.* There is ample evidence that the various circulatory systems in the body with autoregulatory mechanisms are influenced by systemic arterial hypertension. It has been shown that, as an adaptive phenomenon, the range of blood flow autoregulation shifts to a higher level in hypertensive patients than in normal persons, in cerebral circulation (Strandgaard et al, 1973; 1975; Fitch et al, 1975; Bill & Linder, 1976; Edvinsson et al, 1976; Jones et al, 1976; Strandgaard, 1976; Boisvert et al, 1977; MacKenzie etal, 1977) and in coronary circulation (Floras, 1988). This adaptation improves a person's tolerance of high BP but at the same time makes the individual less tolerant of low BP. There are a number of reports in which patients with malignant hypertension, who had their BP lowered precipitously to prevent severe neurological and cardiovascular complications, suffered immediate and permanent blindness or severe visual loss from AION (Cove et al, 1979; Hulse et al, 1979; Pryor et al, 1979; Wetherill, 1979; Taylor et al, 1981). This is fairly good proof that the autoregulatory mechanism in the ONH is deranged in hypertension (Hayreh et al, 1986a). The same holds good for the coronary circulation where a marked fall of BP in hypertensives during sleep can precipitate myocardial ischemia, particularly in patients with obstructive coronary artery disease (Floras, 1988). Alderman et al (1988) found that a large reduction in BP (≥18 mm Hg) caused by treatment of hypertension showed a relationship with development of myocardial ischemia. Thus nocturnal hypotension in hypertensives, as discussed below, could contribute to cardiac, cerebral and ONH ischemia because of defective autoregulation produced by hypertension in these organs.

*Other Causes.* Derangement of the autoregulation in the ONH may be produced by other causes so far unknown - perhaps the aging process, arteriosclerosis and atherosclerosis. We need to learn a lot more on this important subject.

*Our Studies.* Finally, we investigated the state of the blood flow and its autoregulation in the retina and ONH in rhesus monkeys which were old, arteriosclerotic, atherosclerotic (because they had been fed an atherogenic diet for about 13 years), and hypertensive (chronic arterial hypertension produced by a modified Goldblatt's procedure)(Hayreh et al, 1991) Our studies revealed defective autoregulation of blood flow in both the retina and ONH, as well as deranged choroidal blood flow.

Thus, evidence is accumulating from both clinical and experimental studies that although blood flow in the ONH is normally autoregulated, it can become defective under certain conditions, e.g., rise or fall of perfusion pressure beyond the critical levels, old age, arteriosclerosis and atherosclerosis. Eyes with such poor autoregulation are at a much increased risk of developing ischemic damage. Therefore, it would seem that derangement in the autoregulation of the ONH may play an important role in the ischemic damage to the ONH. Pillunat et al (1985a; 1985b; 1986a; 1987) found evidence to suggest defective autoregulation in the ONH in primary open angle glaucoma and low-tension glaucoma. Robert et al (1989) found that the ability of glaucomatous eyes to adjust the blood supply in the optic disc to raised intraocular pressure is significantly ($p<0.001$) reduced compared with normals. Grunwald et al (1984) reported that their results suggest an abnormal autoregulation of macular retinal blood flow in primary open angle glaucoma in man. Thus, we have evidence that in glaucomatous optic neuropathy there is defective autoregulation of the blood flow in the ONH. There are, however, still many unanswered questions on this.

## Vascular Resistance

The resistance to flow is directly related to the length of the tube and the viscosity of the fluid, and inversely related to the fourth power of the radius of the tube. In the vascular system, the main factor normally deciding the vascular resistance is the size of the lumen of the pre-capillary arterioles - the primary resistance vessels.

As mentioned earlier, the blood flow in the ONH is inversely proportional to the vascular resistance in the small vessels of the ONH, so that an increase in resistance would decrease the blood flow in the ONH. The increase in vascular resistance can occur in a variety of systemic, vascular and hematologic conditions, including arterial hypertension, arterio-sclerosis, atherosclerosis and hematologic disorders. In arterial hypertension, although there is a rise in the mean BP, increased peripheral vascular resistance may ultimately in fact reduce the blood flow in the ONH markedly.

Langham (1987), from his pulse amplitude and intraocular pressure studies in primary open angle glaucoma and low-tension glaucoma, found that some patients with visual field loss revealed abnormally high ocular vascular resistances, but some patients having minimal or no visual field loss showed no such increase in resistance. He further stated that ocular hypotensive drugs, such as beta-blockers (e.g. Timolol), have the potential to increase vasoconstrictor tone and ocular vascular resistance in the eye which would suggest that these agents may in fact have a detrimental influence on the ONH blood flow.

In this category also fall the various vasospastic conditions. In vasospasm there is an episodic, reversible narrowing of blood vessels which decreases the blood flow. This can produce variable ischemia, depending upon the extent of narrowing of the vessels. The mechanism of vasospasm is ill-understood. Typical examples of these disorders are migraine and Raynaud's disease. Migraine is associated with decreased cerebral blood flow (Dukes & Vieth, 1964; Edmeads, 1977; Lauritzen et al, 1982; 1983). Attacks of migraine have been reported to cause ischemic infarctions of the brain (Connor, 1962; Dorfman et al, 1979), retina (Connor, 1962; Graveson, 1949; Pearce & Foster, 1965) and optic nerve (McDonald & Sanders, 1971; Weinstein & Feman, 1982), and PCA-ischemia (Connor, 1962; Pasteur et al, 1937). An association between low-tension glaucoma and migraine has been shown (Phelps & Corbett, 1985; Drance et al, 1988; Kitazawa et al, 1989).

## Perfusion

Pressure: perfusion pressure is the most important factor determining the blood flow in the ONH. In the intraocular vascular bed, the perfusion pressure is equal to the mean BP minus the intraocular pressure. A fall in the perfusion pressure in the ONH capillaries, thus, may be due to either a fall in the mean arterial BP or a rise in the intraocular pressure or a combination of both. Such a fall in perfusion pressure, in the presence of defective autoregulation of blood flow, would result in ONH ischemia.

## The Role of Disorders of Systemic Arterial BP in ONH Ischemia

Mean BP evidently depends upon the systemic BP. In the literature systemic arterial hypotension as well as hypertension have been described as associated with the development of AION and of visual loss in glaucoma, particularly in low-tension glaucoma. Thus, both hypotension and hypertension play an important role in ONH ischemic disorders.

*Systemic Arterial Hypotension.* An overall fall in the systemic mean BP would produce a fall in mean BP in the ONH, resulting in low perfusion pressure and reduced blood flow in the ONH. Nocturnal hypotension, a fall of arterial BP during sleep, is now well established by the technique of ambulatory BP monitoring (Floras, 1988; Floras & Sleight, 1983; Kleiman, 1988; Hayreh et al, 1994). Our studies (Hayreh, 1995) have suggested that nocturnal hypotension is of great clinical significance in the pathogenesis of ischemic disorders of the ONH. For example, in our experience with over 600 patients with AION, patients typically discover the visual loss on waking up in the morning, indicating that the neuropathy develops in the vast majority during sleep. The same is true of cerebro-vascular accidents (Caplan et al, 1986; Marsh et al, 1990) and myocardial ischemia (Floras, 1988; Floras & Sleight, 1983; Kleiman et al, 1988). Our studies in patients with AION and glaucomatous optic neuropathy have shown a substantially greater drop in BP during the night than in normal healthy individuals (Hayreh et al, 1995); in our studies the fall was about 27% and 33% in systolic and diastolic BP respectively whereas in normal persons it is reported as 7-13% and 9-18% respectively (Kennedy et al, 1983; Drayer et al, 1984; Wallace et al, 1984).

Instead of or in addition to the systemic arterial hypotension, there may be regional fall in the mean BP in the ocular blood vessels. This may be due to narrowing of the regional arteries. For example, Langham et al (1981) found that with 95 to 100% stenosis of the internal carotid artery the ipsilateral ophthalmic artery systolic BP was 50 mm Hg (33% of the brachial artery systolic BP), as compared to 89 mm Hg (66% of the brachial artery systolic BP) in age matched normal subjects. When the internal carotid artery stenosis was 36.5±7.9%, the ophthalmic artery systolic BP was 70 mm Hg. They found that endarterectomy of the occluded internal carotid artery caused a significant ipsilateral increase in the ophthalmic artery systolic BP and in the intraocular pulse amplitude. This indicates that only severe internal carotid artery stenosis would appreciably reduce the mean BP in the ONH. Similarly, local ocular vascular disease, e.g., narrowing of the ophthalmic artery or of one or more of the PCAs supplying the ONH could be responsible for a fall in the mean BP in the ONH. Our experimental and clinical studies on choroidal and ONH blood flow in health and disease have shown that, contrary to the accepted view, the mean BP in all the major ocular arteries arising from the ophthalmic artery is not always the same. The importance of this is that in the event of a fall of perfusion pressure, the vascular bed supplied by one artery may be affected earlier and more severely than others. I have discussed this elsewhere (Hayreh, 1990; 1985; 1989). This could be further aggravated if atherosclerosis and arteriosclerosis are more severe in one artery than another.

*Systemic Arterial Hypertension.* From the above discussion, it would appear that systemic arterial hypertension should improve the ONH blood flow and protect the ONH from developing ischemia. However, that is not true, because in systemic arterial hypertension the blood flow in the ONH is once again reduced and that could be due to the following 4 mechanisms, at least:

i. In systemic arterial hypertension the basic pathological lesion is almost invariably the narrowing of the terminal arterioles, which increases the peripheral vascular resistance and thereby produces arterial hypertension. Blood flow in the ONH is inversely proportional to the vascular resistance, as mentioned above, so that increased vascular resistance at the level of the terminal arterioles would reduce the blood flow in the ONH capillaries. Our experimental studies have shown that in malignant arterial hypertension, in spite of very high BP, the ONH develops a variable degree of ischemia and AION during the acute phase (Kishi et al, 1985; Hayreh et al, 1986a).

ii. Chronic systemic arterial hypertension causes secondary hypertensive changes in the vessel of the ONH which would render the ONH more susceptible to ischemic damage.

iii. Systemic arterial hypertension would influence the blood flow in the ONH by interfering with the normal autoregulation of blood flow in the ONH, as discussed above. In hypertensives the critical range of blood flow autoregulation shifts to higher levels to adapt to hypertension (Standgaard et al, 1973; 1975; Fitch et al, 1975; Bill & Linder, 1976; Edvinsson et al, 1976; Jones et al, 1976; Strandgaard, 1976; Boisvert et al, 1977; MacKenzie et al, 1977) . Such an adjustment makes the ONH less tolerant to low BP, so that a level of hypotension which would normally be quite safe can result in ONH ischemia. Nocturnal hypotension under such circumstances may play an important role in the pathogenesis of ONH ischemic disorders. Recent ambulatory monitoring of the systemic BP over a 24-hour period has provided new and important information on the BP pattern in hypertensives, particularly in those on anti-hypertensive therapy. This has suggested that in patients with ONH ischemia the real culprit may be not hypertension but nocturnal hypotension. For example, Floras (1988) found that in hypertensives being treated by beta-blockers, during sleep a mean hourly diastolic BP of 50 mm Hg or less was recorded much more frequently - even a mean diastolic BP of 30 mm or less during 7 hours of sleep and diastolic BP between 30 and 40 mm Hg during 10 hours.

## Inter-individual Variation in the Pattern of Blood Supply of the Optic Nerve Head

In addition to the various factors discussed above, I feel this can play an important role in controlling the blood flow in the ONH and in determining the susceptibility, development, location and extent of ischemia of the ONH by the ischemic process, as shown by our studies on AION and glaucomatous optic neuropathy. The subject is discussed at length elsewhere (Hayreh, 1990; 1985; 1989) and following is a very brief summary of that.

Since the PCAs, via the peripapillary choroid or the short PCAs, are the main source of blood supply to the ONH (Hayreh & Perkins, 1968; Hayreh, 1969; 1970; 1974d; 1974e; 1975b; 1981; 1985; 1978; 1989) (Fig.1), it is but natural that variation in their pattern of distribution should influence the pattern of blood supply of the ONH. The following factors related to the PCA-circulation may produce the inter-individual variation:

a. *Variations in number of PCAs supplying an eyeball:* There may be one to five PCAs, though usually two or three, supplying the eyeball and ONH (Hayreh, 1962; 1990).

b. *Variation in the area of the ONH supplied by the various PCAs:* The ONH may be supplied by one, two, three, or rarely even more PCAs, which also vary in their contribution.

c. *Location of the watershed zone(s) between the various PCAs:* A "*watershed zone*" is the border between the areas of distribution of two adjacent end-arteries in a

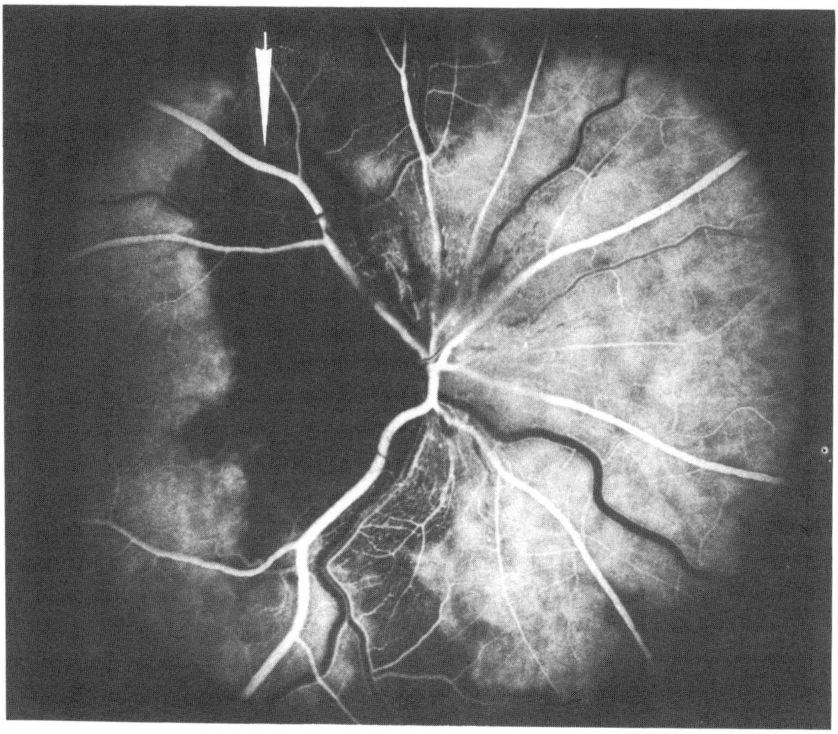

**Figure 2.** Fluorescein fundus angiogram, of right eye of a 60-year-old man with non-arteritic anterior ischemic optic neuropathy, shows non-filling of the watershed zone (indicated by arrows) between the lateral and medial PCAs. (Reproduced from Hayreh, 1985)

tissue. Watershed zones between the various cerebral arteries are a well-known example. The importance of the watershed zone is that in the event of a fall in perfusion pressure in the vascular bed of one or more end-arteries, the watershed zone, being the area of comparatively poorest vascularity, is most vulnerable to ischemia. Our *in vivo* studies (Hayreh & Baines, 1972a; Hayreh, 1974b; 1974c; 1974d; 1974e 1975a; 1975b; 1981; 1985; 1990; Hayreh & Chopdar, 1982) have shown conclusively that the PCAs are end-arteries and that there are watershed zones between them (Fig. 2).

Fig. 3 is a diagrammatic representation of some of the locations of the watershed zone when an eye has two (medial and lateral) PCAs. The watershed zone may be located:

i. temporal to the peripapillary choroid (Fig. 3-B),
ii. passing through the temporal peripapillary choroid (Fig. 3-C),
iii. passing through one or the other part of the optic disc, or the entire optic disc may lie in the watershed zone (Fig. 3-C,D,E),
iv. passing through the nasal peripapillary choroid (Fig. 3-F), or
v. various combinations of the above.

With 3 PCAs, the watershed zone frequently assumes the shape of the letter 'Y', passing through a part of the optic disc; or the entire optic disc may lie in the watershed zone. Figs. 4B-E show diagrammatically the various combinations of watershed zones which can occur when an eye has more than two PCAs. We have seen many of these (Hayreh, 1990;

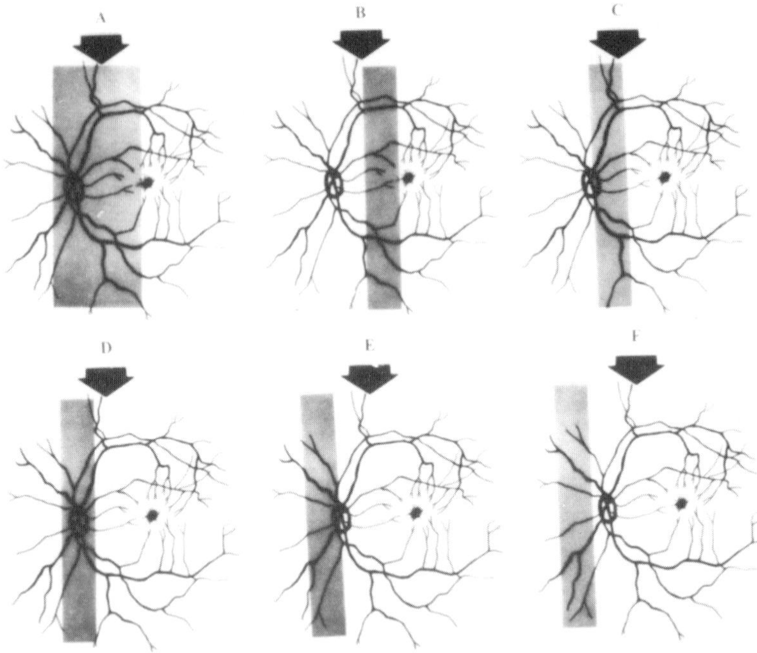


**Figure 3.** Diagrammatic representation of some of the locations of the watershed zone (shaded area) between the medial and lateral PCAs in human eyes. In (A), the shaded area represents the location where the watershed zone may be situated anywhere within this area. (B-F) are some examples of the variations in the location. (Reproduced from Hayreh, SS, 1988 In: Bernstein EF (ed): *Amaurosis Fugax*. New York: Springer-Verlag, pp.1-23)

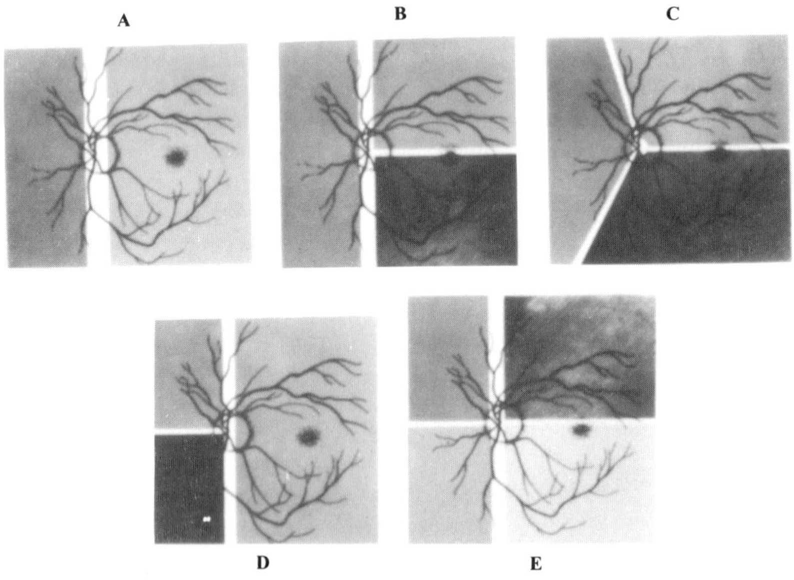


**Figure 4.** Diagrammatic representations of some examples of the locations of the borders between the territories of the various PCAs: (A) - with two PCAs: one medial and the other lateral; (B-D) - with three PCAs in different combinations: (B,C) have one medial and two lateral PCAs, and (D) one lateral and two medial; and (E) with four PCAs: two medial and two lateral. (Reproduced from Hayreh, 1989)

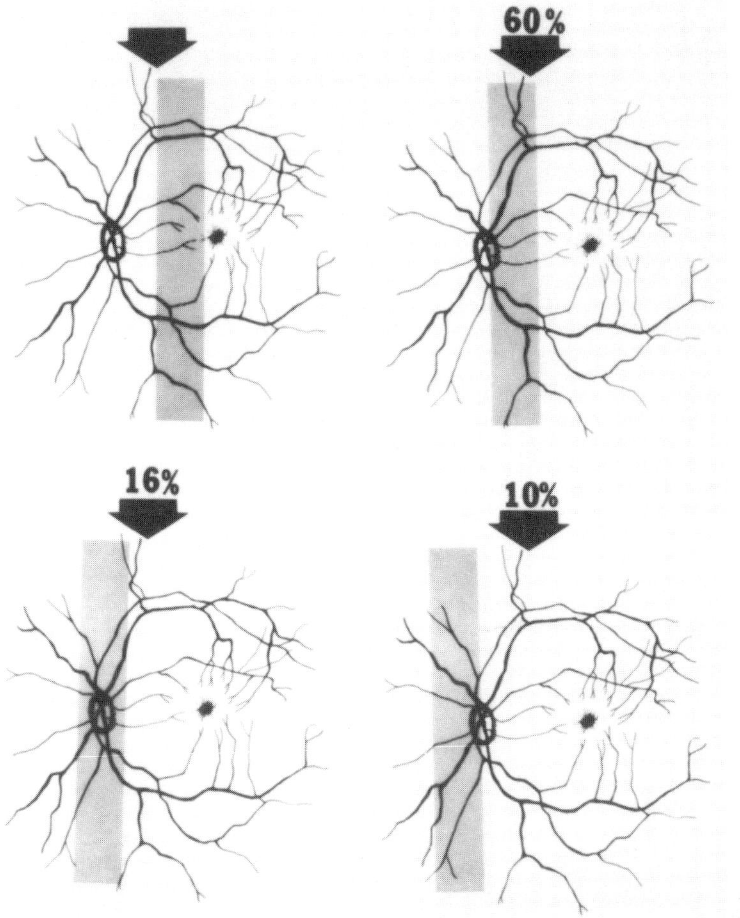


**Figure 5.** Diagrammatic representation of some of the locations of the watershed zone (shaded area) between medial and lateral PCAs and their incidence in human eyes with primary open angle glaucoma and low-tension glaucoma. (Reproduced from Hayreh, 1989).

1985; 1989). When there are 3 or more PCAs, the vertical watershed, instead of being one continuous watershed zone in its entire length (as is the case when there are two PCAs - Figs. 2,3 and 4A) becomes divided into two parts - upper and lower halves (Figs. 4B-E); not only that but horizontal or oblique watershed zones also appear (Fig. 4B-E). The practical implication of this is that watershed zone filling defects are not always present throughout the entire vertical length but may be present only in the upper or lower half.

From this brief account, it is evident that the PCA-watershed zones are usually located close to or in various regions of the optic disc. The exact location of the PCA-watershed zones in relation to the optic disc is extremely important because our studies indicate that the location of the watershed zone determines the vulnerability of the corresponding part of the optic disc to ischemia. The part of the optic disc that lies in the watershed zone is more vulnerable to ischemia than the part that does not. When the entire optic disc lies in the center of a watershed zone, the disc is most vulnerable to ischemia (Figs. 3-D). In our fluorescein fundus angiographic studies on patients with glaucoma, in the eyes where we could outline the watershed zone, we found the incidence of the various locations of the watershed zone as shown in Fig. 5: the most common location being the temporal part of the optic disc - in 60%; this may be the reason for the frequently noted nasal step in glaucoma, and why the temporal island of vision is the last to be lost.

Thus, the location of the PCA-watershed zone in relation to the optic disc is an important factor in determining the site and severity of ischemia of the ONH.

## CONCLUSIONS

It is clear from the above discussion that the blood flow in the ONH is controlled by a large number of factors. These factors, acting in different combinations, produce ONH ischemia. Thus, ONH ischemic disorders are generally multifactorial in nature - a fact not fully comprehended by the vast majority.

## ACKNOWLEDGMENTS

This research was supported by research grants from the National Institutes of Health (EY-1151, EY-1576 and RR-59), and unrestricted grants from Research to Prevent Blindness, Inc., New York, and from the Alcon Research Foundation. *Dr. S.S. Hayreh is a Research to Prevent Blindness Senior Scientific Investigator.*

## REFERENCES

Alderman, M.H., Ooi, W.L., Madhavan, S. and Cohen, H. (1988) Treatment-induced blood pressure reduction and the risk of myocardial infarction, *JAMA* 262:920-924.

Alm, A. and Bill A. (1972) The oxygen supply to the retina. II. Effects of high intraocular pressure and of increased arterial carbon dioxide tension on uveal and retinal blood flow in cats; a study with radioactively labeled microspheres including flow determinations in brain and some other tissues, *Acta Physiol. Scand.* 84:306-319.

Alm, A. and Bill, A. (1973) Ocular and optic nerve blood flow at normal and increased intraocular pressures in monkeys (Macaca irus): a study with radioactively labeled microspheres including flow determinations in brain and some other tissues, *Exp. Eye Res.* 15:15-29.

Bill, A. (1974) Effects of acetazolamide and carotid occlusion on the ocular blood flow in unanesthetized rabbits, *Invest. Ophthalmol. Vis. Sci.* 13:954-958.

Bill, A. (1985) Some aspects of the ocular circulation, *Invest. Ophthalmol. Vis. Sci.* 26:410-424.

Bill, A. and Linder, J. (1976) Sympathetic control of cerebral blood flow in acute arterial hypertension, *Acta Physiol. Scand.* 96:114-121.

Bill, A. and Sperber, G.O. (1987) Blood flow and glucose consumption in the optic nerve: Effects of high intraocular pressure. In:Krieglstein GK, editor. Glaucoma Update III. Berlin Heidelberg: Springer, pp 51-57.

Boisvert, D.J.P., Jones, J.V. and Harper, A.M. (1977) Cerebral blood flow autoregulation to acutely increasing blood pressure during sympathetic stimulation, *Acta Neurol. Scand.* 56 Suppl. 64:48-49.

Caplan, L.R. and Stein, RW. (1986) Stroke: A Clinical Approach. Boston: Butterworths, pp 11.

Connor, R.C. (1962) Complicated migraine: a study of permanent neurological and visual defects caused by migraine, *Lancet* 2:1072-1075.

Cove, D.H., Seddon, M., Fletcher, R.F. and Dukes. D.C. (1979) Blindness after treatment for malignant hypertension, *Br. Med. J.* 2:245-246.

Dollery, C.T., Hill, D.W. and Hodge, J.V. (1963) The response of normal retinal blood vessels to angiotensin and noradrenaline, *J Physiol.* 165:500-507.

Dorfman, L.J., Marshall, W.H. and Deiter, E.R.. (1979) Cerebral infarction and migraine: clinical and radiological correlations, *Neurology* 29:317-322.

Drance, S.M., Douglas, G.R., Wijsman, K., Chulzer, M. and Britton, R.J. (1988) Response of blood flow to warm and cold in normal and low-tension glaucoma patients, *Am. J. Ophthalmol.* 105:35-39.

Drayer, J.I.M., Weber, M.A. and Chard, E.R. (1984) Non-invasive automated blood pressure monitoring in ambulatory normotensive men. In:Weber, M.A.and Drayer, J.I.M., editors. Ambulatory Blood Pressure Monitoring. Darmstadt: Steinkopff Verlag, pp:129-135.

Dukes, H.T. and Vieth, H.G. (1964) Cerebral arteriography during migraine prodrome and headache, *Neurology* 14:636-639.

Edmeads, J. (1977) Cerebral blood flow in migraine, *Headache* 17:148-152.

Edvinsson, L., Owman, C. and Siesjo, B. (1976) Physiological role of cerebrovascular sympathetic nerves in the autoregulation of cerebral blood flow, *Brain Res.* 117:519-523.

Ernest, J.T. (1974) Autoregulation of optic-disk oxygen tension, *Invest. Ophthalmol. Vis. Sci.* 13:101-106.

Ernest, J.T.. (1976) Optic disc blood flow, *Trans. Ophthalmol. Soc. UK.* 96:348-351.

Ernest, J.T. (1977) Optic disk oxygen tension, *Exp. Eye Res.* 24:271-278.

Ernest, J.T. (1979) Autoregulation of blood flow in the distal segment of the optic nerve. In:Krieglstein, G.K., Leydhecker, W, eds. Glaucoma Update. Heidelberg: Springer, pp 93-100.

Fitch, W., MacKenzie, E.T. and Harper, A.M. (1975) Effects of decreasing arterial blood pressure on cerebral blood flow in the baboon; influence of the sympathetic nervous system, *Circ. Res.* 37:550-557.

Floras, J.S. (1988) Antihypertensive treatment, myocardial infraction, and nocturnal myocardial ischaemia, *Lancet* 2:994-996.

Floras, J. and Sleight, P. (1983) Ambulatory monitoring of blood pressure. In:Sleight, P. and Jones, J.V., eds. Scientific foundations of cardiology. London: Heinemann, pp155-164.

Geijer, C. and Bill, A. (1979) Effects of raised intraocular pressure on retinal, prelaminar, laminar, and retrolaminar optic nerve blood flow in monkeys, *Invest. Ophthalmol. Vis. Sci.* 18:1030-1042.

Graveson, G.S. (1949) Retinal arterial occlusion in migraine, *Br. Med. J.* 2:838-840.

Grunwald, J.E., Riva, C.E., Stone, R.A., Keates, E.U. and Petrig, B.L. (1984) Retinal autoregulation in open-angle glaucoma, *Ophthalmology* 91:1690-1694.

Hayreh, S.S. (1958) A study of the central artery of the retina in human beings. [Master of Surgery Thesis]. India: Panjab University.

Hayreh, S.S. (1962) The ophthalmic artery. III. Branches, *Br. J. Ophthalmol.* 46:212-247.

Hayreh, S.S. (1963a) The central artery of the retina - its role in the blood supply of the optic nerve, *Br. J. Ophthalmol.* 47:651-663.

Hayreh, S.S. (1963b) Blood supply and vascular disorders of the optic nerve, *Anal. Inst. Barraquer* 4:7-109.

Hayreh, S.S. (1964) The orbital vessels of rhesus monkeys, *Exp. Eye. Res.* 3:16-30.

Hayreh, S.S. (1969) Blood supply of the optic nerve head and its role in optic atrophy, glaucoma and oedema of the optic disc, *Br. J. Ophthalmol.* 53:721-748.

Hayreh, S.S. (1970) Pathogenesis of visual field defects - role of the ciliary circulation, *Br. J. Ophthalmol.* 54:289-311.

Hayreh, S.S. (1974a) The choriocapillaris, *Graefe's Arch. Clin. Exp. Ophthalmol.* 192:165-179.

Hayreh, S.S. (1974b) Submacular choroidal vascular pattern - experimental fluorescein fundus angiographic studies, *Graefe's Arch. Clin. Exp. Ophthalmol.* 192:181-196.

Hayreh, S.S. (1974c) The long posterior ciliary arteries - an experimental study, *Graefe's Arch. Clin. Exp. Ophthalmol.* 192:197-213.

Hayreh, S.S. (1974d) Anterior ischaemic optic neuropathy - I. Terminology and pathogenesis, *Br. J. Ophthalmol.* 58:955-963.

Hayreh, S.S. (1974e) Anterior ischaemic optic neuropathy - II. Fundus on ophthalmoscopy and fluorescein angiography, *Br. J. Ophthalmol.* 58:964-980.

Hayreh, S.S. (1975a) Segmental nature of the choroidal vasculature, *Br. J. Ophthalmol.* 59:631-648.

Hayreh, S.S. (1975b) Anterior Ischemic Optic Neuropathy. New York:Springer-Verlag.

Hayreh, S.S. (1978) Structure and blood supply of the optic nerve. In:Heilmann K, Richardson KT, editor. Glaucoma: Conceptions of a disease. Stuttgart:Thieme, pp78-96.

Hayreh, S.S. (1981) Anterior ischemic optic neuropathy, *Arch. Neurol.* 38:675-678.

Hayreh, S.S. (1985) Inter-individual variation in blood supply of the optic nerve head. Its importance in various ischemic disorders of the optic nerve head, and glaucoma, low-tension glaucoma and allied disorders, *Doc. Ophthalmol.* 59:217-246.

Hayreh, S.S. (1987) Factors determining the glaucomatous optic nerve head damage. In: Krieglstein GK, editor. Glaucoma Update-III. Heidelberg:Springer-Verlag, pp 40-46.

Hayreh, S.S. (1989) Blood supply of the optic nerve head in health and disease. In:Lambrou GN, Greve EL, editors. Ocular blood flow in glaucoma: Means, methods and measurements. Amstelveen:Kugler & Ghedini, pp 3-54.

Hayreh, S.S. (1990) In vivo choroidal circulation and its watershed zones, *Eye* 4:273-289.

Hayreh, S.S. and Baines, J.A.B. (1972a) Occlusion of the posterior ciliary artery I. Effects on choroidal circulation, *Br. J. Ophthalmol.* 56:719-735.

Hayreh, S.S and Baines, J.A.B. (1972b) Occlusion of the posterior ciliary artery III. Effects on the optic nerve head, *Br. J. Ophthalmol.* 56:754-764.

Hayreh, S.S and Baines, J.A.B. (1973) Occlusion of the vortex veins, *Br. J. Ophthalmol.* 57:217-238.

Hayreh, S.S., Bill, A. and Sperber, G.O. (1991) Metabolic effects of high intraocular pressure in old arteriosclerotic monkeys, *Invest. Ophthalmol .Vis. Sci.* 32:810.

Hayreh, S.S and Chopdar, A. (1982) Occlusion of the posterior ciliary artery - V. Protective influence of simultaneous vortex vein occlusion, *Arch. Ophthalmol.* 100:1481-1491.

Hayreh, S.S and Perkins, E.S. (1968) Clinical and experimental studies on the circulation at the optic nerve head. In:Cant JS, editor. The William MacKenzie Centenary Symposium on the Ocular Circulation in Health and Disease. Kimpton: London, pp 71-86.

Hayreh, S.S. and Perkins, E.S. (1971) The effects of raised intraocular pressure on the blood vessels of the retina and optic disc. In:Amalric P, editor. Proceedings of the International Symposium on Fluorescein Angiography; 1969 Albi. Karger:Basel, pp 323-328.

Hayreh, S.S., Revie, I.H.S. and Edwards, J. (1970) Vasogenic origin of visual field defects and optic nerve changes in glaucoma, *Br. J. Ophthalmol.* 54:461-472.

Hayreh, S.S., Servais, G.E. and Virdi, P.S. (1986a) Fundus lesions in malignant hypertension - V. Hypertensive optic neuropathy, *Ophthalmology* 93:74-87.

Hayreh, S.S., Servais, G.E. and Virdi, P.S. (1986b) Fundus lesions in malignant hypertension - VI. Hypertensive choroidopathy, *Ophthalmology* 93:1383-1400.

Hayreh, S.S. and Walker, W.M. (1967) Fluorescent fundus photography in glaucoma, *Am. J. Ophthalmol.* 63:982-989.

Hayreh, S.S. and Weingeist, T.A. (1980) Experimental occlusion of the central artery of the retina - I. Ophthalmoscopic and fluorescein fundus angiographic studies, *Br. J. Ophthalmol.* 64:896-912.

Hayreh, S.S., Zimmerman, M.B., Podhajsky, P. and Alward, W.L.(1994) Nocturnal arterial hypotension and its role in optic nerve head and ocular ischemic disorders, *Am. J. Ophthalmol.* 117:603-624.

Hulse, J.A., Taylor, D.S.I. and Dillion, M.J. (1979) Blindness and paraplegia in severe childhood hypertension, *Lancet* 2:553-556.

Jones, J.V., Fitch, W., MacKenzie, E.T., Strandgaard, S. and Harper, A.M. (1976) Lower limit of cerebral blood flow autoregulation in experimental renovascular hypertension in the baboon, *Circ. Res.* 39:555-557.

Kennedy, H.L., Horan, M.J., Sprague, M.K., Padgett, N.E. and Shriver, K.K. (1983) Ambulatory blood pressure in healthy normotensive males, *Am. Heart J.* 106:717-722.

Kishi, S., Tso, M.O.M. and Hayreh, S.S. (1985) Fundus lesions in malignant hypertension - II. A pathologic study of experimental hypertensive optic neuropathy, *Arch. Ophthalmol.* 103:1198-1206.

Kitazawa, Y., Shirai, H. and Go, F.J. (1989) The effect of Ca2+-antagonist on visual field in low-tension glaucoma, *Graefe's Arch. Clin. Exp. Ophthalmol.* 227:408-412.

Kleiman, N., Goodman, D., Schechtman, K. and Roberts, R. (1988) (Diltiazem Reinfarction Study.) Lack of diurnal variation in the occurrence of non-Q-wave myocardial infarction: results of a prospective study, J. Am. Coll. Cardiol. 11:27A.

Langham, M.E., Tomey, K.F. and Preziosi, T.J. (1981) Carotid occlusive disease: effect of complete occlusion of internal carotid artery on intraocular pulse/pressure relation and on ophthalmic arterial pressure, *Stroke* 12:759-765.

Langham, M.E. (1987) Ocular blood flow and visual loss in glaucomatous eyes. In:Krieglstein GK, editor. Glaucoma Update III. Heidelberg: Springer, pp 58-66.

Lauritzen, M., Olsen, T.S. and Paulson, O.B. (1982) Regional cerebral blood flow in classic migraine: a possible relationship to the spreading depression of Leao? In:Rose FC, editor. Advances in migraine research and therapy. New York: Raven Press, pp 105-115.

Lauritzen, M., Olsen, T.S., Larsen, N.A. and Paulson, O.B. (1983) Changes in the regional cerebral blood flow during the course of classic migraine attacks, *Ann. Neurol.* 13:633-641.

MacKenzie, E.T., McGerorge, A.P., Graham, D.I., Fitch, W., Edvinsson, L. and Harper, A.M. (1977) Break-through of cerebral autoregulation and the sympathetic nervous system, *Acta. Neurol. Scand.* 56 Suppl 64:48-49.

Marsh, E.E., Bille, J., Adams, H.P., Marler, J.R., Hulbert, J.R., Love, B.B., et al. (1990) Circadian variation in onset of acute ischemic stroke, *Arch. Neurol.* 47:1178-1180.

McDonald, W.I. and Sanders, M.D. (1971) Migraine complicated by ischaemic papillopathy, *Lancet* 2:521-523.

Novack, R.L., Stefansson, E., Hatchell, D.L. (1990) Intraocular pressure effects on optic nerve-head oxidative metabolism measured in vivo, *Graefe's Arch. Clin. Exp. Ophthalmol.* 228:128-133.

Olver, J.M., Splaton, D.J. and McCartney, A.C.E. (1990) Microvascular study of the retrolaminar optic nerve in man: the possible significance in anterior ischaemic optic neuropathy, *Eye* 4:7-24.

Pasteur, V.R., Blamoutier, P., Mawas, L. and Hamburger, J. (1937) Acces de migraine ophtalmique suivis d'une hemorragie retinienne, *Ann. Med.* 42:132-137.

Pearce, J.M.S. and Foster, J.B. (1965) An investigation of complicated migraine, *Neurology* 15:333-340.

Phelps, C.D. and Corbett, J.S. (1985) Migraine and low-tension glaucoma: A case control study, *Invest. Ophthalmol. Vis. Sci.* 26:1105-1108.

Pillunat, L.E., Stodtmeister, R,. Wilmanns, I. and Christ, Th. (1985a) Autoregulation of ocular blood flow during changes in intraocular pressure - Preliminary results, *Graefe's Arch. Clin. Exp. Ophthalmol.* 223:219-223.

Pillunat, L.E., Stodtmeister, R., Wilmanns, I. and Christ, Th. (1985b) New aspects in pressure tolerance of the optic nerve head, *Invest. Ophthalmol. Vis. Sci.* 26 ARVO suppl 223.

Pillunat, L.E., Stodtmeister, R., Wilmanns, I. and Metzner, D. (1986a) Effect of timolol on optic nerve head autoregulation, *Ophthalmologica* 193:146-153.

Pillunat, L.E., Stodtmeister, R., Wilmanns, I. and Christ ,Th. (1986b) Drucktoleranztest des sehnervenkopfes bei okulaerer hypertension, *Klin. Monatsbl. Augenheilkd.* 186:39-44.

Pillunat, L.E., Stodtmeister, R. and Wilmanns, I. (1987) Pressure compliance of the optic nerve head in low tension glaucoma, *Br. J. Ophthalmol.* 71:181-187.

Porsaa, K. (1941) Experimental studies on the vasomotor innervation of the retinal arteries, *Acta Ophthalmol.* Suppl. 18.

Pryor, J.S., Davies,P.D. and Hamilton, D.V. (1979) Blindness and malignant hypertension, *Lancet* 2:803.

Robert, Y. and Maurer, W. (1984) Pallor of the optic disc in glaucoma patients with artificial hypertension, *Doc. Ophthalmol.* 57:203-214.

Robert, Y., Grauwiller, T., Hendrickson, P.H. and Brunner, H.R. (1988) Die autoregulation der papillengefaesse under ihre verhalten unter halothan, *Klin. Monatsbl. Augenheilkd.* 192:117-121.

Robert, Y., Steiner. D. and Hendrickson, P. (1989) Papillary circulation dynamics in glaucoma, *Graefe's Arch. Clin. Exp. Ophthalmol.* 227:436-439.

Russell, R.W.R. (1963) Observations on intracerebral aneurysms, *Brain* 86:425-442.

Singh, S. and Dass, R. (1960a) The central artery of the retina I. Origin and course, *Br. J. Ophthalmol.* 44:193-212.

Singh, S. and Dass, R. (1960b) The central artery of the retina II. Distribution and anastomoses, *Br. J. Ophthalmol.* 44:280-299.

Sossi, N. and Anderson, D.R.. (1983) Effect of elevated intraocular pressure on blood flow: occurrence in cat optic nerve head studies with iodantipyrine I 125, *Arch. Ophthalmol.* 101:98-101.

Sperber, G.O. and Bill, A. (1985) Blood flow and glucose consumption in optic nerve, retina and brain: effects of high intraocular pressure, *Exp. Eye Res.* 41:639-653.

Strandgaard, S. (1976) Autoregulation of cerebral blood flow in hypertensive patients: the modifying influence of prolonged antihypertensive treatment on the tolerance to acute, drug-induced hypotension, *Circulation* 53:720-727.

Strandgaard, S., Jones, J.V., MacKenzie, E.T. and Harper, A.M. (1975) Upper limit of cerebral blood flow autoregulation in experimental renovascular hypertension in the baboon, *Circ. Res.* 37:164-167.

Strandgaard, S., Olesen, J., Skinhoj, E. and Lassen, N.A. (1973) Autoregulation of brain circulation in severe arterial hypertension, *Br. Med. J.* 1:507-510.

Taylor, D., Ramsay, J., Day, S., Dillion, M. (1981) Infarction of the optic nerve head in children with accelerated hypertension, *Br. J. Ophthalmol.* 65:153-160.

Ujiie, K. and Takahashi, M. (1987) Angioarchitecture of anterior segment of the optic nerve. In:Tsuchiya M, Asano M, Mishima Y, Oda M, editors. Microcirculation; - An Update. Proceedings of the Fourth World Congress for Microcirculation, Tokyo, *Amsterdam: Excerpta. Medica.*, 2:85-87.

Wallace, J.M., Thornton, W.E., Kennedy, H.L., Pickering, T.G., Harshfield, G.A., Frohlich, E.D., et al. (1984) Ambulatory blood pressure in 199 normal subjects, a collaborative study. In:Weber MA, Drayer JIM, editors. Ambulatory Blood Pressure Monitoring. Darmstadt: Steinkopff Verlag, pp 117-127.

Weinstein, J.M., Duckrow, R.B., Beard, D. and Brennan, RW. (1983) Regional optic nerve blood flow and its autoregulation, *Invest. Ophthalmol. Vis. Sci.* 24:1559-1565.

Weinstein, J.M. and Feman, S.S. (1982) Ischemic optic neuropathy in migraine, *Arch. Ophthalmol.* 100:1097-1100.

Wetherill, J.H. (1979) Blindness after treatment for malignant hypertension, *Br. Med. J.* 2:550.

# SECTION II

## Retinal Structure, Function and Disease

3

# THE ARCHITECTURE OF FUNCTIONAL NEURAL CIRCUITS IN THE CAT RETINA

Helga Kolb

University of Utah
Departments of Ophthalmology and Physiology and John A. Moran Eye Centre
University of Utah Health Sciences Center
50 North Medical Drive, Salt Lake City, Utah 84132

The cat retina has been a model retina for anatomical, physiological and pharmacological studies to elucidate its functional organization for at least a quarter of a century now. Over most of this time Hisako Ikeda has been at the forefront of research on ganglion cell function in this retina in both normal, abnormal and developmental conditions. Thus, it is fitting that in this paper which is to be amongst contributions from colleagues and friends of Hisako Ikeda I describe my research on the anatomical organization of the cat retina. This research fits very well with her physiological findings and represents our mutual interests in understanding how cat ganglion cell receptive fields are organized.

The retina has been called an "accessible part of the brain" (see Dowling, 1987, for a review) and is indeed a beautiful layered structure of sensory nerve cells and precisely organized pathways of neurons, functioning to receive and partially process messages concerning the visual world. All vertebrate retinas are similarly organized into three layers of neurons and two areas of synaptic neuropil where interactions occur between the neurons of the layers. This is best seen in a radial section of the cat retina (Fig. 1). Light rays have to pass through the whole thickness of the retina before striking the photoreceptors (Fig. 1) and following the translation of the photic stimulation into electrical stimulation, the information flows as electrical signals through a cascade of neurons to have output finally as the ganglion cell receptive field and impulse behaviour. The first area of neuropil is the outer plexiform layer (OPL) (Fig. 1) where connections between photoreceptors, and vertically running bipolar cells and horizontally oriented horizontal cells occur. The second neuropil, the inner plexiform layer (IPL), is more complex judging by its thickness, and functions to interconnect the vertical-information-carrying nerve cells, the bipolar cells, and ganglion cells with laterally running amacrine cells (Fig. 1).

The great spanish anatomist Ramon y Cajal (1892) gave us the first comprehensive description of the different nerve cells that make up the neural layers of the retina. Using the Golgi silver staining technique he described in beautiful drawings many of the bipolar, horizontal, amacrine and ganglion cell types in the mammalian retina. Although he did not stain cat retina per se his drawings of the cells of the ox and dog retinas were considered to

*Basic and Clinical Perspectives in Vision Research*, Edited by Jon Robbins et al.
Plenum Press, New York, 1995

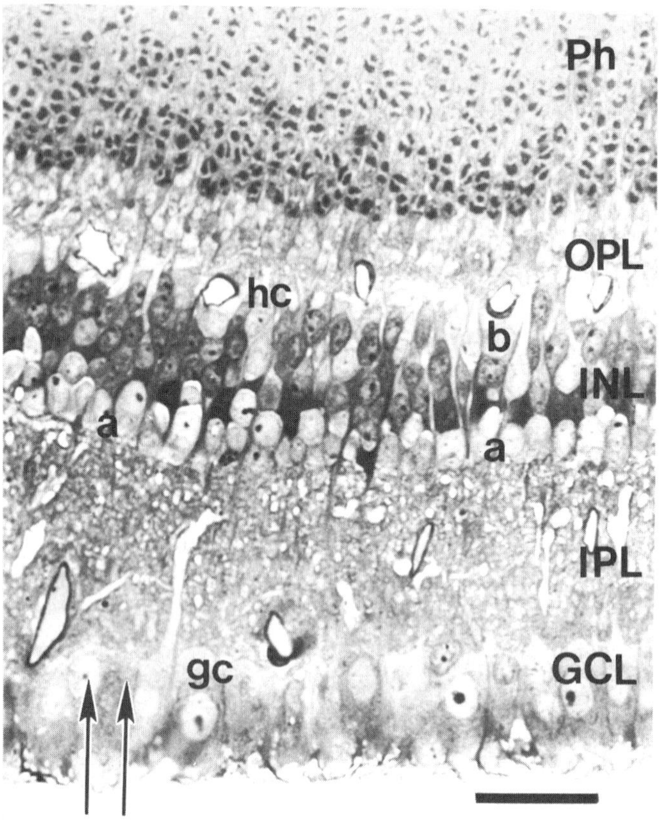


**Figure 1.** Light micrograph of a 1 μm thick radial section through central cat retina. Light (Large arrows) passes from the surface through to the photoreceptors at the very back of the retina. Three layers, photoreceptor layer (Ph), inner nuclear layer (INL) and ganglion cell layer (GCL) contain the neurons, while the two plexiform layers (OPL) and (IPL) are areas of synaptic neuropil. hc, horizontal cells; b, bipolar cells; a, amacrine cells; gc, ganglion cells. Scale bar 25 μm.

be archetypical of those of the cat retina. In the early 1970's I worked in collaboration with Boycott (Boycott & Kolb, 1973) to stain the cat retina with Golgi techniques and later with Famiglietti, Mariani and Nelson we continued further with modifications of these techniques (Famiglietti & Kolb, 1976; Famiglietti & Kolb, 1975; Kolb & Famiglietti 1974; Kolb & Famiglietti 1976; Kolb, Nelson & Mariani; 1981). The development of wholemount Golgi techniques, in several laboratories (Boycott & Kolb, 1973a; Famiglietti, 1983; Kolb, Nelson, & Mariani, 1981; Mariani, 1981; Ogden, 1974; Stell & Witkovsky, 1973), was particularly advantagous for it allowed us to see the complete dendritic tree and intricacies of dendritic morphology of the whole cell in the retina. Some of the cell types seen by Golgi impregnation in wholemounted cat retina are seen in figure 2. It is clear that some cells are much smaller and bushier than others, e.g. cells A4 and β ganglion cell, compared with A13, while others are wider-field and radiate in branching pattern, e.g. A19. The branching characteristics define the different Golgi-stained examples into different classes of cells. In particular the stratification level of their dendrites and axons are important functional classification criteria of nerve cells for it puts their processes into definite sublayers of the neuropils to ensure correct interactions.

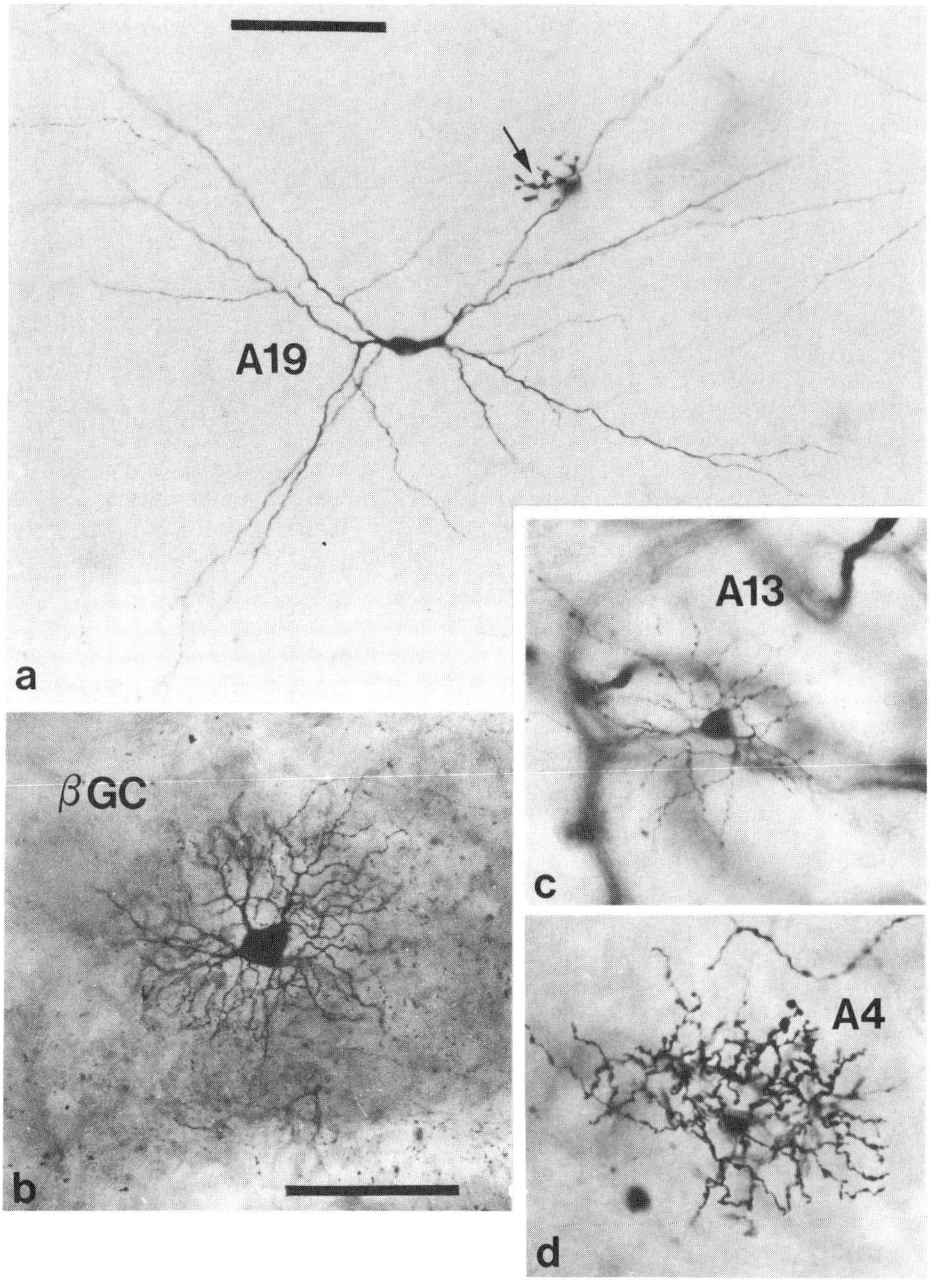


**Figure 2.** Light micrographs of Golgi-impregnated neurons as seen in wholemounts of the cat retina. a) A19 amacrine cell has a large-field and radiate branching pattern. A bipolar cell axon terminal (arrow) branches at the same plane of focus as the dendrites of A19. b) A small beta ganglion cell (βGC) with a complex, tufted branching pattern. c) A13 has a small field, diffuse branching pattern with thin dendrites bearing distinct beads at intervals. d) A4 is a small field cell with a complex, multibranched, appendage-ladened dendritic tree. Scale bars 100 μm.

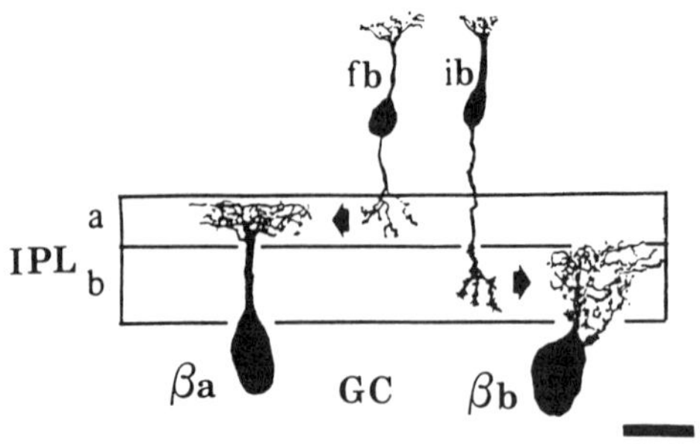


**Figure 3.** Camera lucida drawings of Golgi-impregnated neurons as seen in vertical views of the cat retina. Cone bipolar (fb) has an axon restricted to sublamina a of the IPL and interacts with βa-type ganglion cells (GC). Invaginating cone bipolar (ib) has an axon ending in sublamina b of the IPL and contacts only βb-type ganglion cells (GC). Scale bar 25 μm.

We first appreciated this after an anatomical study of the primate retina (Kolb, 1970) where two types of midget bipolar cells were discovered. The distinction of the two types from each other was made on the ultrastructural characteristics of their contacts with the cone pedicles in the OPL and on their axonal endings at different levels in the IPL. This placed these two types of midget bipolar cells into neuropil of the IPL to interact with two correspondingly different levels of dendritic branching of midget ganglion cells. Thus, with Famiglietti (Famiglietti & Kolb, 1976, Kolb & Famiglietti, 1976) I showed that cat X-type ganglion cells, now known as beta cell types (Boycott & Wässle, 1974), come as paramorphic pairs like the two types of midget ganglion cell in primate retina. One variety (βa type, Fig. 3) branches only in distal IPL (area labeled a) while the other (βb type, Fig. 3) branches only in proximal IPL, (area labeled b) thereby allowing only synaptic interactions between the beta a type and the fb bipolar and between beta b type and the ib bipolar (Fig. 3).

Light microscopy of Golgi-impregnated cat retina allowed us many insights into possible synaptic interactions between neurons in particular pathways but did not unequivocally confirm synapses. Thus, we turned to electron microscopy.

Electron microscopy (EM) enables us to differentiate between many of the neural elements of the retina on cytological criteria. Thus we have been able to distinguish rod photoreceptors from cone photoreceptors by differences in outer segment morphology and differences in structure of their synaptic endings in the OPL. Rod spherules are small compact synaptic endings containing a single synaptic ribbon and a single complex of postsynaptic profiles (Fig. 4b), while cone pedicles have much larger synaptic endings, that contain many synaptic ribbons and invaginated profiles of second order neurons (Fig. 4a and c shows one of the many complexes of synapses per cone pedicle) (Dowling & Boycott, 1966, Kolb, 1970, Missotten, 1965, Rodieck, 1973).

In the IPL, where bipolar cells contact ganglion and amacrine cells, we learnt (Dowling & Boycott, 1966, Kidd, 1962), that bipolar cells can be recognised by the presence of ribbons at their synapses, and amacrine cells by the presence of clusters of synaptic vesicles at dense membrane specializations (Fig. 4d, CB arrow, A open arrow, respectively). Studying of EM images in random sections gives us a good understanding of general recognition features of neurons in the retina but it is difficult to tell exactly which type of neuron we are dealing with. For example we know from the Golgi staining studies that cat retina contains about 70 different types of neurons each with a characteristic field size, stratification and role to play in neurocircuitry (Kolb, Nelson & Mariani, 1981). Thus we needed to turn to techniques that intracellularly stained individual neurons, like Golgi-EM studies (Boycott & Kolb, 1973, Boycott & Kolb, 1973a, Kolb, 1970, Kolb, 1974, Mariani, 1981, Mariani, 1983, Mariani, 1984, Stell, 1967, Stell and Lightfoot, 1975, Stell et al., 1975) or examination of HRP-iontophoresed and intracellularly recorded cell types (Djamgoz & Kolb, 1993, Lasansky, 1978).

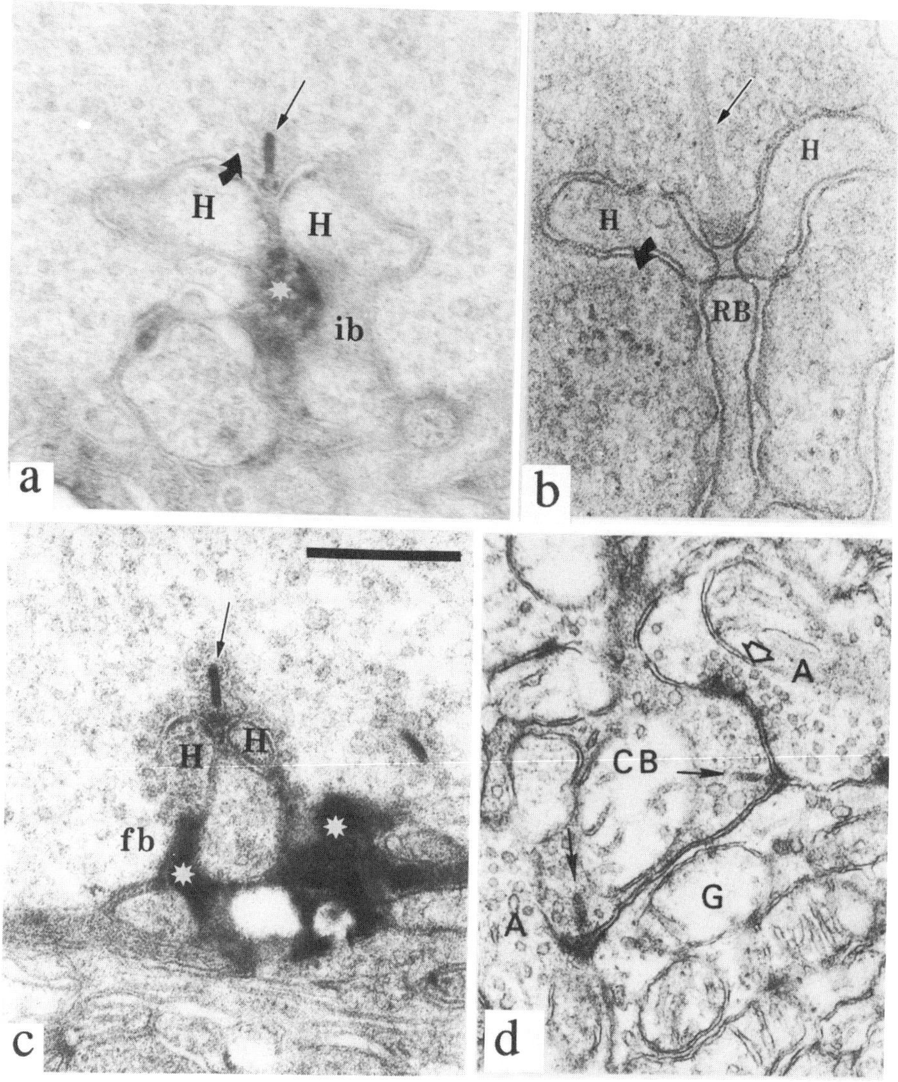


**Figure 4.** Electron micrographs of bipolar contacts in the cat retina. a) An HRP-stained cone bipolar (white star) makes an invaginating (ib) contact to end beneath the ribbon (ribbons arrowed) of the triad synapse in a cone pedicle. b) Rod bipolar (RB) dendrite also ends at an invaginating central position in the rod spherule triad. c) An HRP-stained flat cone bipolar dendrite (fb, white stars) ends at basal junctions on the surface of the cone pedicle away from the synaptic ribbon of the triad. H, is horizontal cell lateral element in a, b and c. d) In the IPL, bipolar cells (CB) are characterized by ribbon synapses to two postsynaptic profiles one of which is often a ganglion cell dendrite (G) and the other an amacrine cell dendrite (A). The amacrine profile frequently makes areciprocal synapse to the bipolar axon terminal (open arrow). Scale bar for all is 0.5 μm.

From these techniques we learned that rod bipolar cells contacted only rod photoreceptors (Fig. 4b) and that cone bipolar cells contacted only cone pedicles (Fig. 4a, c). Furthermore, we saw that the rod bipolar dendrite penetrates the rod spherule to end at the synaptic ribbon in a central invaginating position (Fig. 4b, RB) but that cone bipolar cell dendrites made synaptic contacts with cones in one of two ways. One variety called an

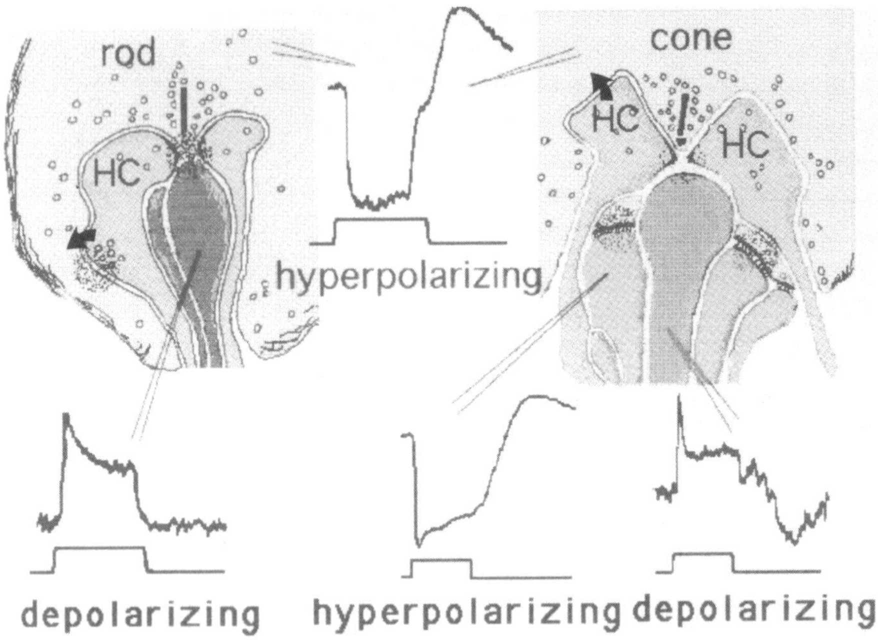


**Figure 5.** Summary diagram of the types of contacts bipolar cells make with rod spherules and cone pedicles with their associated intracellular response to light. Rods and cones give slow hyperpolarizations to light-ON (upward deflection on bar below the light response in all cases). Bipolar dendrites that invaginate the rod or cone to end at the ribbon of the triad give slow depolarizing responses to light. The bipolars that end at basal junctions give slow hyperpolarizing responses like the photoreceptor itself. Horizontal cell dendrites (HC) form lateral elements of the triads. They respond with a hyperpolarization like the photoreceptor (not shown) and feed back information concerning surround responses to the rod or cone as indicated at the curved arrows.

invaginating bipolar type (Fig. 4a), has its dendrites penetrating the pedicle to get close to the synaptic ribbons in the cone pedicle. The other variety called a flat midget bipolar (Fig. 4c, fb) has dendritic terminals ending at basal junctions on the surface of the cone pedicle away from the ribbon. Another important finding was that horizontal cells made synaptic contacts with rods and cones as lateral elements at the synaptic ribbons forming two of the three components of the triad synapse (Fig. 4a, b and c, H).

The arrangement of horizontal cell dendrites on either side of the ribbon synapse in photoreceptors, with bipolar cell dendrites forming central or basal junction position, is an important local circuit in retinal image processing. We know that the photoreceptor neurotransmitter (glutamate) is released in the dark in the vertebrate retina (Bortoff, 1964, Dowling, 1987, Massey, 1990) (for reviews). When a light stimulates the photoreceptor, transmitter release ceases and the photoreceptor responds with a slow hyperpolarizing response (Fig. 5), but the postsynaptic bipolar cells respond with either a sustained hyperpolarization or a sustained depolarization (Fig. 5) (Kaneko, 1970, Slaughter & Miller, 1981, Slaughter & Miller, 1983, Werblin & Dowling, 1969). The horizontal cells also responds, with a slow potential sustained hyperpolarization (Svaetichin & MacNichol, 1958, Tomita, 1965), but a feedback synapse from the horizontal cell to the photoreceptor (Baylor, Fuortes & O'Bryan, 1971) is thought to signal information from a summed network of horizontal cells in the OPL. Thus, this horizontal cell message influences the photoreceptor and the bipolar cell to include a response coming from a surround region of retina (Kaneko, 1970, Naka, 1976, Toyoda, 1972, Werblin and Dowling, 1969). This local circuit provides the bipolar cell with an antagonistic surround to its receptive field center i.e. a hyperpolarizing surround for a depolarizing bipolar cell and a depolarizing surround for a hyperpolarizing bipolar cell.

A similar local circuit at bipolar synapses to amacrine or ganglion cells is also possible at the next stage of retinal processing in the IPL. In figure 4d we see the manner in which rod and cone bipolar cells are presynaptic at ribbon synapses to two postsynaptic elements. Often the postsynaptic amacrine cell profile makes a feed-back (reciprocal) synapse to the bipolar (Fig. 4d, A, open arrow) in the same strategic position the horizontal cell dendrite did at the photoreceptor synapse. Some researchers interpret this to indicate a similar role in surround formation in the IPL (Kolb & Nelson, 1993, Marc, Liu & Muller, 1988).

Understanding such local circuits in the OPL or IPL are the result of EM serial section analyses. By making tracings of serial electron micrographs we can reconstruct portions of retinal cells or even the complete dendritic tree of a cell when it is small enough, to study all the synaptic input and output relationships. By this means over the years I and other colleagues have reconstructed rod and the various cone bipolar axons in the IPL, the small bistratified AII amacrine cell involved in the rod system (see later), A17, A8, A4 and A19 amacrine cells, and portions of alpha and beta ganglion cells in the cat retina (Cohen & Sterling, 1990a, Cohen & Sterling, 1990b, Cohen & Sterling, 1991, Famiglietti & Kolb, 1975, Kolb & Famiglietti, 1976, Kolb & Nelson, 1983, Kolb & Nelson, 1984, Kolb & Nelson, 1985, Kolb & Nelson, 1993, Kolb et al., 1993, McGuire, Stevens & Sterling, 1984, McGuire, Stevens & Sterling, 1986, Sterling, 1983).

Based on what we have learnt from using the anatomical and intracellular staining techniques in the cat retina that have been described above, we now understand much about the wiring and physiological responses of neurons processing the visual message at the retinal level and forming the receptive field organization of cat ganglion cells. In the following sections I shall summarize what we currently know concerning: 1) circuitry of rod pathways underlying night vision, 2) circuits mediating successive contrast otherwise known as ON and OFF pathways, and 3) circuits mediating simultaneous contrast or center/surround organization in the cat retina.

## CIRCUITS OF ROD PATHWAYS UNDERLYING NIGHT VISION

After our Golgi-EM studies of bipolar cell connections with photoreceptors in cat retina (Boycott & Kolb, 1973, Kolb & Nelson, 1983) it became clear that rod pathways through the retina were differently wired than cone pathways. Firstly, rod pathways involve large convergences and divergences of cell contacts. Many rods (30-45) converge information to a single rod bipolar cell. The rod bipolar carries signals to the next stage of processing in the IPL where divergence occurs to several different amacrine cell types, the most important of which are the AII and the A17 cells. Thus, secondly, rod bipolar axons do not contact ganglion cells with direct input synapses but use amacrine cells as intermediaries in the synaptic network (Famiglietti & Kolb, 1975, Kolb & Famiglietti, 1974). About 75,000 rods drive 5000 rod bipolars and 250 AII amacrine cells to get information to one alpha ganglion cell in cat retina (Freed & Sterling, 1988, Kolb & Nelson, 1993, Sterling, Freed & Smith, 1988). Smaller numbers of rods (2100), 140 rod bipolars and 7 AII amacrines converge upon the smaller ganglion cell, the beta cell.

Figure 6 summarizes the physiology and wiring of the rod pathways in the cat retina. Signals from hyperpolarizing rod photoreceptors become sign inverted in the rod bipolar cell. Thus the rod bipolar cells respond with depolarizing slow potentials to light (Fig. 6, top left) (Dacheux & Raviola, 1986). The rod bipolars contact two depolarizing amacrine cell types, the A17s (Fig. 6 upper left) and AIIs (Fig. 6 middle right) (Nelson, 1982, Nelson and

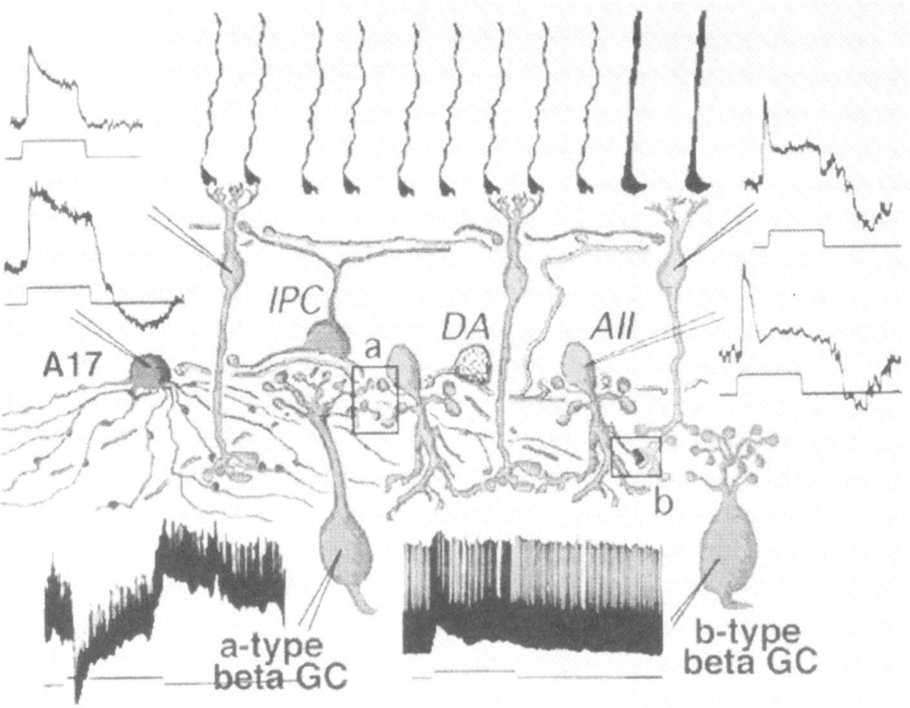


**Figure 6.** Summary diagram of the neurons and their responses involved in rod-driven pathways of night vision in the cat retina. See text for a full description.

Kolb, 1985). Rod signals reach one group of ganglion cell (Fig. 6, a-type beta GC) via chemical synaptic input from the AII amacrine cell lobular appendages (boxed area a, Fig. 6): other types of ganglion cells (Fig. 6, b-type beta GC) receive rod signals, via AII amacrine cell electrical synapses to cone bipolar cells (Famiglietti & Kolb, 1975, Kolb, 1979, Kolb & Famiglietti, 1974, Kolb & Nelson, 1993, Sterling et al., 1987, Sterling, Freed & Smith, 1988). The cone bipolars stimulate the relevent ganglion cell by excitatory chemical synapses (Kolb, 1979). Two other cell types are also influential upon the rod pathways. These are the dopaminergic amacrine (Fig. 6, DA) and the GABAergic interplexiform cell (Fig. 6, IPC) (Kolb et al., 1990, Kolb & West, 1977, Pourcho, 1982, Pourcho & Goebel, 1983). They may modulate and feed back and forth rod signals to further amplify low light sensitivities and reset levels of dark adaptation between the input and output of the system.

## CIRCUITS MEDIATING SUCCESSIVE CONTRAST (ON/OFF PATHWAYS)

Compared with the rod pathways, cone pathways are less convergent in the cat retina. A relatively small number of cones drive cone bipolars and a small number of these have input to a single ganglion cell. As few as 3 cone bipolar cells drive a central beta cell for example (Kolb, 1984, Wässle et al., 1989). Cone pathways are organized as two parallel streams of information directly from the cone photoreceptor to the ganglion cell without intermediary amacrine cells (Kolb & Famiglietti, 1976). This organization allows one channel to provide information to the ganglion cell concerning brighter than background stimuli (the ON-center channel) and the other darker than background stimuli (the OFF-center channel). This is the basis for successive contrast in visual perception.

The origin of these two important ON-center and OFF-center channels, is the type of synaptic contacts cone bipolar cell dendrites make with cone pedicles (Kolb, 1970). Thus, as summarized in figure 5, we found that one type of cone bipolar cell making invaginating contacts responds with an inverted sign of response from the photoreceptor's i.e. as a depolarizing slow potential to light. On the other hand, the type of bipolar cell making basal or flat contacts responds to the same stimulus with a hyperpolarizing slow potential just like the photoreceptor's. The hyperpolarizing types of contacts are the start of OFF-center responses in flat bipolar cells (Fig. 7, fbc, right hand response) and the depolarizing types of contact are the start of ON-center channels in invaginating bipolar cells (Fig. 7, ibc, left hand response). In other words the photoresponse is split into ON and OFF center types at the photoreceptor/cone bipolar contacts in the OPL in the cat, and in all other vertebrate retinas as far as we know at present (Famiglietti & Kolb, 1976, Kolb & Famiglietti, 1976, Nelson & Kolb, 1983, Raviola & Gilula, 1975, Slaughter & Miller, 1981, Slaughter & Miller, 1983).

When the two different cone bipolar cells come to make connections with ganglion cells in the IPL (Fig. 7) probably only straight transmission of the bipolar response occurs without any sign inversion. Therefore, the signal transmitted by the ganglion cell is determined by the type of cone bipolar contacting it. To keep the ON and OFF channels separate through the ganglion cells to the brain, the IPL is arranged into 2 discrete sublaminae, called a and b (see Fig. 3). This ensures that interactions are only allowed between basal-contacting cone bipolars (Fig. 7, fbc) and one set of ganglion cells in sublamina a (Fig. 7, a-type GC), while invaginating-contacting cone bipolar cells (Fig. 7, ibc) can only interact with another set of ganglion cells branching in sublamina b (Fig. 7, b-type GC).

Peter Gouras was the first to suggest that this specificity of bipolar to ganglion cell contacts underlay ON-center and OFF-center midget ganglion cell responses in monkey in 1971 (Gouras, 1971). Then, Ralph Nelson conclusively proved this hypothesis by means of intracellular recording and marking experiments in ganglion cells of cat in 1975, and later reported it in 1978 (Nelson, Famiglietti, and Kolb, 1978). In summary then we can see in figure 7 how information flows in the cone pathways to produce ON-center and OFF-center receptive fields of cat ganglion cells. Cones hyperpolarize to light but two channels, one a depolarizing bipolar and the other a hyperpolarizing bipolar, split the original cone signal into lightness or ON-center and darkness or OFF-center. These sign channels are transmitted directly to ganglion cells architecturally separated to different neuropils of the IPL, producing ganglion cells that are ON-center when they branch in sublamina b, and OFF-center when they branch in sublamina a (Fig. 7).

## CIRCUITS FOR SIMULTANEOUS CONTRAST (CENTER-SURROUND ORGANIZATION)

Information concerning the overall brightness or darkness of the image is of primary importance for visual sensation, however, putting these two informations in simultaneous contrast to each other greatly improves the resolution of the image. In the retina, simultaneous contrast is formed by creating a center-surround organization of a neurons physiological receptive field. The higher acuity or sustained X ganglion cell pathways in the central area of the cat retina are particularly precisely tuned with a center-surround organization (Enroth-Cugell & Robson, 1966, Ikeda & Wright, 1972, Shapley & Perry, 1986).

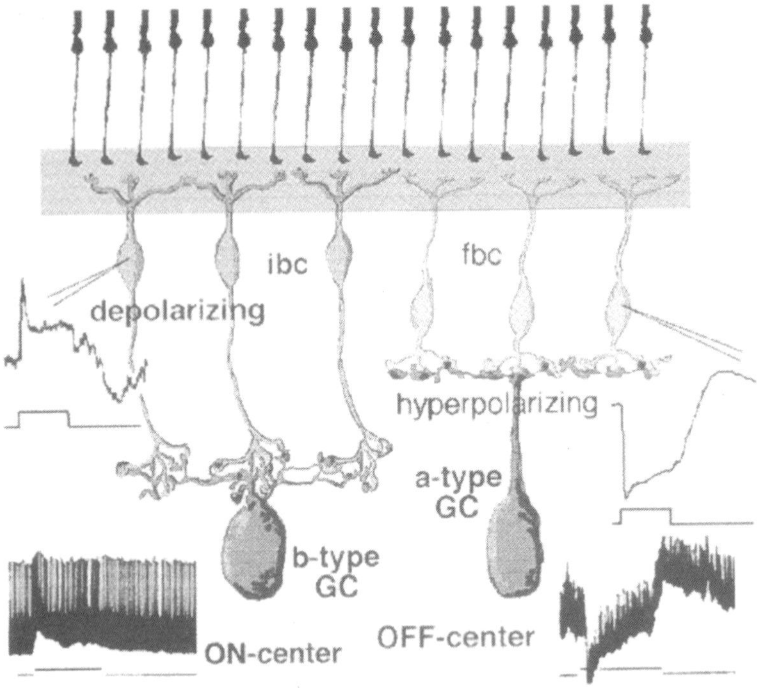


**Figure 7.** Summary diagram of the neurons and their physiological responses to light for the diffuse cone pathways responsible for successive contrast (ON and OFF) pathways in the cat retina. See text for a full description.

It is thought that horizontal cells at the OPL provide, through a mechanism of lateral inhibition, a surround arranged around the receptive field center of the bipolar cell response (Kaneko, 1970, Werblin & Dowling, 1969). The wiring responsible appears to start at the small local circuit at the rod or cone triads illustrated in figures 4 and 5. The architecture for center pathways is created by the cone to bipolar to ganglion cell through-pathways as was discussed above (Fig. 7), while the injection of horizontal cell information from a wider spatial area, than the small group of cones that have direct input to a single cone bipolar, provides a surround to the center: an OFF-surround for the ON-center channel and an ON-surround for the OFF-center channel. In cat retina, surround responses are not as strong in bipolar cells (Nelson & Kolb, 1983) as in some other cold-blooded vertebrates (Kaneko, 1970, Werblin & Dowling, 1969). However, surround responses are very strong in cat ganglion cells and so we have recently suggested that an additional surround is constructed by certain amacrine cells in the IPL in this species (Kolb & Nelson, 1993).

## SUMMARY AND CONCLUSION

The culmination of the anatomical and physiological studies in cat retina, only the broadest of which could be discussed in this paper, can be summarized in the final figure (Fig. 9). A schematized wiring diagram of central cat retina as seen in radial section is depicted. At night, dim light information flows from rods (r) to rod bipolars (rb) to several amacrines, chief amongst them the little bistratified AIIs (AII) and the wide-field A17s (A17).

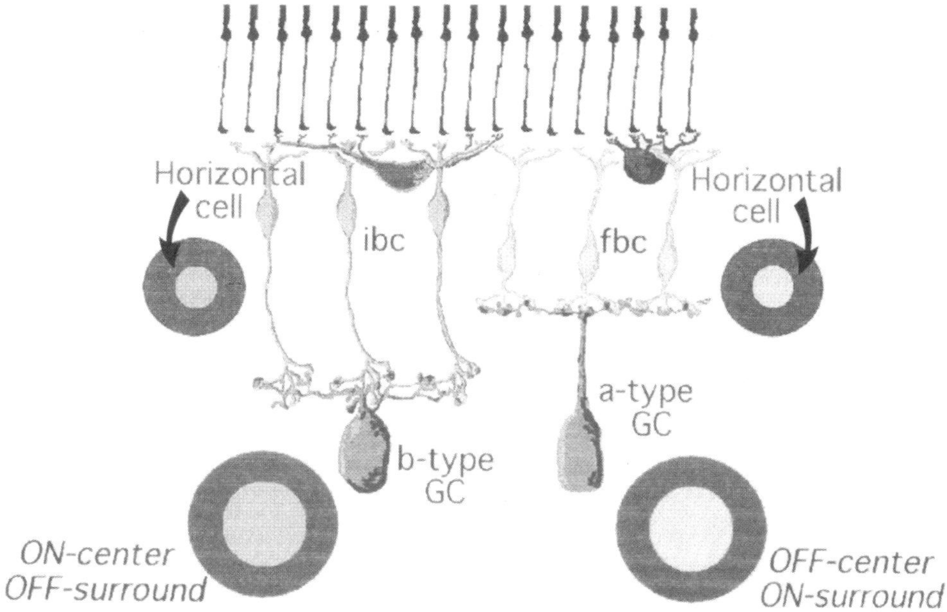


**Figure 8.** Summary diagram of the neurons involved in simultaneous contrast or center-surround organization of the receptive field of ganglion cells in the cat retina. The center response is formed by the cones to bipolar to ganglion cell channels (central spots in receptive field maps) and the concentrically organized antagonistic surround (curved arrows, rings around the central spots of the receptive field maps) is formed by horizontal cells. See text for a full description.

OFF-center alpha and beta ganglion cells get rod-driven input by direct synapses from AII lobular appendages restricted to sublamina a of the IPL. In contrast, ON-center alpha and beta ganglion cells get rod-driven messages in sublamina b through indirect gap-junction mediated ib cone bipolar inputs (ib, orange cell). An interplexiform cell (I, black/white cell) is probably involved in retrograde loops of information transfer between the two synaptic layers of the cat retina and may in some way help adapt the different circuits to different light conditions.

In brighter cone-driven light, diffuse cone pathways consist of cones contacting two parallel channels of bipolar cells (ib and fb) to ON-center and OFF-center ganglion cells. ON- and OFF-center responses of the ganglion cells are determined by specific cone bipolar contacts with cone pedicles, and restriction of synaptic interactions between bipolars and ganglion cells to the correct sublamina of the IPL. Bipolar and ganglion cell synapses in sublamina a of the IPL result in ganglion cells having OFF-center receptive fields. Bipolar and ganglion cell interactions in sublamina b result in ganglion cells with ON-center receptive fields. Spatially opponent surrounds to bipolar cells are added at the OPL level by horizontal cells (Fig. 9, HC) and to the ganglion cells at the IPL level by amacrine cell circuits, possibly involving A6 (A6), A8 (A8) and A13 (A13).

Working on anatomical investigations of the many neurons of the cat retina is particularly satisfying because we regard the knowledge we gain concerning their connections in central retina to be directly applicable to our understanding of extrafoveal human retina. Knowing how specific these wiring patterns are for functional pathways in the adult cat allows us to understand what might be going on in the human retina if some kind of developmental problem, such as strabismus or amblyopia is not corrected early on before

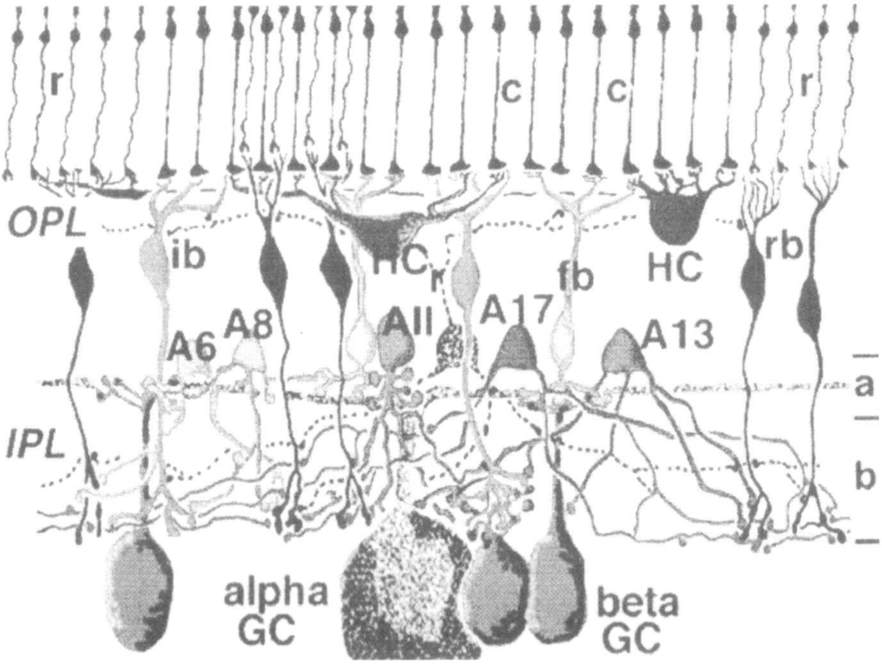


**Figure 9.** Summary diagram of a schematic cat retina with the neurons of the three major pathways described in this paper shown. i.e 1) night vision (rod) pathways, 2) successive contrast (ON and OFF) pathways, and 3) simultaneous contrast (center-surround) pathways. See text for a full description.

the pathways are hard-wired. This is an area that Hisako Ikeda has been so active in, and I think it is clear that our anatomical findings bear great relevance to this area.

## ACKNOWLEDGMENTS

This work has been supported over the years by NEI grants EY03323, EY04845 and grants from the Fight for Sight organization and Research to Prevent Blindness. Peter Ahnelt, University of Vienna, Austria, Patricia Goede, Ophthalmology Department, John A. Moran Eye Center, Utah and Robert McDermott, Supercomputing Institute, Utah gave unstintingly of time and effort to the creation of the videotape presentation from which this paper was generated. The videotape is available from the author.

## REFERENCES

Baylor, D. A., M. G. F. Fuortes, and P. M. O'Bryan (1971) Receptive fields of the cones in the retina of the turtle, *J. Physiol., Lond.* 214:265-294.

Bortoff, A. (1964) Localization of slow potential responses in the Necturus retina, *Vision Res.* 4:627-635.

Boycott, B. B . and Kolb, H. (1973a) The horizontal cells of the rhesus monkey retina, *J. Comp. Neurol.* 148:115-140.

Boycott, B. B . and Kolb, H. (1973) The connection between the bipolar cells and photoreceptors in the retina of the domestic cat, *J. Comp. Neurol.* 148:91-114.

Boycott, B. B . and Wässle, H. (1974) The morphological types of ganglion cells of the domestic cat's retina, *J. Physiol.* 240:397-419.

Cajal, S. R. (1892) The Structure of the Retina. Translated by S.A. Thorpe and M. Glickstein (Springfield, Il., Thomas),1972.

Cohen, E . and Sterling, P. (1990a) Demonstration of cell types among cone bipolar neurons of cat retina, *Phil. Trans. R. Soc. B* 330:305-322.

Cohen, E . and Sterling, P. (1990b) Convergence and divergence of cones onto bipolar cells in the central area of cat retina, *Phil. Trans. R. Soc., B* 330:323-328.

Cohen, E . and Sterling, P. (1991) Microcircuitry related to the receptive-field center of the ON-beta ganglion cell, *J. Neurophysiol.* 65:352-359.

Dacheux, R. F . and Raviola, E. (1986) The rod pathway in the rabbit: a depolarizing bipolar and amacrine cell, *J. Neurosci.* 6:331-345.

Djamgoz, M. B. A . and Kolb, H. (1993) Ultrastructural and functional connectivity of intracellularly stained neurones in the vertebrate retina: Correlative analyses, *Micr. Res. & Tech.* 24:43-66.

Dowling, J. E. (1987) The Retina: an approachable part of the brain. The Belknap Press, Harvard University Press, Cambridge, Massachusetts.

Dowling, J. E. , and Boycott, B.B. (1966) Organization of the primate retina; electron microscopy, *Proc. R. Soc. B* 166:80-111.

Enroth-Cugell, C . and Robson, J.G. (1966) The contrast sensitivity of retinal ganglion cells of the cat, *J. Physiol.* 187:517-552.

Famiglietti, E. V. (1983) 'Starburst' amacrine cells and cholinergic neurons: mirror-symmetric ON and OFF amacrine cells of rabbit retina, *Brain Res.* 261:138-144.

Famiglietti, E. V . and Kolb, H. (1975) A bistratified amacrine cell and synaptic circuitry in the inner plexiform layer of the retina, *Brain Res.* 84:293-300.

Famiglietti, E. V. and Kolb, H. (1976) Structural basis for ON- and OFF-center responses in retinal ganglion cells, *Science* 194:193-195.

Freed, M. A . and Sterling, P. (1988) The ON-alpha ganglion cell of the cat retina and its presynaptic cell types, *J. Neurosci.* 8:2303-2320.

Gouras, P. (1971) The function of the midget system in primate color vision, *Vision Res. Suppl.* 3:397-410.

Ikeda, H . and Wright, M.J. (1972) Receptive field organization of 'sustained' and 'transient' retinal ganglion cells which subserve different functional roles, *J. Physiol.* 227:769-800.

Kaneko, A. (1970) Physiological and morphological identification of horizontal, bipolar and amacrine cells in goldfish retina, *J. Physiol.* 207:623-633.

Kidd, M. (1962) Electron microscopy of the inner plexiform layer of the retina in the cat and the pigeon, *J. Anat.* 96: 179-187.

Kolb, H. (1970) Organization of the outer plexiform layer of the primate retina: electron microscopy of Golgi-impregnated cells, *Phil. Trans. R. Soc. B* 258:261-283.

Kolb, H. (1974) The connexions between horizontal cells and photoreceptors in the retina of the cat: electron microscopy of Golgi-preparations, *J. Comp. Neurol.* 155:1-14.

Kolb, H. (1979) The inner plexiform layer in the retina of the cat: electron microscopic observations, *J. Neurocytol.* 8:295-329.

Kolb, H. (1984) Cone pathways in the mammalian retina. In "Molecular and Cellular Basis of Visual Acuity" (Eds. Hilfer, S.R. and Sheffield, J.B.), Springer-Verlag, New York Inc., pp. 56-78.

Kolb, H. , N. Cuenca, Wang, H.H. and DeKorver, L. (1990) The synaptic organization of the dopaminergic amacrine cell in the cat retina., *J. Neurocytol.* 19:343-366.

Kolb, H . and Famiglietti, E.V. (1974) Rod and cone pathways in the inner plexiform layer of the cat retina, *Science*186:47-49.

Kolb, H . and Famiglietti, E.V. (1976) Rod and cone pathways in the retina of the cat, Invest. Ophthal. 15:935-946.

Kolb, H . and Nelson, R. (1983) Rod pathways in the retina of the cat, *Vision Res.* 23:301-312.

Kolb, H . and Nelson, R. (1984) Neural architecture of the cat retina, *Prog. Ret. Res.* 3:21-60.

Kolb, H . and Nelson, R. (1985) Functional neurocircuitry of amacrine cells in the cat retina. In "Neurocuitry of the Retina: a Cajal Memorial" (Eds. Gallego, A. and Gouras, P.), Elsevier Press, New York, pp. 215-232.

Kolb, H . and Nelson, R. (1993) Off-alpha and off-beta ganglion cells in the cat retina. II. Neural circuitry as revealed by electron microscopy of HRP stains, *J. Comp. Neurol.* 329:85-110.

Kolb, H., Nelson, R., and Mariani, A. (1981) Amacrine cells, bipolar cells and ganglion cells of the cat retina: A Golgi study, *Vision Res.* 21:1081-1114.

Kolb, H., Pflüg, R., DeKorver, L. and Nelson, R. (1993) Synaptic organization of hyperpolarizing amacrine cell A13, *Invest. Ophthal. Vis. Science (Suppl.)* 34:p1153

Kolb, H . and West, R.W. (1977) Synaptic connections of the interplexiform cell in the retina of the cat. *J. Neurocytol.* 6:155-170.

Lasansky, A. (1978) Contacts between receptors and electrophysiologically identified neurones in the retina of the larval tiger salamander, *J. Physiol.* 285:531-542.

Marc, R. E . Liu,W-L. S. and Muller, J.F. (1988) Multiple GABA-mediated surround channels in goldfish retina, Suppl. Invest. Ophthal. Vis. Sci. 29:p. 272.

Mariani, A. P. (1981) A diffuse, invaginating cone bipolar cell in primate retina, *J. Comp. Neurol.* 197:661-671.

Mariani, A. P. (1983) Giant bistratified bipolar cells in the monkey retina, *Anat. Rec.* 206:215-220.

Mariani, A. P. (1984) Bipolar cells in monkey retina selective for cones likely to be blue-sensitive, *Nature* 308:184-186.

Massey, S. C. (1990) Cell types using glutamate as a neurotransmitter in the vertebrate retina, *Prog. Ret. Res.* 9:399-425.

McGuire, B. A.,. Stevens, J.K. and Sterling, P. (1984) Microcircuitry of bipolar cells in cat retina, *J. Neurosci.* 4:2920-2938.

McGuire, B.A., Stevens, J.K. and Sterling, P. (1986) Microcircuitry of Beta ganglion cells in cat retina, *J. Neurosci.* 6:907-918.

Missotten, L. (1965) The ultrastructure of the human retina. Arscia Uitgaven N.V., Brussel.

Naka, K.-I. (1976) Neuronal circuitry in the catfish retina, *Invest. Ophthal.* 15:926-935.

Nelson, R. (1982) AII amacrine cells quicken the time course of rod signals in the cat retina, *J. Neurophysiol.* 47:928-947.

Nelson, R., Famiglietti, E.V. and. Kolb, H. (1978) Intracellular staining reveals different levels of stratification for on-center and off-center ganglion cells in the cat retina, *J. Neurophysiol.* 41:427-483.

Nelson, R . and Kolb, H. (1983) Synaptic patterns and response properties of bipolar and ganglion cells in the cat retina, *Vision Res.* 23:1183-1195.

Nelson, R., and Kolb, H. (1985) A17: a broad-field amacrine cell of the rod system in the retina of the cat, *J. Neurophysiol.* 54:592-614

.Ogden, T. E. (1974) The morphology of retinal neurons of the owl monkey, Aotes, *J. Comp. Neurol.* 153: 399-428.

Pourcho, R. G. (1982) Dopaminergic amacrine cells in the cat retina, *Brain Res.* 252:101-109.

Pourcho, R. G . and Goebel, D.J. (1983) Neuronal subpopulations in cat retina which accumulate the GABA agonist ($^3$H)muscimol: a combined Golgi and autoradiographic study, *J. Comp. Neurol.* 219:25-35.

Raviola, E . and Gilula, N.B. (1975) Intramembrane organization of specialized contacts in the outer plexiform layer of the retina: A freeze-fracture study in monkeys and rabbits, *J. Cell Biol.* 65:192-222.

Rodieck, R. W. (1973) The Vertebrate Retina: principles of structure and function. W.H. Freeman and Company, San Francisco.

Shapley, R . and Perry, V.H. (1986) Cat and monkey retinal ganglion cells and their visual functional roles, *Trends Neurosci.* 9:229-235.

Slaughter, M. M . and Miller, R.F. (1981) 2-amino-4-phosphonobutyric acid: A new pharmacological tool for retina research, *Science* 211:182-184.

Slaughter, M. M . and Miller, R.F. (1983) An excitatory amino acid antagonist blocks cone input to sign-conserving second-order retinal neurons, *Science* 219:1230-1232.

Stell, W. K. (1967) The structure and relationships of horizontal cells and photoreceptor-bipolar synaptic complexes in goldfish retina, *Amer. J. Anat.* 121:401-424.

Stell, W. K . and Lightfoot, D.O. (1975) Color-specific interconnections of cones and horizontal cells in the retina of the goldfish, *J. Comp. Neurol.* 159:473-501.

Stell, W. K., Lightfoot, D.O., Wheeler, T.G. and Leeper, H.F. (1975) Goldfish retina: Functional polarization of cone horizontal cell dendrites and synapses, *Science* 190: 989-990.

Stell, W. K . and Witkovsky, P. (1973) Retinal structure in the smooth dogfish, Mustelus canis: light microscopy of photoreceptor and horizontal cells, *J. Comp. Neurol.* 148:33-46.

Sterling, P. (1983) Microcircuitry of the cat retina, *Ann. Rev. Neurosci.* 6:149-185.

Sterling, P., Cohen, E., Freed, M.A. and Smith, R.G. (1987) Microcircuitry of the ON-beta ganglion cell in daylight, twilight, and starlight, *Neurosci. Res. Suppl.* 6:S269-S285.

Sterling, P., Freed, M.A. and Smith, R.G. (1988) Architecture of rod and cone circuits to the On-beta ganglion cell, *J. Neurosci.* 8:623-642.

Svaetichin, G . and MacNichol, E.F.Jr. (1958) Retinal mechanisms for chromatic and achromatic vision, *Ann. NY. Acad. Sci.* 74:385-404.

Tomita, T. (1965) Electrophysiological study of the mechanism subserving color coding in the fish retina, *Cold Spring Harbor Symp. Quant. Biol.* 30:559-566.

Toyoda, J.-I. (1972) Membrane resistance changes underlying the bipolar cell response in the carp retina, *Vision Res.* 12:283-294.

Wässle, H., Grünert, U., Röhrenbeck, J. and Boycott B.B. (1989) Cortical magnification factor and the ganglion cell density of the primate retina, *Nature* 341:643-646.

Werblin, F. S . and Dowling, J.E. (1969) Organization of the retina of the mudpuppy, Necturus maculosus. II. Intracellular recording, *J. Neurophysiol.* 32:339-355.

4

# RHODOPSIN GENE MUTATIONS CAUSING RETINITIS PIGMENTOSA

## Functional Phenotypes of Codon 23 and Codon 135 Genotypes

Samuel G. Jacobson,[1] Colin M. Kemp,[2] Artur V. Cideciyan,[1] and Jeremy Nathans[3]

[1] Scheie Eye Institute
University of Pennsylvania
51 North 39th Street, Philadelphia, Pennsylvania 19104.
[2] Department of Ophthalmology
University of Miami School of Medicine
Bascom Palmer Eye Institute
Miami, Florida
[3] Departments of Molecular Biology and Genetics and Neurosciences
Howard Hughes Medical Institute
Johns Hopkins University School of Medicine
Baltimore, Maryland

## INTRODUCTION

Major progress in understanding retinitis pigmentosa (RP), a group of hereditary retinal degenerations, occurred with the discovery that some forms of autosomal dominant (ad) RP are caused by mutations in the gene encoding the rod visual pigment, rhodopsin (reviewed in Humphries, et al, 1993). An important question arising from these molecular genetic discoveries is: How does the molecular pathology lead to rod photoreceptor dysfunction and progressive retinal degeneration? A number of different scientific approaches, from molecular to clinical, are being used to try to answer this question. We are studying patients with rhodopsin gene mutations using non-invasive tests of retinal function to try to understand the relationship between the molecular lesions and the clinical phenotype (Jacobson et al, 1991; Kemp et al, 1992a; Jacobson et al, 1994).

In this report, we compare retinal function test results in two groups of adRP patients with rhodopsin mutations: one group with a codon 23 mutation, proline-23-histidine (P23H), and another group representing three different mutations at codon 135: arginine-135-leucine (R135L), arginine-135-tryptophan (R135W) and arginine-135-glycine (R135G). The results indicate that different sites of the mutation lead to different patterns of visual dysfunction and, for these genotypes, there is both intra- and interfamilial consistency of phenotypic

*Basic and Clinical Perspectives in Vision Research*, Edited by Jon Robbins et al.
Plenum Press, New York, 1995

expression (Jacobson et al, 1991; Kemp et al, 1992a). The relationship of our results in patients to those in tissue culture expression systems and in transgenic animals is discussed.

## SUBJECTS AND METHODS

### Subjects

We studied 16 patients with the P23H rhodopsin mutation, representing five different families; and 11 patients with codon 135 mutations: one family with the R135L mutation, one family with the R135G mutation, and three families with the R135W mutation. The rhodopsin mutations in these families were determined in previous molecular genetic investigations (Sung et al, 1991a; Macke et al, 1993). The patients are identified in the text by a mutation code-family number (e.g. P23H-2) and generation and individual numbers (e.g. V/2).

### Visual Function Tests

Brief descriptions of the non-invasive testing techniques are given below; full details have been published (Jacobson et al, 1991; Kemp et al, 1992a; Jacobson et al, 1994; Jacobson et al, 1992; Jacobson et al, 1986; Faulkner & Kemp, 1984).

### Perimetry

Goldmann kinetic perimetry was performed with V-4e and I-4e test targets. Static threshold perimetry was performed at 75 loci throughout the visual field (12° grid) in the dark-adapted (500 and 650 nm stimuli) and light-adapted (600 nm stimulus) states. The results are displayed as gray scales of sensitivity loss for rods (0-54 dB range) and midspectral cones (0-25 dB range). Two-color dark-adapted perimetry was also performed at 31 loci along the horizontal meridian (central 60°; 2° spacing).

### Electroretinography

Rod (blue flash, dark-adapted), mixed cone and rod (white flash, dark-adapted), and cone flicker (white flashes at 29 Hz) ERGs were recorded. Rod b-waves were elicited with increasing intensities of blue light in the dark-adapted state. A curve was fitted to the measured amplitudes using the Naka-Rushton equation.

### Adaptometry

Adaptometry was performed with 460 or 500 nm and 650 nm stimuli after a light exposure expected to bleach 95% of the pigment present. Final recovery of sensitivity is defined as a return to pre-bleach sensitivity levels, which were determined after >3 hours of dark adaptation at sessions without exposure to bright lights.

### Imaging Fundus Reflectometry

The retina was illuminated with a sequence of eight measuring lights (460 to 680 nm) and retinal images were obtained using a high sensitivity TV system. Reflectance values of the light-adapted eye were compared to those from the dark-adapted eye. The spectral variation of the double density changes (differences of the logarithms of the reflectance values obtained in the light- and dark-adapted conditions) provides a measure of the level of visual pigment present in the dark-adapted retina.

## RESULTS

Figures 1 and 2 illustrate that patients with different rhodopsin mutations can show different patterns of visual dysfunction; the data are from patients with the mildest disease expression we have encountered to date in their genotype. The results in Figure 1 are from an 18 year old woman with the P23H mutation. Her fundus appeared normal and kinetic fields are normal. Rod sensitivity is reduced by about one log unit across most of the field while cone sensitivity is within normal limits at all loci tested. The rod (R) b-wave is reduced in amplitude and delayed in timing; mixed (M) cone-rod ERG a- and b-waves are reduced in amplitude; cone (C) flicker is normal in amplitude and slightly delayed in timing. In the rod b-wave amplitude-intensity function (symbols and solid curve), Vmax and K are abnormal (normal mean function shown as dotted curve; bars on axes are normal mean Vmax - 2 s.d. and K ± 2 s.d.). Dark adaptation shows a normal rod-cone break but recovery of rod sensitivity is abnormally prolonged to ~200 minutes (normal function shown as dotted line). Dark adaptometry results from P23H patients at other disease stages (see below) were very similar, thereby suggesting that this rod adaptation abnormality is characteristic of the genotype (Kemp et al, 1992b; 1993).

Figure 2 shows results from a 40 year old man carrying the R135W mutation. There were typical features of retinal degeneration throughout the fundus. The kinetic field is full with V-4e but limited with I-4e. There is no measurable rod sensitivity throughout the field and considerable cone sensitivity loss. ERGs were not detectable to any stimuli. With dark adaptometry, only cone function was measurable. All patients with codon 135 mutations that we have examined to date show no measurable rod function and only cone-mediated function by dark-adapted perimetry throughout their visual field; also, all have non-detectable rod ERGs and barely detectable or non-detectable cone ERGs.

Intra- and interfamilial consistency of phenotype in these genotypes is demonstrated in Figures 3 and 4. Figure 3 shows that in five patients with the P23H mutation, representing two families, the rod-cone break of the dark adaptation function occurs at about 15 minutes and then rod sensitivity drifts back to pre-bleach levels at a much slower rate than normal. Recovery of rod sensitivity in two siblings of P23H-2 are nearly superimposable. In P23H-3, the results of two of three siblings are also similar, reaching pre-bleach levels in ~180 minutes, but the other sibling shows a slower recovery, taking ~300 minutes. The greater degree of abnormality in this patient may reflect the fact that thresholds were measured close to the edge of his remaining area of functioning rods, so that they may have been affected by disease processes apart from the primary lesion.

Patients from the same family with different stages of progression of disease also show intrafamilial consistency of phenotype in the sense that results in more severely affected members appear to reflect progression of patterns in mildly affected members. In Figure 4, two-color dark-adapted perimetric profiles from three members (ages 16-48 yrs) representing two generations of a family with the R135L mutation show no measurable rod function at any of the stages. With increasing severity of disease (top to bottom), the cone-mediated (C) sensitivity becomes further reduced (dotted lines are normal data to 650 nm at cone plateau) and there is decreasing extent of functioning field.

Figure 4 also shows a sequence of dark-adapted perimetric profiles from three generations (ages 18-62 yrs) of a P23H family. There is a progression (top to bottom) from almost normal rod sensitivities to unmeasurable rod function. The most mildly affected member (upper right) has nearly normal rod- (R) and mixed (M) rod- and cone-mediated

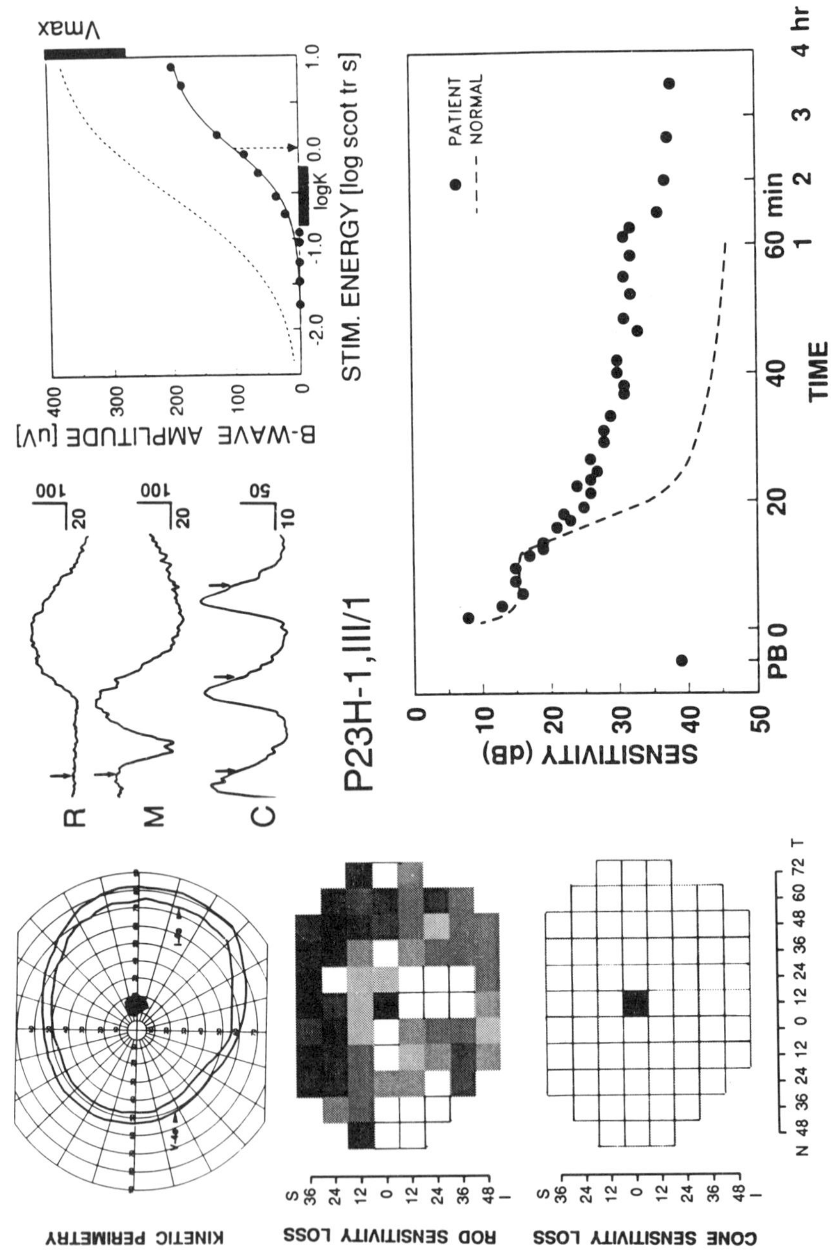


**Figure 1.** Visual function test results in a patient with the P23H rhodopsin gene mutation.

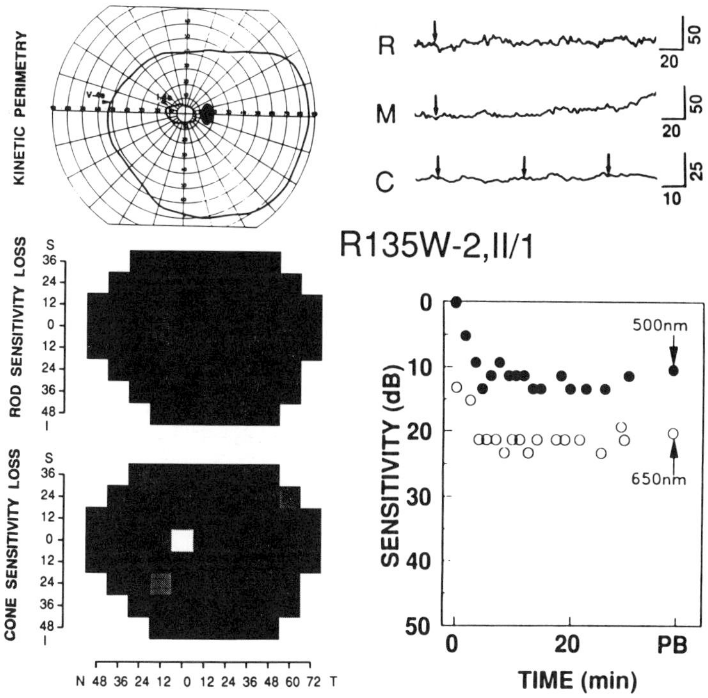


**Figure 2.** Visual function test results in a patient with R135W rhodopsin gene mutation.

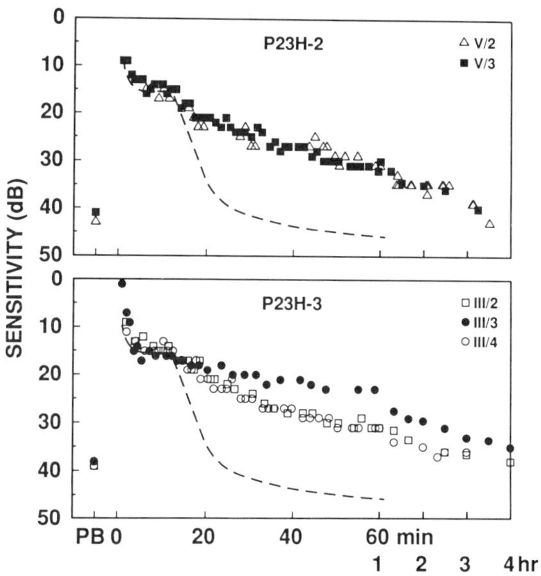


**Figure 3.** Dark adaptation test results in five patients, representing two families, with the P23H rhodopsin gene mutation.

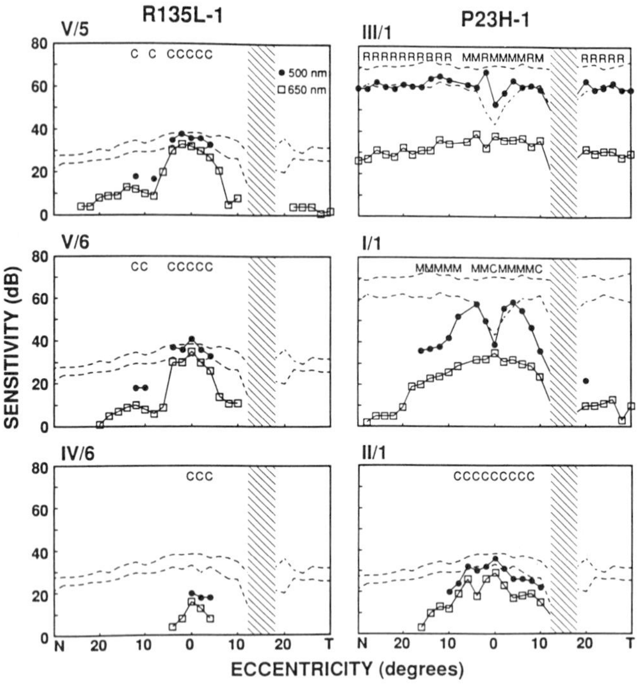


**Figure 4.** Dark-adapted static perimetric profiles in three family members with the R135L rhodopsin gene mutation (left) and three family members with the P23H rhodopsin gene mutation (right).

function (dotted lines are normal data to 500 nm). A family member at a more advanced stage (middle right) has normal rod sensitivity at only a few paracentral loci and rod and cone sensitivity losses that increase with eccentricity. In the most severely affected member (lower right), there is no measurable rod function and cone sensitivity is reduced more than 2 s.d. from the mean normal (for 650 nm at cone plateau) at all loci. The sequences of disease progression in the P23H and R135L families illustrate that there can be convergence of functional phenotypes for some of the genotypes (compare profiles from P23H-1,II/1 and all three R135L patients).

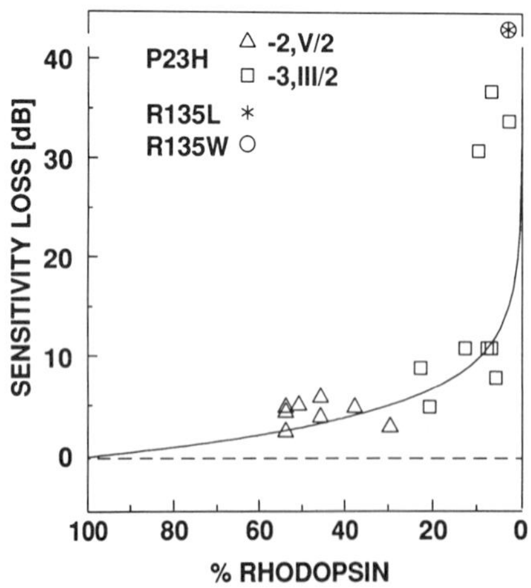


**Figure 5.** Relationship of rhodopsin and rod sensitivity levels in patients with P23H and Rl35L/W rhodopsin gene mutations.

Patients with codon 135 mutations had no measurable rhodopsin. There were measurable levels of pigment in patients with the P23H mutation; double density differences were reduced but the spectra approximated that of normal rhodopsin (Kemp et al, 1992b). Figure 5 shows the relationship between rod sensitivity and levels of rhodopsin. The solid line defines the relationship expected if reduced sensitivity was caused solely by decreased probability of light absorption from reduced levels of rhodopsin; the data points from the patients lie relatively close to this line.

## DISCUSSION

These studies of the phenotype of adRP patients with mutations at codon 23 or codon 135 of the rhodopsin gene indicate that different rhodopsin genotypes can show different patterns of visual dysfunction and, for each genotype, there is intra- and interfamilial consistency of phenotypic expression. A summary description of the phenotype of each genotype is given below.

All three codon 135 mutations showed a similar phenotype. Even patients with the mildest disease expression have no measurable rod-mediated sensitivity across the retina by dark-adapted psychophysics, no detectable rod ERG a- or b-waves, and no measurable rhodopsin. Residual cone function measured with psychophysics and ERG is abnormal in all patients. At more advanced stages, cone function becomes even further reduced. Of interest, all 11 patients gave a history of night blindness from very early in life. Members of another family with the R135L rhodopsin gene mutation were also reported to have this phenotype (Andreasson et al, 1992).

Patients with the P23H mutation at early stages of the disease can be asymptomatic and have only mildly abnormal rod sensitivity across the retina and normal cone sensitivity. Rod ERG a- and b-waves are sizeable. Rhodopsin levels are decreased below normal by amounts which indicate that the reduced ability of the rods to absorb incident light accounts fully for their diminished sensitivity. At intermediate stages, some retinal regions retain mild rod sensitivity loss and no cone sensitivity loss while other regions (e.g. inferior and/or midperipheral retina) progress to more pronounced rod dysfunction and some cone dysfunction. The altitudinal pattern of visual loss associated with this mutation (Kemp et al, 1992a; Heckenlively et al, 1991; Stone et al, 1991) is also seen in other rhodopsin mutations (e.g. Sullivan et al, 1993; Li et al, 1994). Fundus reflectometry in regions including both mild and severe rod dysfunction shows changes in rhodopsin levels that parallel the sensitivity losses. Advanced stages of the disease are characterized by severe visual loss and regions of residual function with little or no rod function, no measurable rhodopsin, abnormal cone function and non-detectable ERGs (Kemp et al, 1992a). All patients, regardless of the extent or severity of their disease, show an abnormally prolonged rod recovery following an intense bleach.

The pathophysiological processes leading from mutation to disease expression are undoubtedly complex (Hargrave & O'Brien, 1991; Fain & Lisman, 1993) and our knowledge of these processes is currently very limited. Some studies, however, have already been performed *in vitro* and in transgenic animals and these permit speculation about the relationship between genotype and phenotype in the codon 135 and 23 mutations. Arginine-135 is at the border between helix C and the second cytoplasmic loop of the rhodopsin molecule, is highly conserved among G protein-coupled receptors, and is involved in binding and activation of transducin (Hargrave & McDowell, 1992; Khorana, 1992). *In vitro* studies of opsins carrying the RP-causing codon 135 mutations suggest that the mutant proteins may be less stable or less efficiently folded than the wild-type protein and are defective in transducin activation (Sung et al, 1991b; 1993; Min et al, 1993; Kaushal & Khorana, 1994); no studies of transgenic animals with these mutations have been reported.

Based on the *in vitro* results we might predict that mildly affected patients with codon 135 mutations (whom, it should be recalled, are heterozygotes) would show rod visual function consistent with having rod outer segments that contain wild-type rhodopsin, and, possibly, also some mutant opsin which is incapable of physiological action; the wild-type protein would be expected to be activated normally by light and to control rod dark adaptation kinetics. We recently found such a pattern of visual function test results (reduced rhodopsin levels and rod sensitivity, sizeable rod a-waves, and relatively normal time course of dark adaptation) in adRP patients with a null mutation of the rhodopsin gene, Q64ter (Jacobson et al, 1994). In the codon 135 mutations, however, the patients' results are not consistent with this prediction but suggest that even at relatively early disease stages there is either little or no photolysable rhodopsin in their rod outer segments, no outer segments, or no rod cells at all. At an advanced stage of this disease, documented in a morphological study of a postmortem donor eye from a 64 year old man with the R135W mutation (father of patient R135W-2,II/1), there is widespread loss of rod (and cone) receptors and the few remaining rod cells lack outer segments (Tucker & Jacobson, 1988). The pathophysiology leading to the early and severe rod cell dysfunction and rod (and cone) degeneration in the codon 135 mutants is unexplained.

Proline-23 is in the intradiscal domain, a region of the rhodopsin molecule considered to have a structural role (Hargrave & McDowell, 1992; Khorana, 1992). *In vitro* studies of the P23H mutant and deletion mutants that include amino acid 23 show that, like the R135 mutants, the proteins are defective in folding and/or stability as evidenced by their low yield, retention in the endoplasmic reticulum, abnormal glycosylation and poor regeneration with 11-*cis*-retinal (Sung et al, 1991b; 1993; Kaushal & Khorana, 1994). Studies of transgenic mouse models of the P23H mutation indicate that the mutant opsin can be produced and transported to the rod outer segment (Olsson et al, 1992; Naash et al, 1993). Based on these results we would predict that patients with the P23H mutation would have rod function consistent with having rod outer segments with reduced levels of rhodopsin that can be activated by light. An abnormality in rhodopsin regeneration, attributable to mutant opsin in the outer segment, might also be expected.

The functional phenotype in patients with the P23H mutation is consistent with this prediction: rod sensitivity levels and rhodopsin measurements indicate that there must be biosynthesis and transport to the outer segment of some rhodopsin (mutant and/or wild-type) and the rod ERG a-wave demonstrates that signal transduction is active, although activation kinetics may not be normal (Hood & Birch, 1994). The rod dark adaptation abnormality provides some evidence that mutant opsin is present in the outer segment. It is of interest in this regard that a transgenic mouse model of this mutation was recently also shown to have abnormally prolonged rod dark adaptation, as measured with the ERG (Goto et al, 1994). Kinetic analysis of the rod adaptation data in our patients suggests that the abnormality is a gross slowing of the rate constant of the slowest reaction controlling rod desensitization, a step loosely identified with the regeneration of rhodopsin (Kemp et al, 1992a; 1993; Lamb, 1990). The P23H mutation could result in a reduction of the ability of the opsin to re-combine with the chromophore, which may relate to the role that the intradiscal domain has been shown to play in protein folding (Khorana, 1992).

In conclusion, our increased understanding of the molecular basis for hereditary retinal degenerations and of the pathophysiology of photoreceptor dysfunction and cell death (Chang et al, 1993; Portera-Cailliau et al., 1994) resulting from the genetic defects makes therapeutic intervention in these disorders a far more hopeful prospect than before.

## ACKNOWLEDGMENTS

Supported in part by Public Health Service research grant EY05627 (SGJ), the National Retinitis Pigmentosa Foundation, Inc. (Baltimore, MD), The Chatlos Foundation, Inc. (Longwood, FL), and the Howard Hughes Medical Institute (Bethesda, MD).

## DEDICATION

To Hisako Ikeda: my teacher, role model and friend.
I am forever grateful for your guidance, inspiration, and generosity.
S.G. Jacobson

## REFERENCES

Andreasson, S., Ehinger, B., Abrahamson, M. and Fex, G. (1992) A six-generation family with autosomal dominant retinitis pigmentosa and a rhodopsin gene mutation (Arginine-135-Leucine), *Ophthalmol. Paed. and Genet.* 13:145-153.

Chang, G-Q., Hao, Y. and Wong, F. (1993) Apoptosis: final common pathway of photoreceptor death in rd, rds and rhodopin mutant mice, *Neuron* 11:595-605.

Fain, G.L. and Lisman, J.E. (1993) Photoreceptor degeneration in vitamin A deprivation and retinitis pigmentosa, *Exp. Eye Res.*57:335-340.

Faulkner, D.J. and Kemp, C.M. (1984) Human rhodopsin measurements using a TV-based imaging fundus reflectometer, *Vision Res.* 24:221-231.

Goto, Y., Peachey, N.S., Ripps, H. and Naash, M.I. (1995) Functional abnormalities in transgenic mice expressing a mutant rhodopsin gene, *Invest. Ophthalmol. Vis. Sci.* 36:62-71.

Hargrave, P.A. and O'Brien, P.J. (1991) Speculations on the molecular basis of retinal degeneration in retinitis pigmentosa. In: Anderson R, LaVail M, Hollyfield J, eds. Proc Stockholm Symposium on Retinal Degenerations. pp 517-528, Boca Raton, FL, CRC Press.

Hargrave, P.A. and McDowell, J.H. (1992) Rhodopsin and phototransduction: a model for G protein-linked receptors, *FASEB J.*6:2323-2331.

Heckenlively, J., Rodriguez ,J.A. and Daiger, S.P. (1991) Autosomal dominant sectoral retinitis pigmentosa. Two families with transversion mutation in codon 23 of rhodopsin, *Arch. Ophthalmol.* 109:84-91.

Hood, D.C. and Birch, D.G. (1994) Rod phototransduction in retinitis pigmentosa: estimation and interpretation of parameters derived from the rod a-wave, *Invest. Ophthalmol. Vis. Sci.* 35:2948-2961.

Humphries, P., Farrar, G.J. and Kenna ,P. (1993) Autosomal dominant retinitis pigmentosa: molecular, genetic and clinical aspects. In: Osborne N, Chader G, eds. Progress in Retinal Research. pp 231-245, Great Britain, Pergamon Press Ltd.

Jacobson, S.G., Kemp ,C.M., Sung, C-H. and Nathans, J. (1991) Retinal function and rhodopsin levels in autosomal dominant retinitis pigmentosa with rhodopsin mutations, *Am. J. Ophthalmol.* 112:256-271.

Jacobson, S.G., Kemp, C.M., Cideciyan, A.V., Macke, J.P., Sung, C-H. and Nathans, J. (1994) Phenotypes of stop codon and splice site rhodopsin mutations causing retinitis pigmentosa, *Invest. Ophthalmol. Vis. Sci.*35:2521-2534.

Jacobson, S.G., Roman, A.J., Cideciyan, A.V., Robey, M.G., Iwata, T. and Inana, G. (1992) X-linked retinitis pigmentosa: functional phenotype of an RP2 genotype, *Invest. Ophthalmol. Vis. Sci.* 33:3481-3492.

Jacobson, S.G., Voigt, W.J., Parel, J-M., et al. (1986) Automated light- and dark-adapted perimetry for evaluating retinitis pigmentosa, *Ophthalmol.* 93:1604-1611.

Kaushal, S. and Khorana, H.G. (1994) Structure and function in rhodopsin. 7. Point mutations associated with autosomal dominant retinitis pigmentosa, *Biochem.* 33:6121-6128.

Kemp, C.M., Jacobson, S.G., Roman, A.J., Sung, C-H. and Nathans, J. (1992a) Abnormal rod dark adaptation in autosomal dominant retinitis pigmentosa with pro-23-his rhodopsin mutation, *Am. J. Ophthalmol.* 113:165-174.

Kemp, C.M., Jacobson, S.G., Cideciyan, A.V., Sung, C.H. and Nathans, J. (1992b) Kinetic analysis of abnormal rod dark adaptation in autosomal dominant retinitis pigmentosa with rhodopsin mutations, *Invest. Ophthalmol. Vis. Sci. Suppl.* 33:1397.

Kemp, C.M., Jacobson, S.G., Cideciyan, A.V. and Wu, S. (1993) Kinetic analyses of rod dark adaptation in normals and in autosomal dominant retinitis pigmentosa with rhodopsin mutations, *Invest. Ophthalmol. Vis. Sci. Suppl.* 34:1074.

Khorana, H.G. (1992) Rhodopsin, photoreceptor of the rod cell, *J. Biol. Chem.* 267:1-4.

Lamb, T.D. (1990) Dark adaptation: a re-examination. In: Hess RF, Sharpe LT, Nordby K, eds. Night Vision. pp 177-222, Cambridge: Cambridge University Press.

Li, Z-Y., Jacobson, S.G. and Milam, A.H. (1994) Autosomal dominant retinitis pigmentosa caused by the threonine-17-methionine rhodopsin mutation: retinal histopathology and immunocytochemistry, *Exp. Eye. Res.* 58:397-408.

Macke, J.P., Davenport ,C.M., Jacobson, S.G. et al.(1993) Identification of novel rhodopsin mutations responsible for retinitis pigmentosa: implications for the structure and function of rhodopsin, *Am. J. Hum. Genet.* 53:80-89.

Min, KC., Zvyaga, T.A., Cypess, A.M. and Sakmar, T.P. (1993) Characterization of mutant rhodopsins responsible for autosomal dominant retinitis pigmentosa. Mutations on the cytoplasmic surface affect transducin activation, *J. Biol. Chem.* 268:9400-9404.

Naash, M.I., Hollyfield, J.G., Al-Ubaidi, M.R. and Baehr, W. (1993) Simulation of human autosomal dominant retinitis pigmentosa in transgenic mice expressing a mutated murine opsin gene, *Proc. Natl. Acad. Sci. USA* 90:5499-5503.

Olsson, J.E., Gorden, J.W., Pawlyk, B.S. et al. (1992) Transgenic mice with a rhodopsin mutation (Pro-23-His); a mouse model of autosomal dominant retinitis pigmentosa, *Neuron* 9:815-830.

Portera-Cailliau,C., Sung, C-H., Nathans, J. and Adler, R. (1994) Apoptotic photoreceptor cell death in mouse models of retinitis pigmentosa, *Proc. Natl.Acad. Sci. USA* 91:974-978.

Stone, E.M., Kimura, A.E., Nichols, B.E., Khadivi, P., Fishman, G.A. and Sheffield, V.C. (1991) Regional distribution of retinal degeneration in patients with the proline to histidine mutation in codon 23 of the rhodopsin gene, *Ophthalmol.* 98:1806-1813.

Sullivan, L.J., Makris, G.S., Dickenson, P., et al. (1993) A new codon 15 rhodopsin gene mutation in autosomal dominant retinitis pigmentosa is associated with sectorial disease, *Arch. Ophthalmol.* 111:1512-1517.

Sung, C-H., Davenport, C.M., Hennessey, J.C., et al. (1991a) Rhodopsin mutations in autosomal dominant retinitis pigmentosa, *Proc. Natl. Acad. Sci. USA* 88:6481-6488.

Sung, C-H., Schneider, B.G., Agarwal, N., Papermaster, D.S. and Nathans, J. (1991b) Functional heterogeneity of mutant rhodopsins responsible for autosomal dominant retinitis pigmentosa, *Proc. Natl. Acad. Sci. USA* 88:8840-8844.

Sung, C-H., Davenport, C.M. and Nathans, J. (1993) Rhodopsin mutations responsible for autosomal dominant retinitis pigmentosa, *J. Biol. Chem.* 268:26645-26649.

Tucker, G. and Jacobson, S. G. (1988) Morphological findings in retinitis pigmentosa with early diffuse rod dysfunction, *Retina* 8:30-41.

5

# THE REGULATORY ROLE OF DOPAMINERGIC PATHWAYS IN THE MAMMALIAN OUTER RETINA

M.W. Hankins* and H. Ikeda

The Vision Research Unit (U.M.D.S.)
The Rayne Institute
St. Thomas' Hospital
London, SE1 7EH, United Kingdom

## INTRODUCTION

This chapter examines the role of dopaminergic pathways in the physiology of the mammalian retina and reviews the current literature regarding this important regulatory retinal pathway. It was my (M.W.H.) good fortune to join Professor Ikeda's group in 1988, where I began to develop *in vitro* preparations of the mammalian retina for intracellular recording. This chapter is illustrated by data from neuropharmacological experiments performed in the Vision Research Unit during the period of 1988-1993.

## ANATOMY OF THE RETINAL DOPAMINE PATHWAY

Dopamine containing neurones have been described in the retinae of many vertebrate species, including human, and appear to be restricted to only two anatomical classes of inner retinal neurone, the dopamine-interplexiform cells (Dowling and Ehinger, 1975;1978) and the dopamine-amacrine cells. Some species possess only one of these classes, though most mammalian retinae express both cell types (see Djamgoz & Wagner, 1992). The dopamine neurones are characterised as a sparse, though regular, population of cells, with typical cell densities which rarely exceed 18-25 cells/mm$^2$ (see Figure 4). This sparse, yet influential group of neurones, has a well defined morphology and connectivity, consistent with a role in the regulation of retinal sensitivity (Witkovsky & Dearry, 1991) . In general terms, the DA-interplexiform cells in mammalian retinae receive synaptic inputs at the innerplexiform layer (IPL) from amacrine and bipolar cells. In response to light stimuli (Godley & Wurtman,

* Current address / address for correspondence- Gunnar Svaetichin Laboratory, Depts. of Biology and Physics, Imperial College Silwood Park, Ascot, Berks. SL5 7PY. U.K.

*Basic and Clinical Perspectives in Vision Research*, Edited by Jon Robbins et al.
Plenum Press, New York, 1995

1988), they release dopamine principally at the outerplexiform layer (OPL) onto horizontal and bipolar cells (Djamgoz & Wagner, 1992). However, dopamine also appears to be released at the IPL onto AII amacrine cells (Hampson et. al., 1992), whilst the distal processes of dopaminergic interplexiform cells reach as far as the outer segments of photoreceptor cells (Nguyen-Legros et. al., 1989).

## THE PHYSIOLOGY AND PHARMACOLOGY OF THE DOPAMINE EFFECTS ON HORIZONTAL CELLS

The post-synaptic effects of dopamine are mediated by membrane receptors, which have broadly been divided into two functional categories, the D1 and the D2 types (Kebabian & Calne, 1979). Recent receptor cloning studies have, however, demonstrated that the D2 receptors in particular are a heterogeneous group comprising a number of discrete receptors including the D2S, D2L, D3 and D4 types (Sibley & Monsma, 1992). Dopamine receptors occur on virtually all morphological classes of retinal neurones, and this is reflected by electrophysiological and biochemical studies in a broad range of species which reveal a multiplicity of dopamine effects on the activity of retinal neurones. The reported range of physiological responses apparently mediated by dopamine in the mammalian neural retina are summarised in Table 1. The range of effects, however, includes a number of responses for which there is no supporting anatomical evidence for direct pre-synaptic contact. This has led to the general hypothesis that some retinal dopamine responses are mediated by the paracrine release of dopamine from DA-interplexiform and/or DA-amacrine cells (Piccolino *et. al.*, 1987).

The first retinal dopamine mediated response to be elucidated was the uncoupling of cone-driven horizontal cells in the teleost fish retina (Negishi & Drujan, 1978), which clearly demonstrated that dopamine was capable of restricting the receptive field properties of horizontal cells. It has since been reported that dopamine has broadly similar effects on the majority of horizontal cell types studied in numerous vertebrate non-mammalian species. We began our studies on the isolated rat retina, in order to examine the physiological role of retinal dopamine at the outerplexiform layer (OPL) in the mammalian system (Hankins and Ikeda, 1991b). The rat retina has been the subject of extensive anatomical and biochemical studies and has been shown to contain all the cellular dopamine components found in higher primate retinae.

## THE EFFECT OF DOPAMINE AND SELECTIVE DOPAMINE ANTAGONISTS ON HORIZONTAL CELLS IN THE RAT RETINA

During the course of these investigations recordings were made from rod-driven horizontal cells using an *in vitro* retina preparation of the pigmented (PVG) rat. We first

**Table 1.** Established effects and sites of action of dopamine at the cellular level in the mammalian retina

| Site of action | Mode of action |
|---|---|
| Photoreceptors | Reduction of the light sensitive pool of cAMP; via D4 receptor (Cohen et al, 1992) |
| Horizontal Cells | Modulates the coupling of rod driven horizontal cells; via D1 receptor (Hankins & Ikeda, 1991b) |
| Amacrine Cells | Affects the coupling of AII amacrines; via D1 receptor (Hampson et al, 1992) |
| Ganglion Cells | Silent DA-synapse, active in early development; via D2 receptors (Ikeda et al, 1986) |

described this preparation in our studies of the non-NMDA excitatory amino acid receptors, which mediate the light-driven inputs to rat horizontal cells (Hankins and Ikeda, 1991a). In brief, the neural retina is isolated from dark-adapted animals and superfused, photoreceptor side upwards, in a recording chamber. In these experiments light stimuli are projected onto the retina from below and intracellular recordings made using glass microelectrodes advanced from the photoreceptor surface. The dark-resting membrane potentials of horizontal cells are in the range of -15 to -35 mV, a relatively depolarised level, which originates from the tonic dark-release of L-glutamate from rod-photoreceptors (Hankins *et. al.*, 1985; Hankins and Ikeda, 1991a). In contrast, the light-driven responses of these cells is characterised by a slow, maintained hyperpolarization known as the S-potential, and this response originates from the light-induced inhibition of L-glutamate release from the photoreceptors.

When dopamine (2 - 10 μM) is applied to the retina it evokes a depolarisation of horizontal cells (2-22mV), which is accompanied by a significant reduction (10 - 60%) in the light evoked responses to a full-field stimulus. The typical dose-dependent effect of dopamine is illustrated in Figure 1A. We also examined the effects of selective dopamine antagonists on these cells. The D1 selective antagonist SCH 23390 (10 - 100 μM) hyperpolarised horizontal cells (5 to 20 mV) and enhanced the responses to full-field light stimuli (Figure 1B), and these effects were observed in all 20 cells examined. In contrast, the D2 type antagonist L-sulpiride (10 - 200 μM) had no effect on any of the 18 cells examined (not illustrated). We also found that the effects of exogenous dopamine (10 μM) were reversed by the co-application of SCH 23390 (Fig. 1C), but not by L-sulpiride. From these experiments we concluded that rod-driven horizontal cells in the rat express D1 but not D2 receptors which respond to exogenous DA, whilst the effects observed with the D1 antagonist alone (Fig. 1B) reveal an endogenous active dopamine input to horizontal cells.

## THE EFFECTS OF DOPAMINE ON HORIZONTAL CELL RECEPTIVE FIELDS

One important characteristic of retinal horizontal cells is their extensive receptive field properties. Single horizontal cells respond to light stimuli which are presented well beyond their individual dendritic field. The origin of this expansive summation lies in the extensive gap-junction coupling between adjacent horizontal cells of a given functional class (Yamada and Ishikawa, 1965; Naka and Rushton, 1967). Previous studies of horizontal cells in the teleost fish had provided evidence that the primary effect of dopamine was upon the radial electrotonic coupling between adjacent horizontal cells (Negishi & Drujan, 1978; Teranishi *et. al.*, 1983). One method for studying this phenomenon is to employ alternating stimuli consisting of a small central spot followed by a surround annulus. The small central spot primarily stimulates the direct dendritic photoreceptor input, whilst the surround annulus stimulates the indirect input through coupled cells. These alternating stimuli are manipulated in intensity, so as to provide roughly equal light responses from the cell, then exogenous dopamine is applied to the perfusate. The results of one such experiment are illustrated in Figure 2. and show that the depolarisation evoked by dopamine is accompanied by a rapid and selective reduction in the light response evoked by the surround annulus, with little or no effect on the direct input evoked by a central spot. From these results we conclude that dopamine applied to horizontal cells modifies their receptive field properties, through a mechanism which involves a reduction in cell-cell coupling. Since horizontal cells provide the surround input to bipolar cells (Kaneko, 1973), dopamine pathways must have an

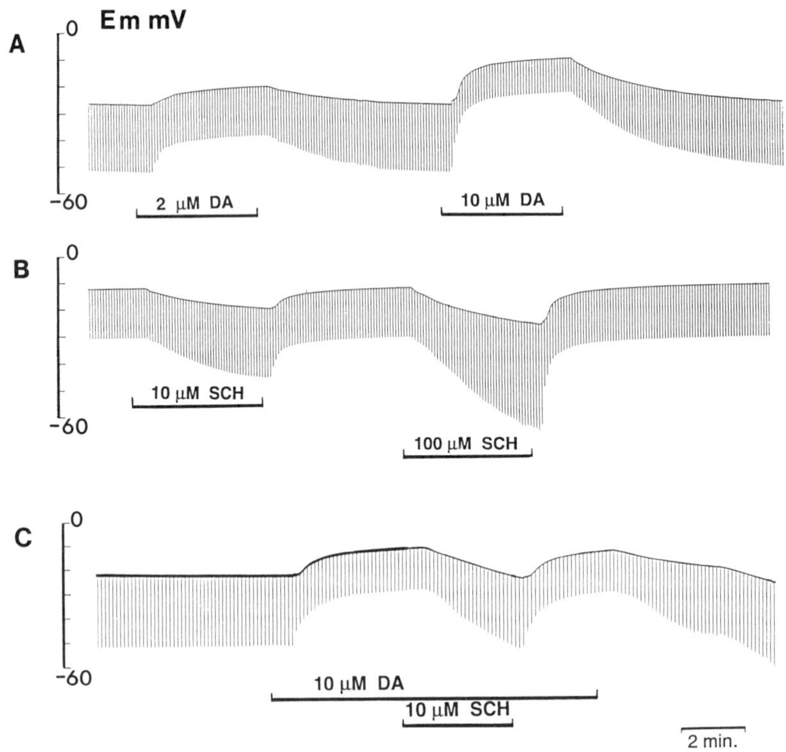


**Figure 1.** The dose dependent effects of dopamine (A) and the D1 receptor selective dopamine antagonist SCH 23390 (B) on the dark-resting membrane potential and the light evoked S-potentials of rod-driven horizontal cells in the normal adult retina. The recordings represent the transmembrane potential (Em, mV) of single cells recorded intracellularly in isolated superfused preparations (Hankins & Ikeda, 1991a). Each downward deflection on the resting membrane potential represents a hyperpolarizing light response to a full-field (4mm dia.) 520nm light stimulus. The drugs, added to the superfusate, are applied at the times indicated by the bars beneath the recordings. (A): 2 μM and 10 μM dopamine (DA) result in dose dependent depolarizations associated with a significant reduction in the full-field light response. The effects being readily reversed following washout. (B): 10 μM and 100 μM SCH 23390 (SCH) evoke dose dependent hyperpolarizations accompanied by a marked enhancement of the full-field light response. (C): The effect of the D1 antagonist SCH23390 (SCH) on the response of horizontal cells to exogenous dopamine (DA). Note that both the depolarization and the reduction in the amplitude of the light response evoked by 10 μM DA are reversed by the subsequent co-application of 10 μM SCH 23390.

important role in the initiation of centre-surround organisation at the first synaptic layer in the mammalian visual pathway.

The effect of dopamine on the receptive fields of horizontal cells can be ascribed to its effect on the coupling conductance. However the origin of the dopamine evoked depolarization of horizontal cells in the rat retina remains somewhat ambiguous. In some non-mammalian species it has be shown that dopamine has an additional effect on the L-glutamate evoked current, which originates from the photoreceptor input to these cells (Knapp and Dowling, 1987). However, we failed to observe such a phenomenon in the rat retina, although we established that the depolarizing effect to dopamine does not persist when all synaptic inputs are blocked by cobalt (Hankins and Ikeda, 1991b)

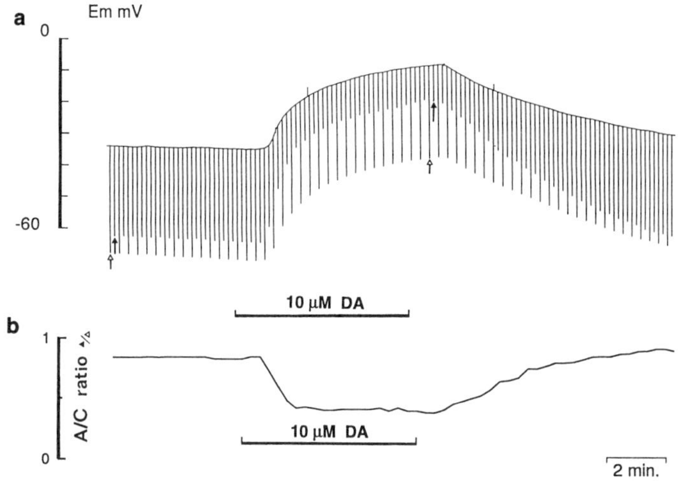


**Figure 2.** The differential effect of dopamine (DA) on the horizontal cell response to peripheral or central light stimuli. (A): In this recording the horizontal cell is stimulated alternately by a small central spot (100μm diameter) and an annulus (150μm inside diameter, 4 mm external diameter) light stimulus, both projected centrally on the recording electrode. Central spot responses are denoted by the open arrow and annulus responses by the filled arrow. Note that the effect of 10 μM dopamine (DA) is to depolarize the cell and selectively reduce the cell response to annular stimulation. (B): A concurrent plot of the annulus/centre (A/C) response ratio reveals the rapid loss in surround sensitivity following the application of dopamine.

## DOPAMINE AND MELATONIN PATHWAYS IN THE MAMMALIAN RETINA. CONTRAREGULATORY SYSTEMS IN LIGHT- AND DARK-ADAPTATION?

Concurrent with our investigations of the retinal dopamine system in the rat, there began a debate on the nature of the functional relationship between retinal dopamine and melatonin pathways. It had, for example, been suggested that retinal dopamine release may represent a critical paracrine signal for slow light adaptation, initiating a number of discrete regulatory processes (See Besharse and Iuvone, 1992), such as those already outlined in Table 1. Furthermore, it also became apparent that retinal dopamine and melatonin pathways may function in contraregulatory roles in the control of such retinal activity (Dubocovich, 1989). In the retinae of numerous species, including man, melatonin synthesis is confined primarily to photoreceptors (Wiechmann and Hollyfield, 1988). The level of melatonin synthesis ultimately governs the level of retinal melatonin release (Redburn and Mitchell, 1989), whilst the rate of synthesis is clearly inhibited by light because n-acetyl transferase (NAT), one of the rate limiting enzymes of melatonin synthesis, is not only inhibited by c-AMP but is also calcium dependent (Iuvone & Besharse, 1986). More recently, it has also been shown in some non-mammalian retinae, that in the absence of normal ambient light cues, the daily rhythm of melatonin synthesis persists (Cahill & Besharse, 1993), which indicates that a circadian oscillator may reside within the retina itself. Functionally, melatonin is principally established as a potent inhibitor of mammalian retinal light-dependent dopamine release (Dubocovich, 1983), whilst, conversely dopamine through its action at photoreceptor D2/D4 receptors (Cohen *et. al.*, 1992) will inhibit melatonin release. The natural conclusion from such findings it that retinal dopamine and melatonin pathways perform contraregulatory roles in retinal light/dark activity. In the mammalian retina, at least, there is strong evidence that the dopaminergic system is chiefly active in the light- or day-period (Godley and Wurtman, 1988; Dong and McReynolds, 1991; Djamgoz and Wagner, 1992). The functional interactions between these retinal regulatory pathways in summarised in Figure 6.

## Studies of the Retinal Dopamine Pathway in an Animal Model of Hereditary Retinal Degeneration, the RCS Rat

The Royal College of Surgeons (RCS) rat is an extensively studied animal model of hereditary retinal dystrophy, in which an autosomal recessive defect is manifested as an apparent reduction in the phagocytosis of rod outersegments. In this model, photoreceptor degeneration begins at around 24 days, and the animal eventually becomes blind at around 60 days (Dowling and Sidman, 1962). Having established some of the essential properties of the retinal dopamine pathway in the normal rat we began to study this pathway in the RCS rat model. Our interest in these studies was initiated by previous work in the Vision Research Unit which showed that melatonin levels were some two fold higher in the dystrophic RCS retina, during the critical time which precedes degeneration (Hawlina *et. al.,* 1992). Since the melatonin and dopamine pathways in the retina are linked (Section 6), we began to look for potential abnormalities in the contraregulatory dopamine system in this model of retinal degeneration.

Experiments were performed in which dopamine and dopamine antagonists were applied to the retinae obtained from dystrophic RCS rats at post-natal day 17-24 (a period prior to retinal degeneration) and in their age-matched controls (Hankins and Ikeda, 1994). The typical results are illustrated in Figure 3, which shows that in both the control and the dystrophic retinae, horizontal cells are depolarized and their light-evoked potentials reduced by the application of dopamine (10 μM). The D1 antagonist, SCH 23390 (10 μM) hyperpolarized the membrane potential and increased the light-evoked potentials of horizontal cells in the control retina, since it antagonises endogenously released dopamine at the D1 receptors on the cells (Fig. 3A). This effect of SCH 23390 on horizontal cells in the control retinae was observed in 13 out of 14 cells studied. In contrast, in the dystrophic RCS retinae horizontal cells consistantly failed to respond to the D1 receptor antagonist even at 10 fold higher concentrations (n = 14), although they responded normally to exogenous dopamine (Fig. 3B). Thus, it appears, under our recording and stimulating conditions, that the endogenous dopaminergic pathway is not functional in the dystrophic retina. We also examined the effects of SCH 23390 on horizontal cells in control retinae recorded in the presence of excess exogenous melatonin (500nM - 1μM) and in these experiments the cells also failed to respond (Fig. 3C). The lack of D1 receptor antagonist action in the presence of melatonin in control retinae was observed in all 6 cells studied. Thus, the responses of the horizontal cells to dopaminergic drugs in the control retinae perfused with excess melatonin are similar to those of cells in the dystrophic retinae.

The apparent total dysfunction of the retinal dopamine pathway in the dystrophic animals led us to examine histologically the DA-containing cells in the dystrophic and control animals. Figure 4 shows epi-fluorescence micrographs visualising catecholamine-containing cells in whole mount retinae obtained from a control rat (top) and a dystrophic rat (bottom) both aged 21 days. The mean cell density counts (± 1 SEM), measured on photomicrographs such as those shown in Figure 4, taken at random locations in 5 control and 5 dystrophic rat retinae were 22 ± 4 cells · $mm^{-2}$ and 24 ± 6 cells · $mm^{-2}$ respectively.

These values were not statistically different. In addition, we could not detect any differences between the shape or size range of catecholaminergic retinal cells between control and dystrophic rats (all aged between 20 - 22 days). For both groups, the longest axis of the labelled cell bodies measured 12 to 18 μm and fluorescent processes were observed extending up to 40 to 50 μm from the cell body. Thus the dopaminergic deficit we have observed in the dystrophic retina cannot be attributed to a cellular dopamine deficiency, a conclusion which is further supported by the finding that dopamine retinal tissue content is normal in the RCS rat (Frucht *et. al.,* 1982).

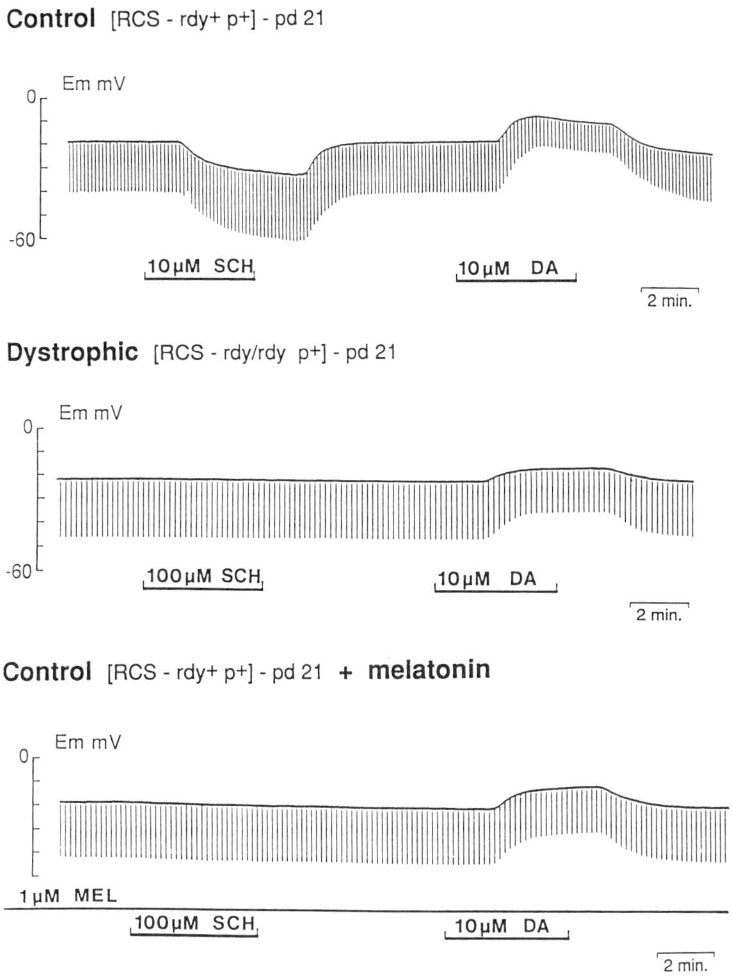


**Figure 3.** The effects of SCH 23390 (SCH, D1 receptor antagonist) and dopamine (DA) on the responses of horizontal cells recorded from control and dystrophic RCS rat retinae. (A): The responses of a horizontal cell from a control RCS rat aged 21 days. (B): The responses from a dystrophic RCS rat aged 21 days (C): Similar responses from a control RCS horizontal cell recorded in the presence of melatonin. Note that, whilst exogenous dopamine depolarized horizontal cells in all three cases, SCH 23390 hyperpolarized only the control RCS cell and not the dystrophic RCS cell, or the control cell recorded in the presence of melatonin. Thus the pharmacological abnormality of the dystrophic RCS cells can be mimicked in control cells by the addition of excess melatonin.

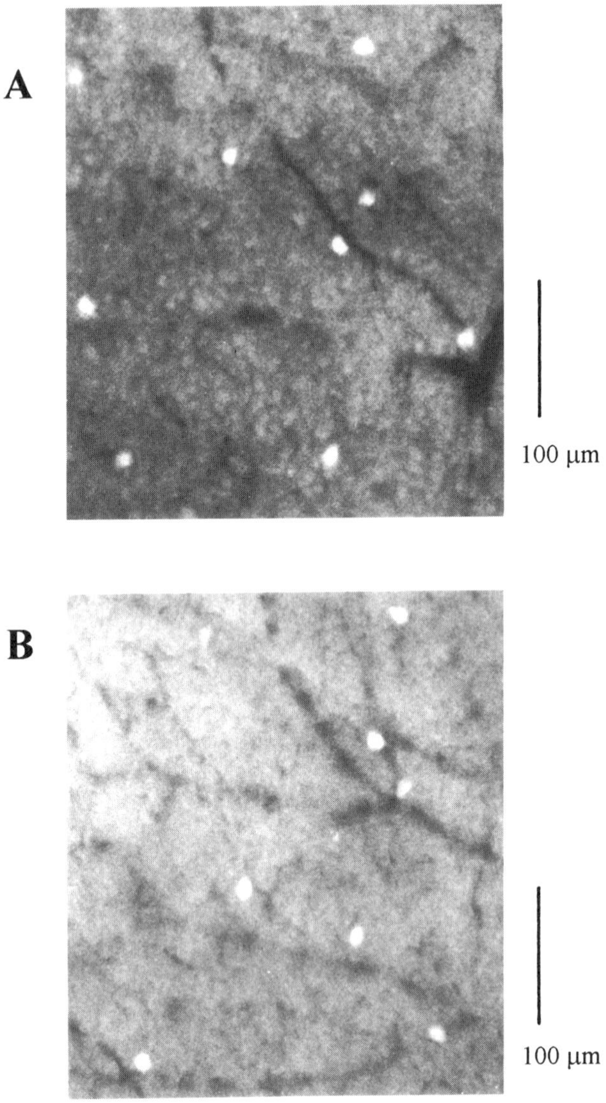


**Figure 4.** Fluorescence photomicrographs of catecholamine-containing cells in whole mount retinae prepared using the FGS technique with no DA analogue preloading. From (A) control and (B) dystrophic RCS rats, both groups aged 21 days. Retinae from both groups showed similar distributions of cell numbers. The scale is provided by the 100 μm markers. Examination of multiple random retinal locations in 6 control and 6 dystrophic retinae revealed no significant differences in the catecholamine cells at the light microscope level.

## Receptive Field Deficits in the RCS Rat Retina

Having established the retinal dopamine deficiency in the dystrophic RCS rat we proceeded to examine whether this deficit might be associated with abnormalities in horizontal cell receptive field properties. This was examined by recording from horizontal cells stimulated by simple small spots of light deflected left and right of the receptive field centre (Fig. 5). These experiments established that the receptive fields of horizontal cells in the dystrophic animals were indeed significantly broader than those observed in the age-matched controls (Fig. 5A). Furthermore, such abnormally broad receptive fields could be replicated in the normal control retinae by the presence of excess exogenous melatonin (> 500 nM; Fig. 5B). These results are consistent with the absence of a functional dopamine system in the dystrophic animals, which leads to abnormally uncoupled horizontal cells with anomalously broad receptive field profiles.

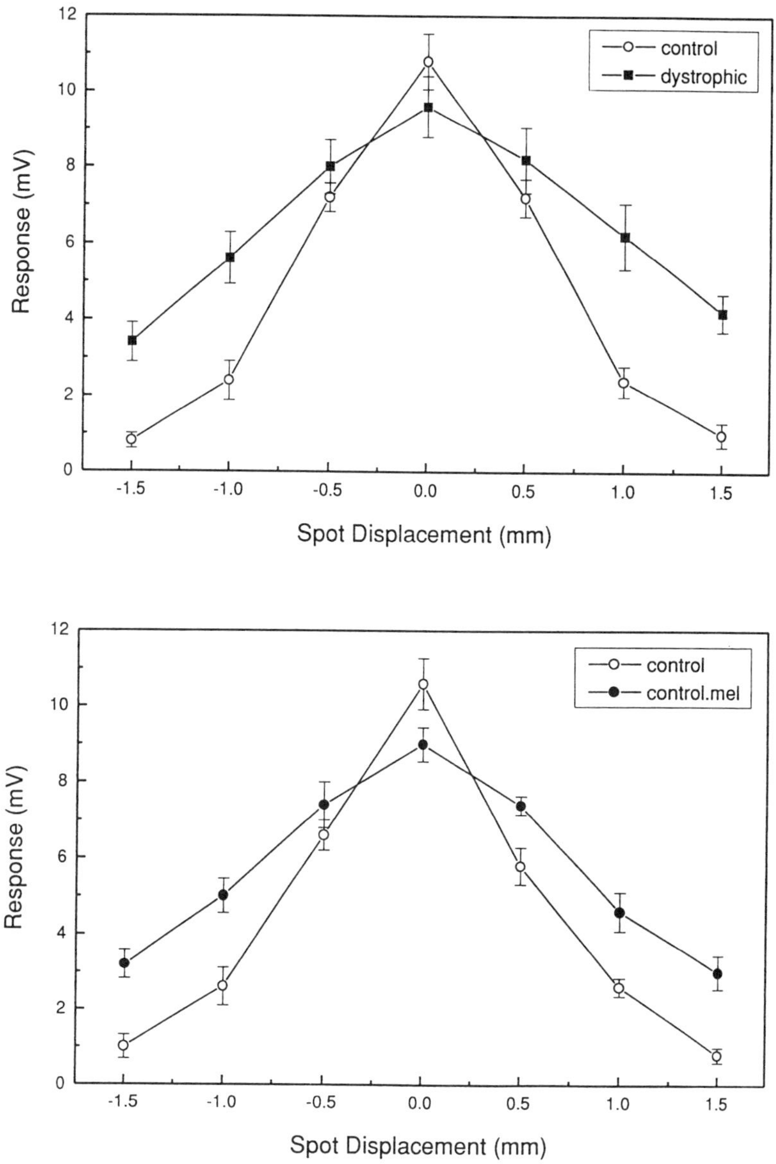


**Figure 5.** Evidence for abnormalities in the receptive field properties of horizontal cells in the dystrophic retina. The extent of horizontal cell receptive fields was assessed by the displacement of a small light spot stimulus (0.5mm dia.) displaced left (-) and right (+) of the recording electrode. In each case recordings were made from 6 cells and the mean amplitude of the resulting light responses (±1 SEM) plotted against spot displacement.(A): The comparison of responses from dystrophic cells with their age-matched controls. Note that the dystrophic cells show a significantly broader receptive field, with larger responses to the stimuli presented at maximal eccentricity. (B): The comparison of the responses of the control group with control cells recorded in the presence of 500 nM melatonin. Note that the presence of melatonin mimics the abnormally broad receptive field responses seen in the dystrophic cells above.

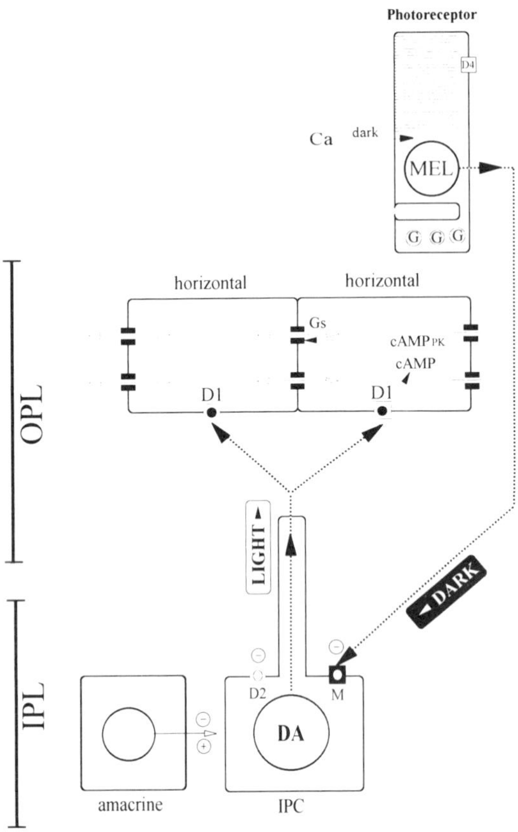


**Figure 6.** Schematic representation of the mammalian retinal dopamine pathway and the relationship with the contraregulatory melatonin pathway. The figure summarises our experimental findings in the rat retina, together with findings from other laboratories as discussed in the text. The level of cyclic-AMP (cAMP) in horizontal cells is regulated through postsynaptic dopamine, D1 type receptors. These receptors are activated primarily via the presynaptic light enhanced release of dopamine (DA) from interplexiform cells (IPCs), although in some species additional paracrine release occurs from DA-amacrine cells. The conductance (Gs) through single gap-junction channels at horizontal cells in the outerplexiform layer (OPL) is modulated principally through cAMP and cAMP-dependent protein kinase (cAMPPK). In this way the DA-IPCs have a direct synaptic effect upon the receptive field properties of retinal horizontal cells, manifested as a reduction in the cell to cell coupling. The DA-IPCs are influenced by a variety of inhibitory and excitatory pathways at the innerplexiform layer (IPL), including: GABA release from amacrine cells and dopamine autoreceptor activation (D2; Harsanyi and Mangel, 1992). Dopamine release is additionally inhibited by the paracrine release of melatonin (MEL) acting at melatonin receptors (M). Whilst the principle photoreceptor transmitter is L-glutamate (G), melatonin synthesis also occurs within the photoreceptors. This proceeds maximally in darkness, but is inhibited by light through a reduction in free calcium (Ca) and a reduction of the D4 receptor coupled changes in c-AMP. In summary, light enhances the synaptic and paracrine retinal dopamine release, whilst darkness inhibits dopamine release via both changes in the synaptic input from amacrine and bipolar cells in the IPL, together with tonic inhibition originating from paracrine retinal melatonin release.

## Conclusions from the RCS Retina Studies

The aetiology of hereditary retinal degeneration in the RCS rat has been shown to be a failure of the RPE to phagocytose the shed outersegment discs of the photoreceptors (Dowling and Sidman, 1962; Bok and Hall, 1971). However, our results suggest that an imbalance of melatonin/dopamine may be an important early factor in the predegenerate RCS retina. We have shown that the high melatonin levels reported in these retinae (Hawlina *et. al.*, 1992), are accompanied by a functional suppression of the retinal dopamine pathway reflected by the observed functional abnormalities of horizontal cells at the outerplexiform layer. Clearly an interesting question is whether this abnormality is an important factor in subsequent retinal degeneration. Interestingly, in support of this notion, melatonin has been shown to accelerate photoreceptor degeneration in the RCS rat (Bubenik and Purtill, 1980). It would thus, be of considerable interest to ascertain whether melatonin antagonists, or indeed dopamine agonists, might either correct these functional abnormalities or perhaps retard retinal degeneration.

## CLINICAL SIGNIFICANCE OF RETINAL DOPAMINE AND THE OPL: PERSPECTIVES

In this chapter we have examined some of the electrophysiological evidence for a functional dopaminergic pathway in the mammalian outerplexiform layer (OPL). This evidence is derived largely from *in vitro* investigations of animal retinae. The functional importance of the OPL is evident since at this, the first synaptic layer of the visual pathway, interactions between photoreceptors, bipolar cells and horizontal cells give rise to retinal centre-surround organisation and hence the initiation of spatial vision. Activity of the retinal dopamine system has a pronounced effect on the receptive field properties of horizontal cells and indeed these cells provide the antagonist surround input to bipolar cells (Kaneko, 1973; Toyoda and Tonasaki, 1978). An important question, therefore, is whether retinal dopamine pathways are functionally expressed in man. The first indication that this may be so originates from vision studies of Parkinson's disease patients, the retinae from whom have been shown to be dopamine deficient (Harnois and DiPaolo, 1990). The functional deficit in contrast sensitivity and visual acuity which has been ascribed to Parkinson's patients (Bodis-Wollner and Yahr, 1978) can be modelled, in part, according to subnormal light-induced uncoupling of horizontal cells at the outerplexiform layer. Further evidence that retinal dopamine pathways have a role in human spatial vision arises from studies of the effects of dopaminergic drugs on normal subjects. For example, it has been shown that the selective dopamine uptake inhibitor, nomifensine, increases contrast sensitivity for spatial frequencies greater than 2 cycles/degree in the normal observer (Domenichi *et al.*, 1985). Subsequent studies, using a range of dopamine agonists, have reported broadly similar results. Therefore once again, it is tempting to associate the spatial sensitivity changes evoked by dopaminergic drugs, with their actions on the coupling of horizontal cells.

We have also provided evidence, derived from animal studies, for a contra-regulatory system involving dopamine and melatonin, which has important implications for the apparent diurnal and adaptational changes that are known to occur in human retinal sensitivity (Birch *et al.*, 1984; Hankins *et al.*, 1995). It is therefore of considerable clinical importance to establish whether such mechanisms are functional in the human retina, and there is indirect evidence for this from a human ERG study on the effects of psoralen drugs, which are known to enhance melatonin release (Souetre *et al.*, 1989). It was reported that 5-methoxypsoralen given to normal subjects caused a pronounced enhancement of retinal b-wave amplitude under photopic conditions and, furthermore, enhanced the b-wave amplitude during the early stages of scotopic dark adaptation. Unfortunately, no studies have yet been performed on the spatial contrast sensitivity of subjects recieving psoralen compounds.

Functional aspects of the role of the retinal dopamine system at the outerplexiform layer of the mammalian retina, together with the relationship with melatonin pathways and the effects of light and darkness have been modelled within the schematic of Figure 6. This is an attempt to incorporate the evidence which we have provided from studies on the rat retinae, together with additional experimental evidence discussed previously . We have primarily investigated the role of dopamine in the regulation of horizontal cell activity, showing how it affects the receptive field properties of these cells. The retinal dopamine system clearly has a role in the regulation of a number of additional discrete mechanisms in the mammalian retina (Table 1). Whilst these individual mechanisms may, at first sight, seem somewhat disparate, they all have potential important roles in the long-term light adaptation of the retina. Whilst primary retinal light and dark adaptation are a photoreceptor phenomenon, it is becoming clear that numerous additional mechanisms prepare the retina for day (light) or night (dark). The experimental evidence so far, suggests that the contraregulatory activity of dopamine and melatonin may integrate these mechanisms and that dopamine is indeed a critical paracrine retinal signal for day (light).

## ACKNOWLEDGMENTS

The experimental work described here was supported by the Special Trustees of St. Thomas' Hospital and the British Retinitis Pigmentosa Society. We are also indebted to Peter de Villiers for technical assistance.

## REFERENCES

Besharse, J.C. and Iuvone, P.M. (1992) Is dopamine a light-adaptive or a dark-adaptive modulator in the retina?, *Neurochem. Int.,* 20:193-199.

Birch, D.G., Bersen, E.L. and Sandberg, M.A. (1984) Diurnal rhythms in the human rod ERG, *Invest. Ophthalmol. Vis. Sci.* 25:236-238.

Bodis-Wollner, I. and Yahr, M. (1978) Measurement of visual evoked potentials in Parkinson's disease, *Brain,* 101:661-671.

Bok, D. and Hall, M.O. (1971) The role of pigment epithelium in the etiology of inherited retinal dystrophy in the rat, *J. Cell Biol.* 14:73-109.

Bubenik, G.A. and Purtill, R.A. (1980) The role of melatonin and dopamine in retinal physiology, *Can. J. Physiol.* 58:1475-1462.

Cahill, G.M. and Besharse, J.C. (1993) Circadian clock functions localised in xenopus photoreceptors, *Neuron* 10:573-577.

Cohen, A.I., Todd, R.D., Harmon, S. and O'Maley, K.L. (1992) Photoreceptors of mouse retinas possess D4 receptors coupled to adenylate cyclase, *Proc. Nat. Acad. Sci. (USA),* 89:12093-12097.

Djamgoz, M.B.A. and Wagner, H.-J. (1992) Localization and function of dopamine in the adult retina, *Neurochem. Int.,* 20:139-191.

Domenichi, L., Trimarchi, C., Piccolino, M., Fiorentini, A. and Maffei, L. (1985) Dopaminergic drugs improve human visual contrast sensitivity, *Human Neurobiol.* 4:195-197.

Dong, C.-J. and McReynolds, J.S. (1991) The relationship between light, dopamine release and horizontal cell coupling in the mudpuppy retina, *J. Physiol.* 440:291-309.

Dowling, J. E. and Ehinger, B. (1975) Synaptic organisation of the amine containing interplexiform cells in the goldfish and cebus monkey retinas, *Science* 188:270-273.

Dowling, J.E. and Ehinger, B. (1978) The interplexiform cell system. I. Synapses of the dopaminergic neurons of the goldfish retina, *Proc. Roy. Soc. Lond. B* 201:7-26.

Dowling, J.E. and Sidman, R.L. (1962) Inherited retinal dystrophy in the rat, *J. Cell Biol.* 14:73-109.

Dubocovich, M.L. (1983) Melatonin is a potent modulator of dopamine release in the retina, *Nature* 306:782-784.

Dubocovich, M.L. (1989) Role of melatonin in the retina, *Prog. Retinal Res.* 8:129-151.

Frucht, Y. Vidauri, J. and Melamed, E. (1982) Light activation of dopaminergic neurones in the rat retina is mediated through photoreceptors, *Brain Res.* 249:153-156.

Godley, B.F. and Wurtman, R.J. (1988) Release of endogenous dopamine from superfused rabbit retina in vitro: effects of light stimulation, *Brain Res.* 452:393-395.

Hampson, E.C.G.M., Vaney, D.I. and Weiler, R. (1992) Dopaminergic modulation of gap-junction permeability between amacrine cells in mammalian retina, *J. Neurosci.* 12:4911-4922.

Hankins, M.W. and Ikeda, H. (1991a) Non-NMDA type excitatory amino acid receptors mediate rod input to horizontal cells in the isolated rat retina, *Vision Res.* 31:609-617.

Hankins, M.W. and Ikeda, H. (1991b) The role of dopaminergic pathways at the outerplexiform layer of the mammalian retina, *Clin. Vision Sci.* 6:87-93.

Hankins, M.W. and Ikeda, H.(1994) Early abnormaities of retinal dopamine pathways in rats with hereditary retinal dystrophy, *Documenta Ophthalmologica.* 86:325-334.

Hankins, M.W., Rowe, J.S. and Ruddock, K.H. (1985) Properties of amino acid binding sites on horizontal cells determined by electrophysiological studies on the isolated roach retina. In: Neurocircuitry of the Retina A Cajal Memorial (Ed. A. Gallego and P. Gouras), Elsevier-N.Y, pp 99-108.

Hankins, M.W., Jones, R.J.M. and Ruddock, K.H. (1995) Diurnal variation in the b-wave component of the human electroretinogram (ERG), *J. Physiol.* 483:42P

Harnois, C. and DiPaolo, T. (1990) Decreased dopamine in the retinas of patients with Parkinson's disease, *Invest.Ophthalmol.Vis. Sci.* 31:2473-2475.

Harsanyi, K. and Mangel, S.C. (1992) Activation of a D2 receptor increases electrical coupling between retinal horizontal cells by inhibiting dopamine release, *Proc. Natl. Acad. Sci.(USA),* 89:9220-9224.

Hawlina, M., Jenkins, H.G., and Ikeda, H. (1992) Diurnal variations in the electroretinographic c-wave and retinal melatonin content in rats with inherited retinal dystrophy, *Documenta Ophthalmologica,* 79:141-150.

Ikeda, H., Priest, T.D., Robbins, J. and Wakakuwa, K. (1986) Silent dopaminergic synapse at feline retinal ganglion cells, *Clin. Vis. Sci.,* 1:25-38.

Iuvone, P.M. and Besharse, J.C. (1986) Dopamine receptor-mediated inhibition of seratonin-N-acetyltransferase activity in the retina, *Brain Res.* 369:168-176.

Kaneko, A. (1973) Receptive field organization of bipolar and amacrine cells in the goldfish retina, *J. Physiol.* 235:133-153.

Kebabian, J.W. and Calne, D.B. (1979) Multiple receptors for dopamine, *Nature* 277:93-96.

Knapp, A.G. and Dowling, J.E. (1987) Dopamine enhances excitatory amino acid-gated conductances in cultured retinal horizontal cells, *Nature* 325:437-439.

Naka, K.-I. and Rushton, W.A.H. (1967) The generation and spread of S-potentials in fish (Cyprinidae), *J. Physiol.* 192:437-461.

Negishi, K. and Drujan, B.D. (1978) Effects of catecholamine on horizontal cell membrane potential in the fish retina, *Sensory Processes* 2:388-395.

Nguyen-Legros, J., Savy, C., Martin-Martinelli, E., and Yelnik J. (1989) Distribution and spatial organisation of dopaminergic interplexiform cells in the rat retina. In: Weiler R. and Osborne N. Eds. Neurobiology of the inner retina. Berlin: Springer-Verlag. pp65-75.

Piccolino, M., Witkovsky, P. and Timarchi, C. (1987) Dopaminergic mechanisms underlying the reduction of the electrical coupling between horizontal cells of the turtle retina induced by D-amphetamine, bicuculline and veratridine, *J. Neurosci.* 7:2273-2284.

Redburn, D.A. and Mitchell, C.K. (1989) Darkness stimulates rapid synthesis and release of melatonin in rat retina, *Vis. Neurosci.* 3:391-403.

Sibley, D.R. and Monsma, F.J. (1992) Molecular biology of dopamine receptors, *Trends Pharmacol. Sci.* 13:61-69.

Souetre, E., De Galeani, B., Gastaud, P., Salvati, E. and Darcourt, G. (1989) 5-methoxypsoralen increases the sensitivity of the retina to light in humans, *Eur. J. Clin. Pharmacol.* 36:59-61.

Teranishi, T., Negishi, K. and Kato, S. (1983) Dopamine modulates S-potential amplitude and dye-coupling between external horizontal cells in carp retina, *Nature* 301:243-246.

Toyoda, J.-I. and Tonasaki, K. (1978) Effect of polarization of horizontal cells on the on the ON-center bipolar cell of the carp retina, *Nature* 276:399-400.

Wiechmann, A.F. and Hollyfield, J.G. (1988) HIOMT-like immunoreactivity in the vertebrate retina: a species comparison, *Exp. Eye Res.* 49:1079-1095.

Witkovsky, P. and Dearry, A. (1991) Functional roles of dopamine in the vertebrate retina, *Prog. Retinal Res.* 11:247-292.

Yamada, E. and Ishikawa, T. (1965) Some observations on the fine structure of the vertebrate retina, *Cold Spring Harb. Symp. Quant. Biol.* 30:383-392.

# SPATIO-CHROMATIC SIGNALLING IN THE VERTEBRATE RETINA

Mustafa B. A. Djamgoz and Renata Petruv

Neurobiology Group
Biology Department
Imperial College of Science, Technology and Medicine
London SW7 2BB, United Kingdom

## INTRODUCTION

The optical image focussed on the photoreceptor layer of the retina has distributed within it a variety of informative features that represent visual conditions in the surrounding environment. One of these features is colour. In fact, the functional organization of the vertebrate visual system is such that the colour of a given 'local' stimulus is not perceived just according to the wavelengths of light reflected from that point but is also influenced by wavelengths emitted from surrounding areas. This spatio-chromatic phenomenon leads to "colour constancy" whereby the colour of an object appears unchanged under different spectral illumination conditions (Land, 1959). Colour constancy has been demonstrated by both psychophysics and electrophysiology in vertebrates as diverse as fish and primates (Zeki, 1980; 1995; Ingle, 1985).

The chromatic world of fish is particularly complex (Levine & MacNichol, 1982; Loew & McFarland, 1990). First, the wavelengths of available light change with depth of water. This dependance is different for sea vs. fresh-water and changes further as the quality and the quantity of suspended chemical (e.g. plant) material vary with season at a given location. Second, fish can change their habitats by migration during development or reproduction. Third, even in a given body of water, at a given depth, the wavelength of light incident upon the retina will depend upon viewing angle. Thus, terrestrial chromatic conditions are made even more complex when light penetrates water.

A considerable amount of work has been done to elucidate the ability of retinal neurones to signal chromatic information in a spatially distributed manner. This article reviews the spatio-chromatic organization of retinal neuronal receptive fields with an emphasis upon (i) fish retinal neurones; and (ii) cone-driven signals. Since spatio-chromatic signalling is likely to be a dynamic process reflecting the variability in stimulus conditions, further emphasis is placed upon the 'plasticity' of receptive fields and underlying neuro-chemical control mechanisms.

*Basic and Clinical Perspectives in Vision Research,* Edited by Jon Robbins et al.
Plenum Press, New York, 1995

## CONE PHOTORECEPTORS

At the sampling stage, spatio-chromatic signalling is affected by the relative densities of 'red-' (R-) 'green-' (G-), 'blue-' (B-), even ultraviolet-sensitive (UV-) cones and their distribution in the plane of the retina. In fish, cones are often arranged in precise and orderly mosaics, for example as "rows", "squares" or "rectangles" over the whole retina (Engström, 1963; Wagner, 1990). In primates, on the other hand, a regular hexagonal mosaic of cones (with a centre - centre separation of 2 - 3 μm) is restricted to the fovea (Wässle & Boycott, 1991).

Intracellular recordings in retinae of turtle (Baylor et al., 1971) and pikeperch (Burkhardt, 1977) have shown that cones summate their light-evoked hyperpolarizing responses over retinal areas of 50 - 250 μm. These findings indicated that cones of a given spectral identity are coupled and interact with one another in a summative manner. Subsequent work demonstrated that cones of different spectral classes also interact in a spatially organized way.

There are two spatially distinct means by which cone - cone interactions can take place.

### Short-Range: Close Apposition

Short-range connections between pairs of cones can occur by means of close apposition or fine processes called "fins". Gap junctions which may mediate low-resistance coupling between cone pairs have been observed in carp (Witkovsky et al., 1974) and goldfish (Stell, 1980).

The physiology of cone - cone interactions by means of 'close apposition' has been studied in the cases of both twin cones and double cones. Twin cones comprise morphologically similar cones with spectrally identical photopigments. In the pikeperch retina, the majority of cones are twin cones that contain photopigments absorbing maximally at 605 nm. Intracellular injection of the fluorescent marker procion (or Lucifer) yellow (P/LY) into one member of given twin cones revealed the dye to remain confined to that member, thereby suggesting lack of substantial dye-coupling between members of twin cones (Burkhardt et al., 1980). However it should be noted that P/LY does not always cross electrotonically communicating gap junctions (e.g. Warner & Lawrence, 1982). In fact, electrical coupling between members of twin cones has been demonstrated in the striped bass retina (Miller & Karenbrot, 1993). The situation may be different in double-cones which are made up of red- and green-sensitive (principal/larger and accessory/smaller, respectively) members, comprising some 50% of the total cone population in cyprinid fish retinae (Scholes, 1975; Stell & Harosi, 1976; Downing et al., 1986). Marchiafava (1985) and Marchiafava et al. (1985) studied cell coupling in isolated double cones of the tench retina. The action spectrum of the double cone was found to be broader than that of single cones, thereby suggesting significant coupling of the two members. This could be an effective means of broadening the spectral sensitivity of a cone mechanism within a short spatial range.

### Long-Range: Telodendritic Contacts

Telodendria are fine neurites that arise from the synaptic terminals of photoreceptors (rods and cones) and extend for distances of 10 - 20 μm, thus mediating 'long' range interactions. Telodendria have been seen in retinae of a variety of vertebrates, from fish to human, and are thought to be a common feature of cones (Schultze, 1866; Cajal, 1893; Stell, 1980; Normann et al., 1984; Kraft & Burkhardt, 1986; Ohtsuka & Kawamata, 1990). Most work has, in fact, been carried out upon cones. Up to some 10 telodendria may arise from a given cone, and a given telodendrion may contact several cone pedicles.

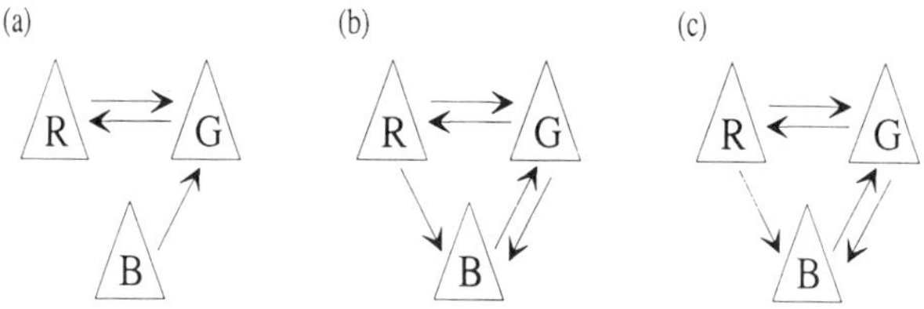


**Figure 1.** The chromaticity and directionality of cone - cone telodendria (indicated by arrows) in retinae of (a) rudd, (b) goldfish and (c) turtle. R, G, B represent red-, green- and blue-sensitive cones. Based upon findings as follows: (a) - Scholes (1975); (b) - Stell (1980); (c) - Ohtsuka & Kawamata (1990).

The chromaticity and directionality of cone - cone telodendritic connections have been described for the cyprinid fish, rudd (Scholes, 1975) and goldfish (Stell, 1980), and the turtle (Ohtsuka & Kawamata, 1990); similar patterns of connectivity were observed (Fig. 1). Reciprocal connections between R- and G-cones occurred in all three species. Goldfish and turtle possessed similar two-way connections between G- and B-cones, whilst this was one way (B $\rightarrow$ G) in rudd. Finally, R $\rightarrow$ B (one-way) telodendria were present in goldfish and turtle, but not in rudd. In perch (walleye) retinae, telodendritic contacts were observed between the two main (orange- and green-sensitive) cone populations present. As regards the nature of the contacts made by telodendria at target cone pedicles, both invaginating (ribbon-related) and basal junctions have been found.

The physiology of telodendritic transmission has been studied in retinae of fish (Kraft & Burkhardt, 1986) and turtle (Normann et al., 1984). Computer simulations of electrotonic spread along fish (walleye) cone telodendria suggested that bidirectional coupling would be significant, although electrophysiological measurements indicated that the intrinsic spectral sensitivities of the cones were not affected. Convincing data for spectral effects involving excitatory interactions between R- and G-cones via their telodendria in the turtle retina were presented by Normann et al. (1984). In a different study, cone telodendria were suggested to be the input elements in the negative feed-back from horizontal cells in the retina of the monochromatic catfish (Sakai & Naka, 1986).

## Modulation of Photoreceptor Coupling

At present, there is no evidence for modulation of cone - cone coupling by light-dark adaptation or putative neuromodulators. On the other hand, rod - cone coupling has been shown to be affected by background light via change of membrane (rather than junctional) conductance (see Barnes, 1995, for short review). Cone mosaics also show considerable structural change: In salmon and trout, for example, the mosaic changes during development (Lyall, 1957). In some cases, a subsct of cones may disappear completely, presumably due to changing chromatic conditions, as in the case of UV-cones in brown trout (Bowmaker & Kunz, 1987). Browman and Hawryshyn (1992) showed that the disappearance of ultraviolet photosensitivity in rainbow trout could be induced by thyroxine treatment.

# HORIZONTAL CELLS

A clear case of spatio-chromatic signalling occurs in horizontal cells (HCs) since these cells are intimately concerned with processing of both spectral and spatial information (Fig. 2). The ability to generate chromatically specific signals results from the cells'

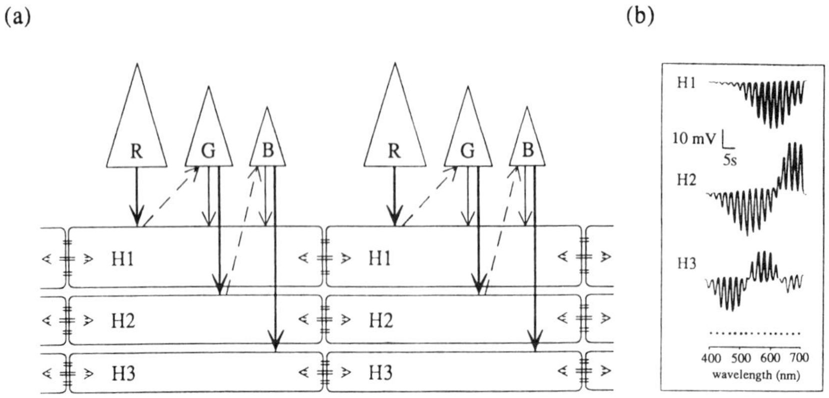


**Figure 2.** (a) Schematic diagram showing the spatio-chromatic organization of horizontal cells in cyprinid fish based mainly on findings by Stell et al. (1975); Downing & Djamgoz (1989); Greenstreet & Djamgoz (1994b). H1 type horizontal cells receive a major excitatory input (as indicated by the thick solid arrows) from red-sensitive (R-) cones and minor excitatory inputs (thin solid arrows) from green- (G-) and blue- (B-) sensitive cones. H1 cells form an inhibitory feed-back pathway (as shown by the dashed arrows) on to G-cones, but also to B- and back on to R-cones (not indicated). H2 type horizontal cells receive a major excitatory input from G-cones and feed-back on to B-cones, which form the major feed-forward cone input to H3 cells. In addition, all three types of horizontal cells are coupled via gap junctions, each forming a separate syncytium across the outer retina which can signal specific spatio-chromatic information (as indicated by the dashed double-headed arrows). (b) Light evoked horizontal cell responses in the roach retina. Red-sensitive luminosity, L/H1-type; biphasic/chromaticity Cb/H2-type and triphasic/chromaticity Ct/H3-type responses are shown. Sixteen different spectral stimuli (20 nm intervals) of near-equal quantum content were presented in sequence (dots at bottom of figure) to cover the spectral range from 400 to 700 nm (inclusive).

connectivity with different subsets of cones (reviewed recently by Djamgoz et al., 1995a). Spatial integration relies upon the cells' syncytial characteristics involving coupling via gap junctions (reviewed recently by Hankins, 1995). Whilst HCs in all vertebrate species studied show some degree of electrotonic coupling, the cone connectivity patterns vary considerably from species to species. On the whole, lower vertebrates (e.g. fish, reptiles) possess a chromatically hierarchical ("cascade") system of feed-forward and feed-back interactions amongst cones and HCs. On the other hand, the chromatic selectivity of cone - HC connectivity appears to be rather variable in mammals.

In the cyprinid fish retina, cone-connected H1, H2 and H3 HCs generating mono-phasic/luminosity, biphasic/chromaticity (red-depolarizing/green-hyperpolarizing, i.e. $R^+G^-$) and triphasic/chromaticity (red-hyperpolarizing/green-depolarizing/blue- hyperpolarizing, i.e. $R^-G^+B^-$) type responses, respectively, form separate syncytia (Fig. 2). Current injection experiments suggested that the coupling of the three lateral systems of HCs are independent with insignificant cross-talk amongst them (Toyoda & Fujimoto,1983).

Spatio-chromatic signalling by the H1 HC network in the carp retina has some novel features. It has been found that spatial summation of responses generated by long wavelength (LW) stimuli is much broader compared with the effect of short wavelength (SW) stimuli (Teranishi et al., 1982; Lipetz & Kaneko, 1984; Kamermans et al., 1989). As regards the cellular mechanisms responsible for this effect, H1 HCs receive the majority of their synaptic drive from R-cones, whilst a minority input is provided by a SW-sensitive (probably B-) cone mechanism. The latter has been suggested to involve a 'conductance- decreasing' receptor, similar to the case of ON-centre bipolar cells, with 2-amino-4- phosphonobutyrate (APB) as an agonist (Yasui et al., 1990; Djamgoz et al., 1993). Thus, the light-evoked changes in the cells' input impendances (due to suppression of cone transmitter release) are in opposite directions with SW stimuli causing a conductance increase (Yamada et al., 1991). Consequently, a resistance shunt is created in the sub-synaptic membrane and this is thought to truncate the spread of light-evoked activity generated by SW stimuli.

## Modulation of Spatio-chromatic Signalling by the HC Network

Spatio-chromatic signalling by the HC network is extremely dynamic since both 'spatial' and 'chromatic' properties show considerable light/dark adaptation dependent plasticity. All three types of HC show some change of spectral character as a result of light/dark adaptation. The effect of light adaptation on H1 HCs is to 'sharpen' the spectral response, mainly by suppressing SW sensitivity (Djamgoz & Yamada, 1992). The chromatic difference in the receptive field size of H1 HCs has also been found to be 'plastic', being enhanced by light adaptation (Petruv et al., 1993). As regards H2 HCs, a type of "hidden C-type response" was described originally by Djamgoz and Ruddock (1978b). Subsequent work showed that the $R^+G^-$ type responses normally generated by H2 type HCs in light-adapted retinae become monophasic with the red-sensitive component gradually disappearing during dark adaptation (reviewed by Wagner & Djamgoz, 1993). The change in the H3 ($R^-G^+B^-$) type HC responses during dark adaptation has recently been elucidated by Greenstreet (1994). Interestingly, the plasticity of the H3 HCs was found to be spectrally selective, the triphasic responses becoming biphasic ($B^-G^+$) only. The behaviour of the H2 and H3 HCs during dark adaptation could be explained by selective suppression of the negative feed-back pathway from H1 HCs onto G-cones. The results on H2 and H3 HCs are consistent with behavioural data on goldfish showing that dark adaptation (Neumeyer & Arnold, 1989), bicuculline (Wietsma & Spekreijse, 1991) and ethambutol (Spekreijse et al., 1991) suppress red-green but not blue-green discrimination.

A substantial body of experimental evidence suggests that the electrotonic coupling of rod- and/or cone-connected HCs (in all vertebrate classes studied) also changes with adaptation (Hankins 1995). On the whole, coupling is weak in the light-adapted state, and becomes stronger during dark adaptation. Further changes may occur after 'prolonged' (>2 h) dark adaptation (Mangel & Dowling, 1985). Spread of signal (i.e. strength of coupling) in the HC syncytium is described by the length constant, $\lambda$ which is proportional to $[g_c/g_m]$ where $g_c$ is the junctional (coupling) conductance and $g_m$ denotes the conductance of the plasma membrane; in turn, $g_m$ comprises synaptic and non-synaptic components. Thus, a change in signal spread can follow from a change in $g_c$ and/or $g_m$, and both types of effect have been suggested to occur (Piccolino et al., 1984; Umino et al., 1991).

The HC gap junction (i.e. $g_c$) is well known to be modulated by dopamine (DA) acting via D1 receptors and protein kinase A, with cAMP as the second messenger (Dowling, 1989; Witkovsky & Dearry, 1991; Djamgoz & Wagner, 1992; Hankins, 1995). Thus, DA mimics the effect of light adaptation (Shigematsu & Yamada, 1988). However, more recent data suggest that light adaptation can cause uncoupling of HCs independently of DA (Baldridge & Ball, 1991). The nitric oxide (NO) system has also been found to be involved in modulating HC coupling (DeVries & Schwartz, 1989; Miyachi et al., 1990; Miyachi & Nishikawa, 1994). Petruv et al. (1993) have shown that application of NO donors to dark-adapted retina induces chromatic difference, as described earlier, in the receptive field size of H1 HCs normally seen in the light-adapted retina. This effect could involve $g_m$ in the form of a cGMP-gated cation channel (Petruv et al., 1993). Nitric oxide may also control other light adaptive changes in the retina such as retinomotor movements and HC spinules (Greenstreet & Djamgoz, 1994a)

## BIPOLAR CELLS

Since colour vision is spatially organized, a light stimulus has to strike spectrally different photoreceptors covering the dendritic area and peripheral regions within a given receptive field. Otherwise, no colour-opponent processing of spatially distributed visual signals can occur. This process is initiated by the bipolar cells (BCs).

Bipolar cells can firstly be classified into two types according to their basic electrophysiological response patterns (Werblin & Dowling, 1969; Kaneko, 1970; Schwartz, 1974). ON-centre cells respond with depolarization to a small central spot of illumination, but with hyperpolarization to surround illumination. OFF-centre cells respond to the same stimuli with opposite polarities, respectively. Thus, with almost no exception, BCs show a receptive field organization consisting of concentric and mutually antagonistic centre and surround zones. Evidence indicates that the centre response reflects direct photoreceptor input, whereas the surround response is mediated by HC activity (Werblin & Dowling, 1969; Kaneko, 1970; Werblin, 1974; Saito & Kujiraoka, 1988). It has been found in the carp retina that BCs of the same morphological and functional type are electrically coupled enabling spatial summation of signals (Saito & Kujiraoka, 1988). This coupling makes the receptive field centres of these neurones much larger than their dendritic trees as observed in goldfish (Kaneko, 1973), dogfish (Ashmore & Falk, 1980) and carp retinae (Saito & Kujiraoka, 1982). However, BC coupling is not nearly as extensive as that of HCs and thus constitutes only a limited syncytium, the functional implications of which are not yet clear.

Bipolar cell responses not only have opponent organization spatially (centre vs. surround), but also spectrally (long vs. medium wavelength) and these aspects combine to give spatio-chromatically organized receptive fields (Fig. 3). Since the BC response to central illumination is from direct photoreceptor input, colour coding in the receptive field centre depends on the type(s) of cones connected. Scholes (1975) examined teleost (rudd) BCs and their photoreceptor connectivity with respect to spectrally distinct cone types, and found the following types of cone-connected BCs:

1. Selective cone BCs: These are 'filamentous' neurones with wide dendritic trees which connect only to cones. Different subtypes synapse exclusively with one or other of the chromatic pathways, though variants were seen for G- or B-, but not for R-cones. This group corresponds to the "small" bipolar cells of Cajal (1893).
2. Mixed cone BCs: These connect predominantly to cones, though some invaginate the spherules of rods as well. Different BCs within this class were found to contact combinations of different chromatic types of cones. The most frequently seen type synapsed with R- and G-cones (and with rods).

Another class of BC (termed "rod" BCs) has dendrites invaginating the spherules of rods and the pedicles of R-cones. There are two distinct morphological subtypes of these BCs having large bulbous synaptic terminals in the two sublaminae of the inner plexiform layer.

The chromatic organization of BC receptive fields was also studied electrophysiologically in goldfish and carp retinae by Kaneko (1973) and Kaneko and Tachibana (1981, 1983). Two types of colour-opponent BCs were found, as shown in Figure 3.

1. Single-colour-opponent BCs: These cells are assumed to get their centre response from R-cones (and rods), and receive same-polarity inputs from R- and G-cones in the receptive field surround (Fig. 3a). The surround responses may come from biphasic chromaticity/H2 type HCs. The two subtypes of this kind of receptive field are red-ON-centre / red- and green-OFF surround; and red-OFF-centre / red- and green-ON-surround. Similar spectrally opponent BCs have been recorded in the turtle retina (Yazulla, 1976).

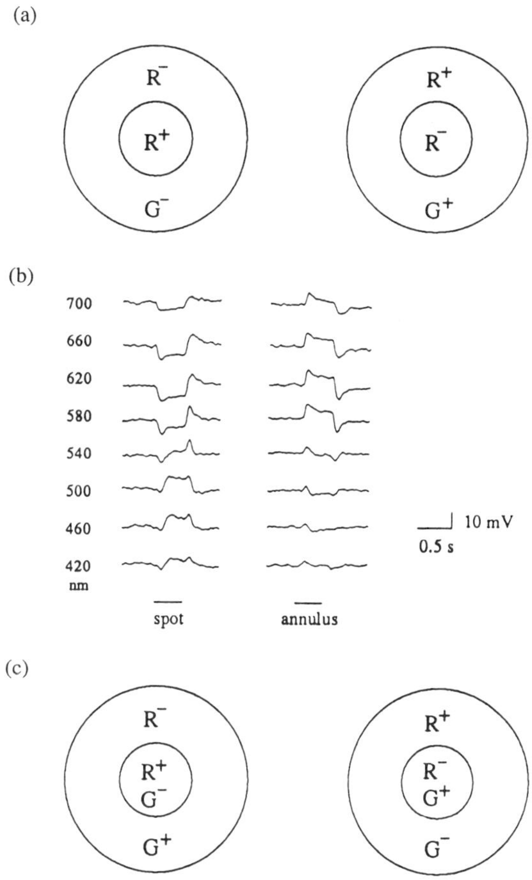


**Figure 3.** Chromatically antagonistic (a) single- and (b & c) double-colour-opponent receptive fields of bipolar cells in the cyprinid fish retina. R and G represent inputs from red- and green-sensitive cones, respectively. Adapted from Kaneko (1973) and Kaneko and Tachibana (1981, 1983). (b) Light-evoked bipolar cell responses in the spectral range 420-700 nm showing double colour opponency in the centre (spot) and surround (annulus) of the receptive field (modified from Kaneko & Tachibana, 1981).

2. Double-colour-opponent BCs: These are colour-opponent in both the centre and surround regions of the receptive fields Fig. 3b, c). Two types of double-opponent cells have been found in carp retina (Kaneko & Tachibana, 1983). One type has a centre response comprising hyperpolarization to red light and depolarization to blue/green light; its surround response is depolarization to red light and hyperpolarization to blue/green light. The other type is its exact mirror image for both centre and surround. These BCs are thought to be the origin of double-colour-opponent receptive fields found earlier in ganglion cells (Daw, 1968).

In addition, some BC receptive fields (probably corresponding to the "rod" BCs) were found to be "non colour-coded". The centre and surround regions in these receptive fields are well balanced and both are driven by rods and R-cones. The surround antagonism in these cells is due to luminosity/H1 type HC input. Similar BCs have been found in turtle by Richter and Simon (1975) and Yazulla (1976). Richter and Simon (1975) found that red backgrounds enhanced the relative sensitivity of OFF-centre BCs to green flashes suggesting that these BCs receive input from red-sensitive members of double cones as well as single R-cones.

In the primate (including human) retina the spatio-chromatic organization of BCs is somewhat different. There appears to be no double-colour-opponent processing in the retina; this occurs in higher cortical centres. Each cone in the fovea and perifoveal area of monkey and human are individually connected to two midget BCs, one for the ON- and the other for the OFF-centre pathway (Kolb, 1970; Gouras, 1971). As a result, midget BCs inherit that photoreceptor's spectral sensitivity (forming the receptive field centre) and convey this chromatic signal to the inner retina. How the opponent surrounds of midget BCs are generated is unclear. It has not yet been possible to record electrophysiologically from BCs in the primate retina to ascertain whether HCs provide the surround response of the midget BCs, although electroretinographic data would indicate that some degree of spectral opponency occurs at the outer plexiform layer of monkey retinae (Sperling, 1986). In mammalian (cat and rabbit) retinae, the antagonistic surrounds of BCs may not be so strongly developed and amacrine cells may have a bigger role in generating the surround fields of ganglion cells, compared with non-mammalian retinae (Nelson & Kolb, 1983; Dacheux & Raviola, 1986). However, Mangel (1991) subsequently presented convincing evidence that HCs do contribute to the receptive field surround of ganglion cells in the rabbit retina.

Single-opponent cells in the visual system may provide the basis for successive colour contrast phenomena whilst double-opponent cells may play a major role in the detection of coloured images by enhancing simultaneous colour contrast (Daw, 1973). In teleost fish, the conversion from simple to double spectral opponency occurs in the retina, while in primates and humans this happens in the brain.

### Modulation of Bipolar Cell Receptive Fields

As already mentioned, BCs of the same physiological type are electrically coupled. Application of DA was found to have little effect on the central receptive field sizes of BCs in the carp retina, indicating that DA does not modulate BC coupling (Yamada & Saito, 1988). However, Dowling and Naka (Dowling, 1986) noted that following DA application to the catfish retina, no antagonistic surround was observed in BC responses. This was due to the strong uncoupling effect of DA on HCs. No antagonistic surround in the BC response was observed because the receptive field size of the HC had decreased to a size smaller than the entre field of the BC. Dopamine, therefore, does not appear to change the basic receptive field organization of BCs; instead it alters the balance of centre and surround responses (Hedden & Dowling, 1978).

It is known that reorganization of double-colour-opponent receptive fields occurs during light/dark adaptation in ganglion cells of fish whereby in the dark-adapted state these cells lose their colour opponency (Adams & Afandor, 1971; Beauchamp & Daw, 1972; Raynauld et al., 1979). It is not known, however, whether the same phenomenon is apparent in double- (or single-) colour-opponent BCs.

## AMACRINE CELLS

Both sustained and transient amacrine cells (ACs) in goldfish have been shown to summate their signals over a retinal area of at least 2.5 mm in diameter (Kaneko, 1973). The lateral spread of signals in the transient (ON-OFF) AC layer of the roach retina was found to be much more sensitive to local $CoCl_2$ application than the spread of signals in the HC layer (Djamgoz & Ruddock, 1978a). This has been attributed to chemical coupling of the ACs. Consistently with this suggestion, Witkovsky and Dowling (1969) have shown that serial, chemical synapses are found amongst AC processes in carp retina. Naka and Christensen (1981), on the other hand, have provided electrophysiological and ultrastructural evidence that ACs of catfish may also couple through electrotonic (gap) junctions. Gleason et al. (1993) studied synaptic transmission between isolated pairs of chick GABAergic ACs in culture and found both chemical and electrical coupling. It is not yet certain whether the electrotonically-coupled ACs *in vivo* all use the same transmitter or different transmitters. Biochemical release

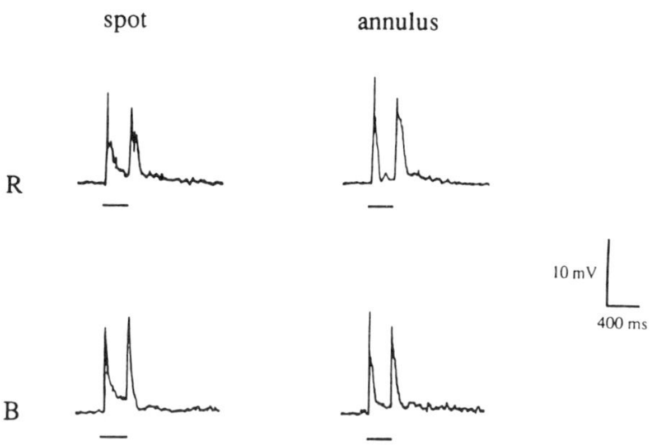


**Figure 4.** Spatio-chromaticity of light-evoked responses of a transient amacrine cell in the carp retina. Responses to central (spot; 0.5 mm diameter) illumination showed colour opponency in the ON and OFF phases of stimulation (as indicated by the short bars) to red (R; 619 nm) and blue (B; 454 nm) light. Peripheral (annulus; 0.4 mm/1 mm inner/outer diameter) illumination evoked an opponent ON-OFF response to blue light only. Photon flux, around 108/photons/m2/sec.

experiments have shown that there are excitatory and inhibitory synaptic interactions between chemically unlike ACs. For example, putative enkephalinergic ACs inhibit the GABAergic cells in goldfish retina (Djamgoz et al., 1981). In turn, the latter and the glycinergic cells are inhibitory to cholinergic ACs in rabbit retinae (Neal, 1983). In the same retina $\alpha$-MSH has been shown to excite the GABAergic cells (Bauer & Ehinger, 1980).

There are conflicting reports about centre-surround antagonism in AC receptive fields. Kaneko (1973; goldfish) and Murakami and Shimoda (1977; carp) did not observe centre-surround antagonisms in ACs. Toyoda et al. (1973) reported, however, that some one-half of the sustained units encountered in the inner nuclear layer of the carp retina had such an organization. In catfish, Naka and Ohtsuka (1975) and Naka et al. (1975) initially maintained that ACs had monophasic receptive fields. Subsequently, Davis and Naka (1980) showed that a large, antagonistic surround field existed. This seems to be true also in the roach retina, where using simple spot and annulus shaped test stimuli, we have observed that some centre-surround antagonism also exists in the maintained component of both sustained and transient units (M.B.A. Djamgoz, unpublished observations).

The chromatic organization of ACs in the cyprinid fish retina was first described by Kaneko (1973). The sustained units of the goldfish were found to be colour-coded similarly to the biphasic chromaticity (H2) type HCs, but they were hyperpolarizing to long, and depolarizing to short wavelength lights. A mirror image of this response pattern was found subsequently by Djamgoz et al. (1990). Mitarai et al. (1978) described some sustained responses with chromatically biphasic receptive fields in the carp retina. ON-OFF transient ACs of the carp also appear to be concerned with spatio-chromatic signalling whereby the chromatic balance of the 'ON' and the 'OFF' transients differs in the central vs. peripheral parts of the receptive fields (Fig. 4).

## Modulation of AC Coupling

At present, very little is known about modulation of AC coupling (or spatio-chromatic signalling by ACs) in vertebrate retinae. Possible effects of light/dark adaptation have not been studied. Most of what is known concerns the effect of DA on dye-coupling. In the mammalian (rabbit) retina, DA was found to modulate the gap junction permeability between ACs. In particular, the dye-coupling of the AII ACs was reduced by DA acting via D1 receptors (Hampson et al., 1992; Mills & Massey, 1995). In contrast, Teranishi and Negishi (1994) showed that DA had no effect on the dye-coupling of ACs in the carp retina, although under the same experimental conditions used, HCs were uncoupled.

## GANGLION CELLS

The basic ganglion cell (GC) receptive field arrangement is an antagonistic centre-surround organization similar to BCs. Light-evoked GC activity comprises action potentials superimposed on slow potentials. The spike train evoked in response to stimulation of the receptive field centre by a small spot of light can be transient or sustained. In turn, sustained responses can be ON or OFF, whilst transient responses can be ON, OFF or ON-OFF; opposite response patterns can be generated by stimulation of corresponding surround zones. These GC responses strongly reflect the properties of the input neurones (BCs and ACs). Additionally, HCs play an important role in the formation of receptive field surrounds of BCs, and therefore contribute to the centre-surround receptive field organization of retinal GCs (Kaneko, 1973; Toyoda & Tonosaki, 1978; Mangel, 1991).

In non-mammalian species, little is yet known concerning the bipolar connectivity to GCs and the role of ACs in receptive field organization at the GC level. BC and AC inputs to GCs have been studied most extensively in the catfish retina. BC to GC transmission was shown to be sign conserving and restricted to similar-polarity members of the ON and the OFF systems (Naka, 1977). In contrast, Witkovsky and Stone (1983) showed in *Xenopus* retina that the chromatic-BC to colour-coded GC synapse is sign-inverting. Lasater (1982) and Sakai and Naka (1987) divided ON-centre and OFF-centre GCs each into "simple" and "complex" cells. Receptive fields of simple GCs reflect mainly linear input(s) from physiologically corresponding BCs, while complex GCs receive substantial input also from sustained ACs of similar physiological type. The greater the AC inputs to GCs, the more complex the receptive field characteristics of the GCs (Dubin, 1970; Dowling, 1979).

Investigations into the spatial properties of goldfish GCs have shown that they possess the same response characteristics as GCs in other species, such as cat (Enroth-Cugell & Robson, 1966; Hochstein & Shapley, 1976). In particular, goldfish GCs could also be classified as X-, Y- or W-like based on the linearity of spatial summation within their receptive fields (Levine & Shefner, 1979; Bilotta & Abramov, 1989a).

Ganglion cell fields were studied extensively in goldfish retina by Daw (1968) and three types unit of were found:

1. Type P GCs: These cells are colour-opponent only in the receptive field centre. They constituted only about 5% of the GC population studied by Daw (1968). Similar units were found by Wagner et al. (1960, 1963).
2. Type O GCs: The receptive fields of these cells are colour-opponent both in the centre and surround regions. They constituted about 50 % of the GC population and seemed ideal for simultaneous colour contrast. Some type O units possessed a narrow intermediate annular zone (surrounding the centre) in which the centre response to green light was unopposed by red due to a slightly larger central field for the green.
3. Type Q GCs: These cells were found to have a response component driven by G-cones that was masked by red-driven input.The hidden green-sensitive response could be isolated by using high intensity stimulation or chromatic adaptation with long wavelength light. The underlying receptive field organization was essentially double-colour-opponent, as in type O GCs.

Subsequent studies showed that GCs of the goldfish retina also possess inputs from B-cones (and rods). Spekreijse et al. (1972) proposed that the blue-sensitive response component was only present in the centres of the receptive fields, alongside a green-sensitive component and had the same sign as the red-sensitive response. Beauchamp and Daw (1972), however, found that while the B-cone input was always found together with a green-sensitive component, it did not necessarily have the same response sign as the red-driven input, and that the B-cone input could occur in both the centre and the periphery of concentrically organized receptive fields. It has subsequently been shown that most goldfish GCs receive input from all three cone types as regards both centre and surround mechanisms, even in cells that superficially appear to possess only the red-sensitive component (Mackintosh et al., 1987).

In frogs, chromatic signal processing at the GC layer has been shown to be even more complex: Directional selectivity, chromatic composition of the stimulus and angular velocity appeared to be interrelated (Grüsser-Cornhels & Langeveld, 1985).

Attempts to relate spatial and spectral properties in the primate retina have examined the response of GCs to luminance contrast. As for BCs, primates only have single-colour-opponent GCs in the retina. In the monkey, double-colour-opponent GCs, comparable to those in the goldfish retina are found only in higher cortical centres (Thorell et al., 1984). Kaplan and Shapley (1986) found that spectrally non-opponent GCs of the macaque monkey were very sensitive to stimulus contrast, whereas spectrally opponent cells were relatively insensitive to contrast. Similar findings were reported for goldfish GCs; spectrally non-opponent cells were more sensitive to contrast than colour-opponent cells (Bilotta & Abramov, 1989b). This may not be a surprising finding since spectrally opponent cells provide colour information and may be less concerned with intensity variations or luminance contrast. These functional properties, for both the monkey and goldfish, do not correspond to the X and Y pathways found in the cat. Bilotta and Abramov (1989b) showed in goldfish that the spatial summation class of GCs (X-, Y- or W-like) is independent of its spectral properties. Each class contains cells that provide colour information (spectrally opponent) and cells that do not (spectrally non-opponent). Also, each spatial summation class contains red-ON centre, red-OFF centre, and red-ON/OFF centre cells. Similarly, van Dijk and Spekreijse (1984) found that about 30% of double-opponent GCs in carp were linearly opponent and that these cells could show both strong and weak colour opponency. Weak opponency was seen by simulation with small spots and low intensities where the R-cone component dominated the centre responses of the cells; strong opponency was observed for large spots and high intensities. It would appear, therefore, that spatial summation, R-cone centre response type, and spectral opponency are functionally independent.

In the primate retina, colour-coded GCs fall into two major physiological classes: (i) red/green opponent cells, which receive antagonistic input from G- and R-cones, and (ii) blue/yellow opponent cells, which receive input from B-cones, opposed by combined G- and R-cone inputs (De Monasterio & Gouras, 1975; Gouras & Zrenner, 1981a & b; Gouras & Eggers, 1982, 1983) . The neural mechanisms underlying colour opponency in the primate retina are not clearly understood. The cone midget BC midget GC chain of retinal neurones is assumed to be the R/G opponent tonic system (Perry et al., 1984). This is because the midget GCs dominate in the fovea, have the smallest receptive fields and respond with spatially opponent, spectrally pure centre and surround responses just as expected for the red- or green-ON-centre, red- or green-OFF-surround, or green- or red-OFF-centre and green- or red-ON-surround cells driven by the single cone midget BC chain. Recently, it has been shown that the GCs responsible for carrying B-cone messages in a spectrally coded manner are of a distinct bistratified type, named the "blue-ON" cells (Dacey & Lee, 1994). Thus, an anatomically distinct pathway that conveys B-cone signals to the brain is thought to exist. In addition, the morphology of this blue-ON GC suggests a novel hypothesis for the retinal circuitry underlying the blue/yellow opponent response (Dacey & Lee, 1994).

Spectrally opponent GCs of the peripheral retina in the macaque monkey do not show overt colour opponency as do cells in the fovea (Zrenner & Gouras, 1983). These peripheral GCs are usually dominated by long wavelength-sensitive cone input and are, therefore, spectrally non-opponent although there is some concealed colour opponency, as in the case of goldfish GCs. Spectrally non-opponent GCs of the goldfish are dominated by the long wavelength-sensitive cone input (Spekreijse et al., 1972) but under chromatic adaptation reveal hidden colour opponency (Abramov & Levine, 1972; Mackintosh et al., 1987).

## Modulation of Spatio-chromatic Signalling in Ganglion Cell Receptive Fields

The effect of adaptation on the basic receptive field organization of GCs was studied initially in the cat retina by Barlow et al. (1957) who showed that as the mean luminance dropped to scotopic levels, the surround response disappeared, so that the cell's response became dominated by its centre mechanism. However, Enroth-Cugell and Lennie (1975) recording directly from cat retinal GCs and Kaplan et al. (1979) recording at the level of the lateral geniculate nucleus found a wide variation in the effects of mean luminance on the centre-surround balance. Kaplan et al. (1979) showed that the surround did not always disappear in the dark and in some cases the surround mechanism was still able to antagonize the centre response following dark adaptation. In goldfish BCs, modulation of the relative strengths of the centre and surround mechanisms was shown to be due to the uncoupling of HCs by DA (Hedden & Dowling, 1978). Similarly, Thier and Alder (1984) reported that DA reduced the strength of the antagonistic surround in all GC types tested in the cat retina.

Adams and Afandor (1971) measured GC responses at varying levels of adaptation in the intact goldfish eye and found that GCs lost their double colour opponency under dark adaptation conditions and became rod-driven (Fig. 5). The sign of the rod input was always the same as that of the red-driven response in the same part of the receptive field. These findings were confirmed by Beauchamp and Daw (1972) and Raynauld et al. (1979).

Schellart et al. (1984) showed that GABA induced a loss of spectral antagonism in the receptive field centre of all OFF-centre single-colour-opponent GCs studied in the goldfish retina. This effect was completely reversed on addition of picrotoxin. In the presence of GABA, a green spot caused a stronger OFF-response than a red spot of equal size and quantal flux, whereas without GABA the 'green' response was weaker and of opposite sign. The spectral sensitivity of these OFF-centre GCs in the presence of GABA showed that the sensitivity in the 'middle' wavelength range was due to input from a G-cone mechanism. The response observed, therefore, was due to a second green-sensitive process which was synergistic to the red-sensitive mechanism in the receptive field centre.

Bicuculline was found to have little effect on the spatial profiles of OFF-centre cells, but greatly simplified the 'complex' receptive fields of ON-centre cells in the catfish retina (Lasater & Lam, 1984b). These results indicated that GABAergic AC input to OFF-GCs simply supplies a tonic inhibitory input in the catfish retina. However, for ON-cells the GABAergic AC is a major source of input which helps to shape the receptive field (Lasater & Lam, 1984a). This input could come directly from ACs or it could be derived indirectly from BCs.

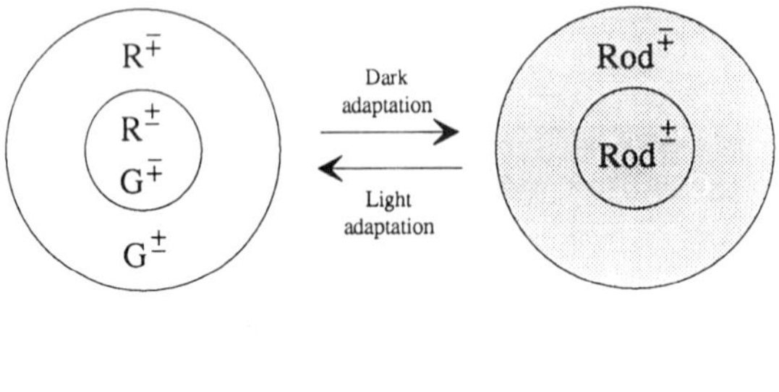


**Figure. 5.** Reorganization of double-colour-opponent ganglion cell receptive fields during light-dark adaptation in the goldfish retina. In the dark-adapted state (indicated by shading) the units lose their colour opponency and become rod-driven. The sign of the rod input is the same as that of the red-sensitive cone input in a given part of the receptive field. R and G represent inputs from red- and green-sensitive cones, respectively. (+) and (-) denote ON and OFF responses, respectively. Based on the findings of Adams and Afandor (1971).

## EFFERENT INPUT

It is now well established that efferent fibres originating in the brain enter the retina via the optic nerve and impinge upon specific neurones (e.g. Cowan, 1970; Stell et al., 1987). In cyprinid fish, these efferents arise from the olfactory bulb and, at least some of them, synapse upon the dopaminergic interplexiform cells (Zucker & Dowling, 1987). Since DA is involved in modulating both spatial and chromatic properties of HCs, an indirect effect upon spatio-chromatic signalling would, therefore, be expected. The effect of eliminating the efferent pathway on absolute visual sensitivity was tested psychophysically in goldfish, but no effect was found (Davis et al., 1988). However, the effect, if any, on spatio-chromatic signalling has not been investigated.

## CONCLUSIONS

The available data, as reviewed in this article, clearly show that all major neuronal types in the retina have an involvement in signalling spatio-chromatic information. This occurs in a number of different ways, e.g. colour-coded, centre-surround antagonistic receptive fields in BCs and GCs, and a hierarchical system of colour-opponent responses in HC syncytia. Importantly, spatio-chromatic signalling is a dynamic process dependent upon light/dark adaptational state of the retina. On the whole, colour-opponent responses (at least in HCs and GCs) are weak or non-existent in dark-adapted retinae, and become potentiated by light adaptation. A substantial body of evidence has shown that DA is a mediator of light adaptive changes (Djamgoz & Wagner, 1992). However, more recent work suggests that NO could comprise another neurochemical mechanism controlling spatio-chromatic signalling by retinal neurones (Petruv et al., 1993). Interestingly, the two mechanisms may not function independently since NO donors have been found to reduce depolarization-induced DA release in the mammalian retina (Djamgoz et al., 1995b).

## REFERENCES


Abramov, I. & Levine, M.W. (1972) The effects of carbon dioxide on the excised goldfish retina, *Vision Res.* 12:1881-1895.

Adams, A.J. & Afandor, A.J. (1971) Ganglion cell receptive field organization at different levels of light adaptation, *Am. J. Optom.* 48:889-894.

Ashmore, J.F. & Falk, G. (1980) Responses of rod bipolar cells in the dark adapted retina of the dogfish, *Scyliorhinus canicula*, *J. Physiol.* 300:115-150.

Barlow, H.B., Fitzhugh, R., & Kuffler, S.W. (1957) Change of organization in the receptive fields of the cat's retina during dark adaptation, *J. Physiol.* 137:338-354.

Barnes, S. (1995) Photoreceptor synaptic output: neurotransmitter release and photoreceptor coupling. In Neurobiology and Clinical Aspects of the Outer Retina (Djamgoz M.B.A., Archer, S. and Vallerga, S. eds.). Chapman & Hall, London. pp. 133-153.

Baldridge, W.H. & Ball, A.K. (1991) Background illumination reduces horizontal cell receptive-field size in both normal and 6-hydroxydopamine lesioned goldfish retinas, *Visual Neurosci.* 7:441-450.

Bauer, B. & Ehinger, B. (1980) Action of alpha-MSH on the release of neurotransmitters from the retina, *Acta Physiol. Scand.* 108:105-107.

Baylor, D.A., Fourtes, M.G.F & O'Bryan, P.M. (1971) Receptive field of cones in the retina of the turtle, *J. Physiol.* 214:65-214.

Beauchamp, R.D. & Daw, N.W. (1972) Rod and cone input to single goldfish optic nerve fibres, *Vision Res.* 12:1201-1212.

Bilotta, J. & Abramov, I. (1989a) Spatial properties of goldfish ganglion cells, *J. Gen. Physiol.* 93:1147-1169.

Bilotta, J. & Abramov, I. (1989b) Spatio-spectral properties of goldfish retinal ganglion cells, *J. Neurophysiol.* 62:1140-1148.

Bowmaker, J.K. & Kunz, Y.W. (1987) Ultraviolet receptors, tetrachromatic colour vision and retinal mosaics in the brown trout (*Salmo trutta*): age-dependent changes, *Vision Res.* 27:2101-2108.

Browman, H.I. & Hawryshyn, C.W. (1992) Thyroxine induces a precocial loss of ultraviolet photosensitivity in rainbow trout (*Oncorhynchus mykiss*), *Vision Res.* 32:2303-2312.

Burkhardt, D.A. (1977) Responses and receptive field organization of cones in perch retina, *J. Neurophysiol.* 40:53-62.

Burkhardt, D.A., Hassin, G., Levine, J.S. & MacNichol, E.F. Jr. (1980) Electrical responses and photopigments of twin cones in the retina of the walleye, *J. Neurophysiol.* 309:215-228.

Cajal, S.R. (1893) La retiné des vertébrés, *La Cellule* 9:17-257.

Cowan, M.W. (1970) Centrifugal fibres to the avian retina, *Br. Med. Bull.* 26:112-118.

Dacey, D.M. & Lee, B.B. (1994) The 'blue-on' opponent pathway in primate retina originates from a distinct bistratified ganglion cell type, *Nature* 367:731-735.

Dacheux, R.F. & Raviola, E. (1986) The rod pathway in the rabbit retina: a depolarizing bipolar and amacrine cell, *J. Neurosci.* 6:331-345.

Davis, G.W. & Naka, K-I. (1980) Spatial organization of catfish retinal neurons. I. Single-and random-bar stimulation, *J. Neurophysiol.* 43:807-831.

Davis, R.E., Kyle, A. & Klinger, P.D. (1988) Nervus terminalis innervation of the goldfish retina and behavioural visual sensitivity, *Neurosci. Lett.* 91:126-130.

Daw, N.W. (1968) Colour-coded ganglion cells in the goldfish retina. Extension of their receptive fields by means of new stimuli, *J. Physiol.* 197:567-592.

Daw, N.W. (1973) Neurophysiology of colour vision, *Physiol. Rev.* 53:571-611.

De Monasterio, F.M. & Gouras, P. (1975) Functional properties of ganglion cells of the rhesus monkey retina, *J. Physiol.* 251:167-195.

DeVries, S.H. & Schwartz, E.A. (1989) Modulation of an electrical synapse between solitary pairs of catfish horizontal cells by dopamine and second messengers, *J. Physiol.* 44:351-375.

van Dijk, B.W. & Spekreijse, H. (1984) Linear color opponency in carp retinal ganglion cells, *Vision Res.* 24:1865-1872.

Djamgoz M.B.A. & Ruddock K.H. (1978a) Properties of amacrine cell responses recorded from isolated fish retinae, *Neurosci. Lett.* 7:89-93.

Djamgoz, M.B.A. & Ruddock, K.H. (1978b) Changes in structure and electrophysiological function of retinal neurones induced by laser irradiation, *Neurosci. Lett.* 7:251-256.

Djamgoz, M.B.A., Stell, W.K., Chin, C.A. & Lam, D.M.K. (1981) An opiate system in the goldfish retina, *Nature* 292:620-623.

Djamgoz, M.B.A., Spadevecchia, L., Usai, C. & Vallerga, S. (1990) Light-evoked response pattern variability and morphological characterization of amacrine cells in goldfish retina, *J. Comp. Neurol.* 301:171-190.

Djamgoz, M.B.A. & Wagner, H-J. (1992) Localization and function of dopamine in the adult vertebrate retina, *Neurochem. Int.* 20:139-191.

Djamgoz, M.B.A & Yamada, M. (1992) Dopamine and light adaptation sharpen the spectral response of H1 horizontal cells in carp retina, *Neurosci.Res. Comm.* 10:149-153.

Djamgoz, M.B.A., Furukawa, T., Yamada, M. and Yasui, S. (1993) Short-wavelength signal transmission to H1 horizontal cells in the isolated carp retina: involvement of APB receptor activated by light adaptation, *J. Physiol.* 467:353P.

Djamgoz, M.B.A., Wagner, H-J & Witkovsky, P. (1995a) Photoreceptor-horizontal cell connectivity, synaptic transmission and neuromodulation. In Neurobiology and Clinical Aspects of the Outer Retina. (Djamgoz M.B.A., Archer S.N. and Vallerga S. eds.). Chapman & Hall, London. pp. 155-193.

Djamgoz, M.B.A., Cunningham, J.R., Hutson, P.H., Murray, F. & Neal, M.J. (1995b) Nitric oxide inhibits depolarization-induced release of endogenous dopamine in the rabbit retina. *Neurosci. Lett.* 198:1-4.

Dowling, J.E. (1979) Information processing by local circuits: the vertebrate retina as a model system. In The Neurosciences, Fourth Study Program (Schmitt F.O. and Worden F.G. eds.). MIT Press, Cambridge, Mass. pp. 163-181.

Dowling, J.E. (1986) Dopamine: a retinal neuromodulator?, *Trends Neurosci.* 9:236-240.

Dowling, J.E. (1989) Neuromodulation in the retina: the role of dopamine, *The Neurosciences* 1, 35-43.

Downing, J.E.G., Djamgoz, M.B.A. & Bowmaker, J.K. (1986) Photoreceptors of a cyprinid fish, the roach: morphological and spectral characteristics, *J. Comp. Physiol. A* 159:859-868.

Downing, J.E.G. and Djamgoz, M.B.A. (1989) Quantitative analysis of cone photoreceptor- horizontal cell connectivity patterns in the retina of a cyprinid fish: electron microscopy of functionally-identified and HRP-labelled horizontal cells, *J. Comp. Neurol.* 289:537-553.

Dubin, M. (1970) The inner plexiform layer of the vertebrate retina: a quantitative and comparative electron microscopic analysis, J. Comp. Neurol. 140, 479-506.

Engström, A. (1963) Cone types and cone arrangements in teleost retinae, *Acta Zool.* 42:179-243.

Enroth-Cugell, C. & Lennie, P. (1975) The control of retinal ganglion cell discharge by receptive field surrounds, *J. Physiol.* 247:51-578.

Enroth-Cugell, C. & Robson, (1966) The contrast sensitivity of retinal ganglion cells of the cat, *J. Physiol.* 87:517-552.

Gleason, E., Borges, S. & Wilson, M. (1993) Synaptic transmission between pairs of retinal amacrine cells in culture, *J. Neurosci.* 13:2359-2370.

Gouras, P. (1971) The function of the midget cell system in primate colour vision, *Vision Res. Suppl.* 3:397-410.

Gouras, P. & Zrenner, E. (1981a) Color coding in primate retina, *Vision Res.* 21:1591-1598.

Gouras, P. & Zrenner, E. (1981b) Color vision: a review from a neurophysiological perspective, *Prog. Sensory Physiol.* 1:139-179.

Gouras, P. & Eggers, H. (1982) Ganglion cells mediating the signals of blue sensitive cones in primate retina detect white-yellow borders independently of brightness, *Vision Res.* 22:675-679.

Gouras, P. & Eggers, H. (1983) Responses of primate ganglion cells to moving spectral contrast, *Vision Res.* 23:1175-1182.

Greenstreet E.H. (1994) Neuronal connectivity and plasticity in the outer retina of cyprinid fish. PhD Thesis. University of London.

Greenstreet E.H. & Djamgoz M.B.A. (1994a) Nitric oxide induces light-adaptive morphological changes in retinal neurones, *NeuroReport* 6:109-112.

Greenstreet, E.H. & Djamgoz, M.B.A. (1994b) Triphasic chromaticity-type horizontal cells selectively contact short wavelength-sensitive cone photoreceptors in the retina of a cyprinid fish, *Rutilus rutilus*, *Proc. R. Soc. Lond. B* 256:227-230.

Grüsser-Cornhels, U. & Langeveld, S. (1985) Velocity sensitivity and directional selectivity of frog retinal ganglion cells depend on chromaticity of moving stimuli, *Brain Behav. Evol.* 27:165-185.

Hampson, E.C.G.M., Vaney, D.I. & Weiler, R. (1992) Dopaminergic modulation of gap junction permeability between amacrine cells in mammalian retina, *J. Neurosci.* 12:4911-4922.

Hankins, M.W. (1995) Horizontal cell coupling and its regulation. In Neurobiology and Clinical Aspects of the Outer Retina. (Djamgoz M.B.A., ArcherS.N. & Vallerga S. eds.). Chapman & Hall. London. pp. 195-220.

Hedden, W.L. & Dowling, J.E. (1978) The interplexiform cell system. II. Effects of dopamine on goldfish retinal neurones, *Proc. R. Soc. Lond. B* 201:27-55.

Hochstein, S. & Shapley, R.M. (1976) Quantitative analysis of retinal ganglion cell classifications, *J. Physiol.* 262:237-264.

Ingle, D.J. (1985) The goldfish as a retinex animal, *Science* 227:755-758.

Kamermans, M., van Dijk, B.W., Spekreijse, H. & Zweypfenning, R.C.V.J. (1989) Lateral feedback from monophasic horizontal cells to cones in carp retina. I. Experiments, *J. Gen. Physiol.* 93:681-694.

Kaneko, A. (1970) Physiological and morphological identification of horizontal, bipolar and amacrine cells of goldfish retina, *J. Physiol.* 207:623-633.

Kaneko, A. (1973) Receptive field organization of bipolar and amacrine cells in the goldfish retina, *J. Physiol.* 235:133-153.

Kaneko, A. & Tachibana, M. (1981) Retinal bipolar cells with double colour-opponent receptive fields, *Nature* 293:220-222.

Kaneko, A. & Tachibana, M. (1983) Double colour-opponent receptive fields of carp bipolar cells, *Vision Res.* 23:381-388.

Kaplan, E., Marcus, S. & So, Y.T. (1979) Effects of dark adaptation on spatial and temporal properties of receptive fields in cat lateral geniculate nucleus, *J. Physiol.* 294:561-580.

Kaplan, E. & Shapley, R.M. (1986) The primate retina contains two types of ganglion cells, with high and low contrast sensitivity, *Proc. Natl. Acad. Sci. USA* 83:2755-2757.

Kolb, H. (1970) Organization of the outer plexiform layer of the primate retina: electron microscopy of Golgi-impregnated cells, *Phil. Trans. R. Soc. Lond. B* 258:261-283.

Kraft, T.W. & Burkhardt, D.A. (1986) Telodendrites of cone photoreceptors: Structure and probable function, *J. Comp. Neurol.* 249:13-27.

Land, E.H. (1959) Color vision and the natural image. Parts I and II., *Proc. Natl. Acad. Sci. USA* 45:115-129 and 636-644.

Lasater, E.M. (1982) Spatial receptive fields of catfish retinal ganglion cells, *J. Neurophysiol.* 48:823-835.

Lasater, E.M. & Lam, D.M.K. (1984a) The identification and some functions of GABAergic neurones in the distal catfish retina., *Vision Res.* 24:497-506.

Lasater, E.M. & Lam, D.M.K. (1984b) The identification and some functions of GABAergic neurones in the proximal retina of the catfish, *Vision Res.* 24:875-881.

Levine, M.W. & Shefner, J.M. (1979) X-like and not X-like cells in goldfish retina, *Vision Res.* 19:95-97.

Levine, J.S. & MacNichol, E.F. Jr. (1982) Color vision in fishes, *Sci. Am.* 246:108-117.

Lipetz, L.E. & Kaneko, A. (1984) Receptive field properties of the photopic luminosity horizontal cell of carp retina, *Vision Res.* 24:1947-1950.

Loew, E.R. & McFarland, W.N. (1990) The underwater environment. In The Visual System of Fish (Douglas R.H. & Djamgoz M.B.A. eds.). Chapman & Hall, London. pp. 1-43.

Lyall, A.H. (1957) Cone arrangement in teleost retinae, *Q. J. Microsc. Sci.* 98:189-201.

Mackintosh, R.M., Bilotta, J. & Abramov, I. (1987) Contributions of short-wavelength cones to goldfish ganglion cells, *J. Comp. Physiol.* A 161:85-94.

Mangel, S.C. (1991) Analysis of the horizontal cell contributionto the receptive field surround of ganglion cells in the rabbit retina, *J. Physiol.* 442:211-234.

Mangel, S.C. & Dowling, J.E. (1985) Responsiveness of receptive field size of carp horizontal cells are reduced by prolonged darkness and dopamine, *Science* 229:1107-1109.

Marchiafava, P.L. (1985) Cell coupling in double cones of the fish retina, *Proc. R. Soc. Lond. B* 226:211-215.

Marchiafava, P.L., Strettoi, E. & Alpigiani, V. (1985) Intracellular recording from single and double cone cells isolated from the fish retina (*Tinca tinca*), *Exp. Biol.* 44:173-180.

Miller, J.L. & Karenbrot, J.I. (1993) Phototransduction and adaptation in rods, single cones and twin cones of the striped bass retina: a comparative study, *Visual Neurosci.* 10:653-667.

Mitarai, G., Goto, T. & Takagi, S. (1978) Receptive field arrangement of colour-opponent bipolar and amacrine cells in the carp retina, *Sensory Processes* 2:375-382.

Miyachi, E-I., Murakami, M. & Nakaki, T. (1990) Arginine blocks gap junctions between retinal horizontal cells, *NeuroReport* 1:107-110.

Miyachi, E-I. & Nishikawa, C. (1994) Blocking effect of L-arginine on retinal gap junctions by activating guanylate cyclase via generation of nitric oxide, *Biogen. Amines* 10:459-464.

Murakami, M. & Shimoda, Y. (1977) Identification of amacrine and ganglion cells in the carp retina, *J. Physiol.* 265:801-818.

Naka, K-I. (1977) Functional organization of catfish retina, *J. Neurophysiol.* 36:502-518.

Naka, K-I. & Ohtsuka, T. (1975) Morphological and functional identifications of catfish retinal neurons. II. Morphological identification., *J. Neurophysiol.* 38:72-91.

Naka, K-I., Marmarelis, P.X. & Chan, R.Y. (1975) Morphological and functional identifications of catfish retinal neurons. III. Functional identification, *J. Neurophysiol.* 38:92-131.

Naka, K-I. & Christensen, B.N. (1981) Direct electrical connections between transient amacrine cells in the catfish retina, *Science* 214:462-464.

Neal, M.J. (1984) Cholinergic mechanisms in the vertebrate retina, *Prog. Retinal Res.* 2:191-212.

Nelson, R. & Kolb, H. (1983) Synaptic patterns and response properties of bipolar and ganglion cells in the cat retina, *Vision Res.* 23:1183-1195.

Neumeyer, C. & Arnold, K. (1989) Tetrachromatic colour vision in the goldfish becomes trichromatic under white adaptation light of moderate intensity, *Vision Res.* 29:1719-1727.

Normann, R.A., Perlman, I., Kolb, H., Jones, J. & Daly, S.J. (1984) Direct excitatory interactions between cones of different spectral types in the turtle retina, *Science* 224:625-627.

Ohtsuka, T. & Kawamata, K. (1990) Telodendrinal contact of HRP-filled photoreceptors in the turtle retina: pathways of photoreceptor coupling, *J. Comp. Neurol.* 292:599-613.

Perry, V.H., Oehler, R. & Cowey, A. (1984) Retinal ganglion cells that project to the dorsal lateral geniculate nucleus in the macaque monkey, *Neurosci.* 12:1101-1123.

Petruv, R., Furukawa, T., Yasui, S. & Djamgoz, M.B.A. (1993) Sodium nitroprusside, a nitric oxide donor, generates chromatic difference in the receptive field size of H1 horizontal cells in isolated retinae of carp, *J. Physiol.* 473:163P.

Piccolino, J., Neyton, J. & Gerschenfeld, H.M. (1984) Decrease of gap junction permeability induced by dopamine and cyclic adenosine d':5'-monophosphate in horizontal cells of turtle retina, *J. Neurosci.* 4:2477-2488.

Raynauld, J.P., Laviolette, J.R. & Wagner ,H-J. (1979) Goldfish retina: a correlate between cone activity and morphology of the horizontal cell in cone pedicules, *Science* 204:1436-1438.

Richter, A. & Simon, E.J. (1975) Properties of centre-hyperpolarizing, red-sensitive bipolar cells in the turtle retina, *J. Physiol.* 248:317-334.

Saito, T. & Kujiraoka, T. (1982) Physiological and morphological identification of two types of on-centre bipolar cells in the carp retina, J. *Comp. Neurol.* 205:161-170.

Saito, T. & Kujiraoka, T. (1988) Characteristics of bipolar-bipolar coupling in the carp retina, *J. Gen. Physiol.* 91:275-287.

Sakai, H.M. & Naka, K-I. (1986) Synaptic organization of the cone horizontal cells in the catfish retina., *J. Comp. Neurol.* 245:107-115.

Sakai H.M. & Naka K-I. (1987) Signal transmission in the catfish retina. IV. Transmission to ganglion cells, *J. Neurophysiol.* 58:1307-1328.

Schellart, N.A.M., van Acker, H.F. & Spekreijse, H. (1984) Influence of GABA and the spectral and spatial coding of goldfish retinal ganglion cells, *Neurosci. Lett.* 48:31-36.

Scholes, J.H. (1975) Colour receptors and their synaptic connections, in the retina of a cyprinid fish, *Phil. Trans. R. Soc. Lond. B* 270:61-118.

Schultze, M. (1866) Zur anatomie und physioloigie der retina, *Arch. Mikr. Anat.* 2:175-286.

Schwartz, E.A. (1974) Responses of bipolar cells in the retina of the turtle, *J. Physiol.* 236:211-224.

Shigematsu, Y. & Yamada, M. (1988) Effects of dopamine on spatial properties of horizontal cell responses in the carp retina, *Neurosci. Res. Suppl.* 8:S69-S80.

Spekreijse, H., Wagner, H.G. & Wolbarsht, M.L. (1972) Spectral and spatial coding of ganglion cell responses in goldfish retina, *J. Neurophysiol.* 35:73-86.

Spekreijse, H., Wietsma, J.J. & Neumeyer, C. (1991) Induced color blindness in goldfish: a behavioral and electrophysiological study, *Vision Res.* 31:551-562.

Sperling H.G. (1986) Spectral sensitivity, intense spectral light studies and the colour receptor mosaic of primates. Vision Res. 26:1557-1571.

Stell W.K. (1980) Photoreceptor-specific synaptic pathways in goldfish retina: a world of colour, a wealth of connections. In Colour Vision Deficiencies V (Verriest, G. ed.). Adam Hilger, Bristol. pp. 1-14.

Stell, W.K. & Harosi, F.I. (1976) Cone structure and visual pigment content in the retina of goldfish, *Vision Res.* 16:647-657.

Stell, W.K., Lightfoot, D.O., Wheeler, T.G & Leeper, H.F. (1975) Goldfish retina: functional polarization of cone horizontal cell dendrites and synapses, *Science* 190:989-990.

Stell, W.K., Walker, S.E. & Ball, A.K. (1987) Functional-anatomical studies on the terminal nerve projection to the retina of bony fishes, *Ann. N.Y. Acad. Sci.* 519:80-96.

Teranishi, T. & Negishi, K. (1994) Double-staining of horizontal and amacrine cells by intracellular injection with lucifer yellow and biocytin in carp retina, *Neurosci.* 59:217-266.

Teranishi, T., Kato, S. & Negishi, K. (1982) Lateral spread of S-potential components in the carp retina, *Exp. Eye Res.* 34:389-399.

Thier, P. & Alder, V. (1984) Action of iontophoretically applied dopamine on cat retinal ganglion cells., *Brain Res.* 292 109-121.

Thorell L.G., De Valois R.L. & Albrecht D.G. (1984) Spatial mapping of monkey V1 cells with pure colour and luminance stimuli, *Vision Res.* 24:751-769.

Toyoda, J-I., Hashimoto, H. & Ohtsu, K. (1973) Bipolar-amacrine transmission in the carp retina, *Vision Res.* 13:295-307.

Toyoda, J-I. & Fujimoto, M. (1983) Analyses of neural mechanisms mediating the effect of horizontal cell polarization., *Vision Res.* 23:1143-1150.

Toyoda, J-I. & Tonosaki, K. (1978) Effect of polarization of horizontal cells on the on-center bipolar cell of carp retina, *Nature* 276:399-400.

Umino, O., Lee, Y.L. & Dowling, J.E. (1991) Effects of light stimuli on the release of dopamine from interplexiform cells in the white perch retina, *Visual Neurosci.* 7:451-458.

Wagner, H-J. (1990) Retinal structure of fishes. In The Visual System of Fish. (Douglas R.H. & Djamgoz M.B.A. eds.). Chapman & Hall, London. pp. 109-157.

Wagner, H.G., MacNichol, E.F. Jr. & Wolbarsht, M.L. (1960) The response properties of single ganglion cells in the goldfish retina, *J. Gen. Physiol.* 43:43-62.

Wagner, H.G., MacNichol, E.F. Jr. & Wolbarsht, M.L. (1963) Functional basis for 'on'-centre and 'off.'-centre receptive fields in the retina, *J. Opt. Soc. Am.* 53:66-70.

Wagner, H-J. & Djamgoz, M.B.A. (1993) Spinules: a case for retinal synaptic plasticity, *Trends Neurosci.* 16:201-206.

Warner, A.E. & Lawrence, P. (1982) Permeability of gap junctions at the segmental border in insect epidermis, *Cell* 28:243-252.

Wässle, H. & Boycott, B.B. (1991) Functional architecture of the mammalian retina, *Physiol. Rev.* 71:447-480.

Werblin, F.S. (1974) Control of retina sensitivity. II. Lateral interactions at the outer plexiform layer, *J. Gen. Physiol.* 63:62-87.

Werblin, F.S. & Dowling, J.E. (1969) Organization of the retina of the mudpuppy, *Necturus maculosus*. II. Intracellular recording, *J. Neurophysiol.* 32:339-355.

Wietsma, J.J. & Spekreijse, H. (1991) Bicuculline produces reversible red-green color blindness in goldfish, as revealed by monocular behavioral testing., *Vision Res.* 31:2101-2107.

Witkovsky, P. & Dowling, J.E. (1969) Synaptic relationships of the plexiform layers of carp retina, *Z. Zellforsch. Mikrosk. Anat.* 100:60-82.

Witkovsky, P., Shakib, M. & Ripps, H. (1974) Interreceptoral junctions in the teleost retina, *Invest. Ophthalmol. Visual Sci.* 13:996-1009.

Witkovsky, P. & Stone, S. (1983) Rod and cone inputs to bipolar and horizontal cells of the Xenopus retina, *Vision Res.* 23:1251-1258.

Witkovsky, P. & Dearry, A. (1991) Functional roles of dopamine in the vertebrate retina, *Prog. Retinal Res.* 11:247-292.

Yamada, M. & Saito, T. (1988) Effects of dopamine on bipolar cells in the carp retina, *Biomed. Res.* 9, *Suppl.* 2:125-130.

Yamada, M., Djamgoz, M.B.A., Low, J.C., Furukawa, T. & Yasui, S.(1991) Conductance decreasing cone-output to H1 horizontal cells in carp retina, *Neurosci. Res. Suppl* 15:S51-S65.

Yasui, S., Yamada, M. & Djamgoz, M.B.A. (1990) Dopamine and 2-amino-4-phosphonobutyrate differentially affect spectral responses of H1 horizontal cells in carp retina, *Exp. Brain Res.* 83:79-84.

Yazulla, S. (1976) Cone input to bipolar cells in the turtle retina, *Vision Res.* 16:737-744.

Zeki, S. (1980) The representation of colours in the cerebral cortex, *Nature* 284:412-418.

Zeki, S. (1995) Behind the scene: An exploration of the visual brain (Ferrier Lecture), *Proc.R. Soc. Lond. B.* (In press).

Zrenner, E. & Gouras, P. (1983) Cone opponency in tonic ganglion cells and its variation with eccentricity in rhesus monkey retina. In Colour Vision: Physiology and Psychophysics (Mollon J.D. and Sharpe L.T. eds). Academic Press, London. pp. 211-223.

Zucker, C.L. & Dowling, J.E. (1987) Centrifugal fibers synapse on dopaminergic interplexiform cells in the teleost retina, *Nature* 330:166-168.

7

# ON-BIPOLAR CELLS, VISUAL SENSITIVITY AND THE b-WAVE

Gertrude Falk and Richard A. Shiells

Department of Physiology
University College
London WC1E 6BT, United Kingdom

## INTRODUCTION

It has been known for some time from psychophysical observations that the visual system can perform as a near-ideal detector of light, capable of detecting the absorption of a few photons. This high sensitivity is generated in the retina; rods respond to single photons and there is a large amplification at the synapse with some bipolar cells (Ashmore & Falk 1976; 1980). The absorption of a photon within the receptive field of a ganglion cell of a dark-adapted retina has a high probability of eliciting extra impulses in that ganglion cell (Barlow, Levick & Yoon 1971).

Great progress has been made within the past 25 years in the understanding of the visual signals generated and their transmission at different levels in the retina. Many of the discoveries were novel and were not anticipated in the light of the then current ideas about how the nervous system operates. As a result the canvas has become much richer. For example, the discovery that photoreceptors hyperpolarise in a graded manner in response to light overturned the classical view that signal transmission is by action potentials generated as a result of depolarisation. The mechanism of phototransduction, in terms of a biochemical cascade leading to the generation of a second messenger, became the model for cellular signal transduction in a wide variety of tissues.

It was known, from single-unit recording of action potentials in ganglion cells and optic nerve, that the retina is organised into ON- and OFF-pathways. These parallel pathways have their origin in the outer retina at the synapses which photoreceptors make with two types of bipolar cells, the ON-bipolar cell which depolarises and the OFF-bipolar cell which hyperpolarises in response to central illumination of their receptive fields.

## THE GLUTAMATE RECEPTORS OF BIPOLAR CELLS

The consequence of rod hyperpolarisation with light is a decrease in the release of transmitter, glutamate, from their terminals where glutamate is released in darkness at a relatively high rate. Glutamate, acting on specialised receptors of bipolar cells, directly or

*Basic and Clinical Perspectives in Vision Research*, Edited by Jon Robbins et al.
Plenum Press, New York, 1995

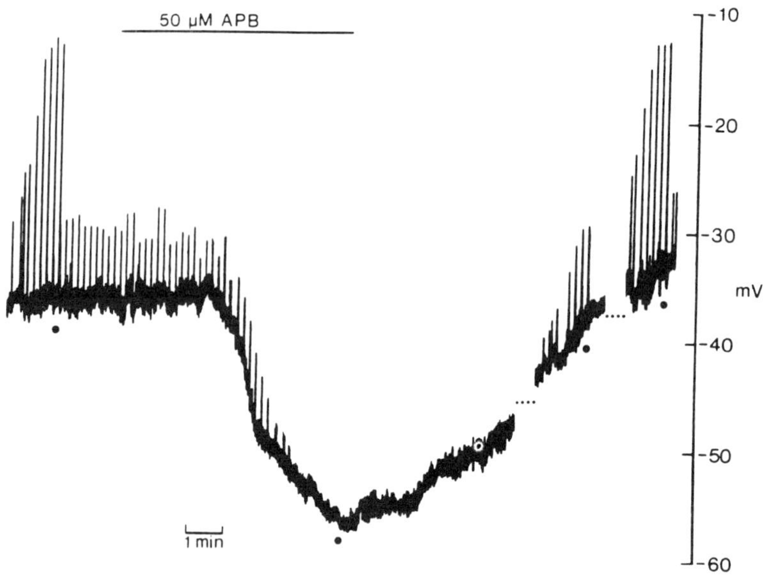


**Figure 1.** The action of APB on a retinal ON-bipolar cell. Intracellular recording of membrane potential of an on-bipolar cell in the virtually all-rod retina of the dogfish in response to flashes of light and to APB. The upward deflections are depolarising responses to light every 10 s. The stepwise increases in response (shown by the dot under the trace) are due to stepwise increases in light intensity, up to a maximum response. When APB was added to the superfusate, the cell hyperpolarised after a delay; due to diffusion, and no longer responded to light. The hyperpolarisation was associated with an increase in membrane resistance as synaptic ion channels close (modified from Shiells et al., 1981).

indirectly controls the opening or closing of bipolar cell ionic channels which are permeable to cations (see Falk, 1989 for review). The glutamate receptor of OFF-bipolar cells, when combined with the neurotransmitter in darkness, directly opens the channels. In the light the channels close and the OFF-bipolar cell hyperpolarises, a signal which, when transmitted to ganglion cells, inhibits the firing of action potentials in the optic nerve.

On the other hand, ionic channels of ON-bipolar cells are closed in the darkness under the influence of glutamate and open in the light when glutamate levels fall. The opening of the channels leads to depolarisation of ON-bipolar cells and finally firing of ON-ganglion cells, hence the terminology used in describing the two kinds of bipolar cells. The glutamate receptors of ON-bipolar cells are specialised and belong to a class of receptors, the metabotropic receptors, which do not directly control the gating of ion channels, but which instead control the operation of intracellular biochemical events giving rise to second messengers. The glutamate receptor of ON-bipolar cells has been cloned (Nakajima et al, 1993) and can be selectively activated by 2-amino-4-phosphonobutyrate (APB), an analogue of glutamate (Shiells, Falk & Naghshineh, 1981; Slaughter & Miller, 1981). Like glutamate, APB causes the closure of cation channels of ON-bipolar cells and hyperpolarisation of the cell (Fig 1). In the presence of APB the light responses of ON-bipolar cells and the b-wave of the ERG are blocked, as is the ON-pathway of the visual system (Schiller, Sandell & Maunsell, 1986).

## THE cGMP CASCADE OF ON-BIPOLAR CELLS

The mechanism by which glutamate closed the channels had been a long standing problem until recently. The discovery of a biochemical cascade of reactions linked to the glutamate receptor was made possible by the technique of whole-cell patch clamping (Shiells & Falk, 1990; Nawy & Jahr, 1990). Contact with the cell interior is made by means of a fine pipette which permits the cell interior to be manipulated and any ionic currents generated as a result, measured. It was soon apparent that when the patch pipette only contained inorganic salts, albeit appropriate to the cell interior, responses to light began to diminish and soon disappeared, a phenomenon known as "washout" due to the loss of diffusible molecules by equilibration with the pipette solution. Light responses could be maintained if guanosine triphosphate (GTP) was included in the patch pipette. Why should GTP be important? GTP could be converted to a cyclic compound, cGMP, by guanylate cyclase. GTP can also be bound to proteins called G-proteins which are hydrolysing enzymes or GTPases. Transducin in rods is such a G-protein.

## The G-protein of ON-Bipolar Cells

In order to investigate the role of GTP, experiments were done in which a non-hydrolysable form of GTP, GTP-γ-S was added to the patch-pipette and so was incorporated into the cell. The effect of GTP-γ-S is illustrated in Fig. 2.

GTP-γ-S caused suppression of an inward current present in darkness, hence the upward change in the trace. The change in current was accompanied by a decrease in membrane conductance due to the closure of ion channels. These observations were important in showing that on-bipolar cells contained a G-protein and that the persistent activation of the G-protein by GTP-γ-S lead to the closure of ionic channels which were involved in the light response of the cell. Put another way, the action of glutamate was mimicked by activation of the G-protein. Like transducin in rods, the G-protein-protein was sensitive to both pertussis and cholera toxins (Shiells & Falk, 1992a). Was the action of glutamate linked to the G-protein and if so, was there a second messenger as in phototransduction ?

## The Intracellular Transmitter, cGMP

The possibility of an intracellular messenger mediating the action of glutamate was explored (Shiells & Falk, 1990). When cGMP was included in the patch pipette an inward current was elicited due to an increase in membrane conductance (Fig 3a). That the effect of cGMP was on the light-modulated channels was evident from the fact that, at higher concentrations, the light responses were markedly reduced, as might be expected if cGMP opened such channels and, at elevated cGMP levels, fewer channels remained to be opened

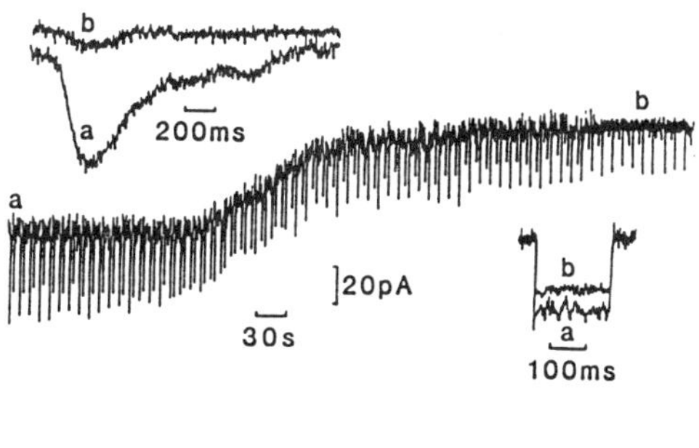


**Figure 2.** The effect of GTP-γ-S via the patch pipette on an ON-bipolar cell. Because the cell is voltage-clamped, the records show changes in membrane current where, by convention, an inward current (which would lead to the depolarisation of an unclamped cell) is shown as a downward deflection. The upper inset shows the timecourse of the inward current in response to a single brief flash of light on an expanded timescale (record a) before GTP-γ-S diffused into the cell and the much diminished light response under the action of GTP-γ-S (record b). The middle record shows, on a slow sweep, the effect of GTP-γ-S on the dark current. There is a delay in the onset of effect, which is due to the time for equilibration of the patch-pipette contents with the cell interior. The downward traces are periodic current responses to light flashes alternating with 1 mV voltage steps to test membrane resistance (the latter shown on an expanded time sweep in the lower inset) (from Shiells & Falk, 1990).

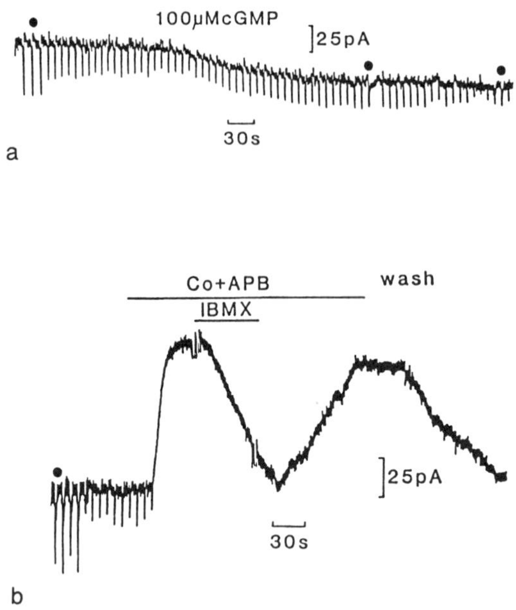


**Figure 3.** The opening of ion channels of on-bipolar cells by intracellular cGMP. (a) The level of cGMP was raised by including 100 µM in the patch-pipette. The delay in the effect (an inward current shown as a downward deflection) is the time for equilibration of the patch-pipette contents with the cell interior. The periodic downward deflections are the current responses to brief light flashes. (b) The action of the phosphodiesterase inhibitor IBMX in the superfusate on whole-cell current after synaptic blockade by $Co^{2+}$ and APB (modified from Shiells & Falk, 1990).

during the light response. The effect was specific for cGMP as the membrane-permeable analogue of cAMP was without action (Shiells & Falk, 1990). If cGMP were the second messenger, one would expect a phosphodiesterase (PDE) to be present. The effect of the PDE inhibitor isobutyl-methyl-xanthine (IBMX) is illustrated in Fig. 3b.

In order to exclude any effect on photoreceptors, it was necessary to block synaptic transmission; hence the use of $Co^{2+}$ to block transmitter release and APB to activate the on-bipolar cell glutamate receptors causing channel closure. IBMX application caused channels to open as evidenced by an increase in both inward current and membrane conductance. The most reasonable interpretation of these results is that inhibition of PDE leads to a rise in cGMP levels and the opening of cation channels.

## The Guanylate Cyclase of ON-Bipolar Cells

A rise in cGMP following PDE inhibition implies that there is a cGMP synthesising enzyme. Two principal forms of guanylate cyclase are known, one a particulate, membrane-bound form found in retinal rods and the kidney, the other a soluble, iron-containing form activated by nitric oxide and inhibited by methylene blue or ferricyanide. It was found that the guanylate cyclase of ON-bipolar cells was the NO-sensitive form (Shiells & Falk, 1992b) and confirmed by molecular biological studies of the inner nuclear layer (Barnstaple & Ahmad, 1992).

## Overview of the Linkage of cGMP to the Glutamate Receptor

The proposed sequence of events is summarised in Fig 4.

The mechanism of transduction at the synapse of rods with ON-bipolar cells has many similarities to phototransduction, whereby light-activated rhodpsin acts on the G-protein transducin, which then activates PDE thereby reducing the level of the second messenger cGMP leading to the closure of ion channels and hyperpolarisation of rods. In place of rhodopsin read glutamate receptor for ON-bipolar cells. Both transducin and the ON-bipolar

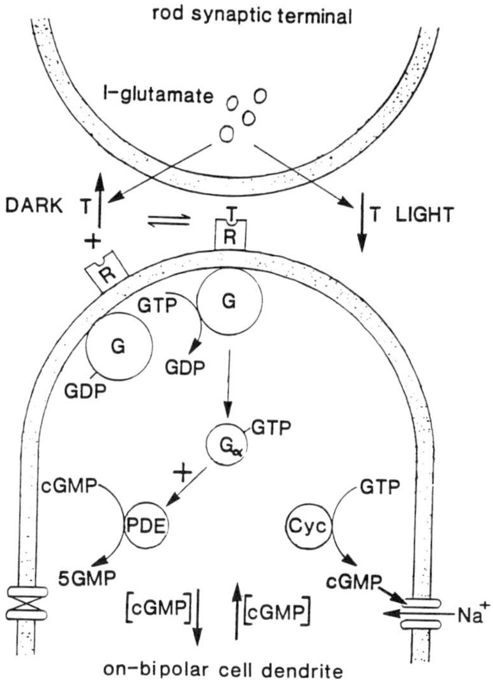


**Figure 4.** The glutamate receptor-linked cGMP cascade. When the receptor binds glutamate, an exchange of GTP for GDP bound to the G-protein is initiated. The GTP-bound form activates the enzyme cGMP PDE. The fall in cGMP levels results in the closure of cation membrane channels in the dark. In the light, rods hyperpolarise reducing the voltage-dependent release of glutamate. Consequently, the sequence of events just described is reversed, cGMP levels rise and ion channels which are directly gated by cGMP open causing the ON-bipolar cells to depolarise.

cell G-protein are sensitive to pertussis and cholera toxins. However, while formally familiar, there are differences in the two systems at the molecular level. The cGMP synthesising enzymes (guanylate cyclase) are very different entities with different control mechanisms (Shiells & Falk, 1992b). The cGMP-gated channels are similar, but not identical (Shiells & Falk, 1992c).

## AMPLIFICATION AT THE SYNAPSE

The very great sensitivity of the dark-adapted human eye is due to amplification in phototransduction and in transmission of rod signals at the synapse with ON-bipolar cells. Evidence for a synaptic voltage gain of 100-200 in the vertebrate retina was first obtained by Ashmore & Falk (1976; 1980) from intracellular recordings, which is in accord with the greater sensitivity of the b-wave compared with the a-wave (see later section). How this amplification of rod signals comes about could not be explained until recently, with the discovery of the cGMP-cascade linked to the glutamate receptor.

How does the cascade generate gain? The single-photon sensitivity of rods arises because an activated rhodopsin molecule can in turn activate many transducin molecules (some 500) and via the phosphodiesterase hydrolyse many more cGMP molecules, controlling ion channels. It can be seen that a similar cascade linked to the glutamate receptor of the ON-bipolar cell could, in principle, provide a second stage of amplification. A quantitative model of voltage gain at the synapse has recently been published (Shiells & Falk, 1994; 1995) showing that the cGMP cascade linked to the glutamate receptor can account for the observed high gain in the transmission of rod signals to ON-bipolar cells. It is important to note that the large voltage gain is not simply due to the multiplication in the ratio of synaptic channels to glutamate receptors bound, but an essential feature is that the binding leads to

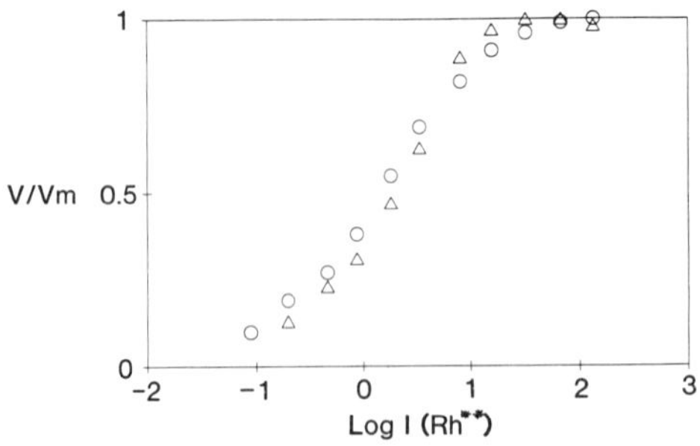


**Figure 5.** Comparison of the peak of the b-wave and the response of an on-bipolar cell as a function of log light intensity Rh**, the number of photoisomerisations per rod. Peak amplitudes are normalised to the respective maximal values (Vm) with bright flashes. Recordings from an isolated superfused sector of dogfish eyecup.

the *closure* of the channels. Had it been the case that channels were opened as a result of glutamate binding, the maximum voltage gain would be severely decreased. The reason for the loss in gain in this case is that the open channels shunt out voltage changes (see Falk, 1989, for a detailed quantitative explanation). A receptor-linked cascade where channels open would formally be equivalent to an increase in apparent affinity of receptor binding which would not increase maximum gain but would only shift the operating point at the synapse. What had been regarded in the past as somewhat arbitrary and perverse, a glutamate receptor involved in the closure of ion channels is seen to be a significant evolutionary adaptation. Given that photoreceptors and bipolar cells differentiate from a common precursor, it will be for future research to unravel the transformation in molecular expression of phototransduction and synaptic transduction via the cGMP cascade.

# CLINICAL IMPLICATIONS

## Relation of ON-Bipolar Cell Responses to the b-Wave

It is generally agreed that the ERG b-wave is due to the extracellular current generated as a result of depolarisation of Muller cells by K ions (Kline, Ripps & Dowling, 1985; Newman & Odette, 1984). The principle source of K ions is in the outer plexiform layer so that it was logical to presume that it was due to ON-bipolar cell activity, a presumption given weight by the fact that the selective block of ON-bipolar cell responses also abolished the b-wave (Shiells et al, 19981; Knapp & Schiller, 1984; Stockton & Slaughter, 1989).

How closely does the b-wave mirror the activity of ON-bipolar cells? To answer this question, simultaneous measurements were made with an intracellular electrode of on-bipolar cell responses to light flashes of varying intensities whilst recording the transretinal ERG. As illustrated in Fig 5, there is an excellent parallelism in the relative amplitude changes with light of ON-bipolar cell responses and the b-wave.

As shown in Fig. 6, the b-wave and ON-bipolar cell responses to flashes at low light intensities have the same timecourse. At high light intensities (not illustrated) the b-wave outlasts the bipolar cell depolarisation, no doubt reflecting the more prolonged release of K ions in proximal layers due to repetitive action potentials in amacrine and ganglion cells. The

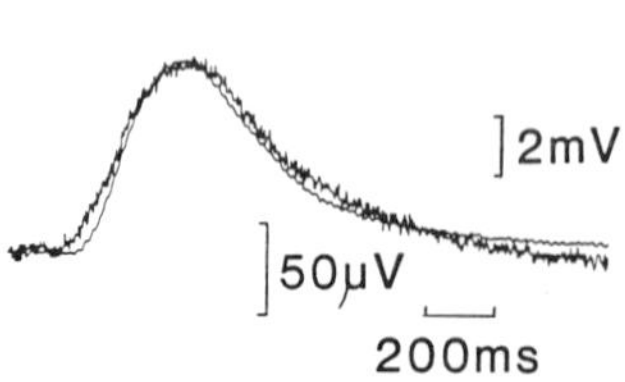


**Figure 6.** Comparison of the time course of the b-wave with the response of ON-bipolar cells to dim light flashes scaled to the same peak and superimposed.

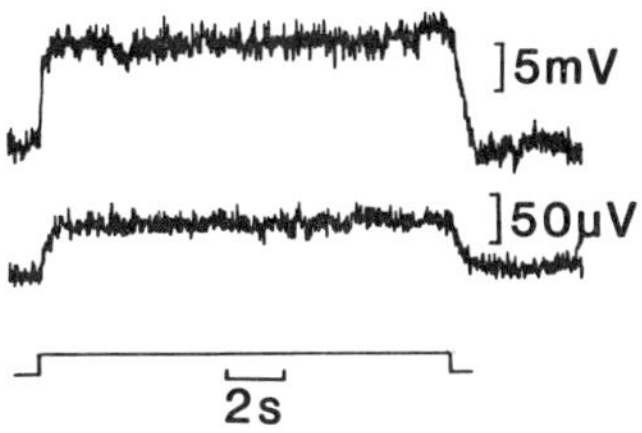


**Figure 7.** The DC component of the ERG and the depolarisation of an on-bipolar cell of the dogfish retina in response to a step of dim light. The duration of the step is indicated in the lowest record.

discrete localisation principally to the outer plexiform layer and secondarily to the inner plexiform layer rather than more uniformly through the inner retina (Newman & Odette, 1984) suggests the hypothesis that K ion efflux largely through cGMP-gated channels of ON-bipolar cells restricted to the outer plexiform layer rather than through voltage-gated K channels.

## The DC Component of the ERG

The DC component of the ERG is illustrated in the upper record of Fig. 7. It can be elicited in pure form by dim light of duration exceeding about 200ms. As shown by the simultaneously recorded response of an ON-bipolar cell (lower record), the DC component simply reflects ON-bipolar activity.

As would be expected, the DC component disappears when ON-bipolar cell responses are blocked by APB. Obviously, the observations on the b-wave, the DC component and the response of ON-bipolar cells have important implications for the various forms of night-blindness, including the recently discovered autoimmune disease, melanoma associated retinopathy.

## ACKNOWLEDGMENTS

Original work has been supported by the Medical Research Council, the Wellcome Trust and Rank Prize Funds. This Chapter is dedicated to Professor Hisako Ikeda.

## REFERENCES

Ashmore, J.F. and Falk, G. (1976) Absolute sensitivity of rod bipolar cells in a dark-adapted retina, *Nature* 263:248-249.

Ashmore, J.F. and Falk, G. (1980) Responses of rod bipolar cells in the dark-adapted retina of the dogfish, Scyliorhinus canicula, *J. Physiol.* 300:115-150.

Barlow, H.B., Levick, W.R. and Yoon, M. (1971) Responses to single quanta of light in retinal ganglion cells of the cat, *Vision Res. Suppl.* 3:87-102.

Barnstaple, C.J. and Ahmad, J. (1992) Molecular cloning and differential laminar expression of soluble and particular guanylate cyclases in rat retina, *Invest. Ophthalmol. Vis. Sci.* 33 p1104.

Falk, G. (1989) Signal transmission from rods to bipolar and horizontal cells: a synthesis, *Prog. Retinal Res.* 8:255-279.

Kline, R.P., Ripps, H. and Dowling, J.E. (1985) Light induced potassium fluxes in the skate retina, *Neuroscience* 14:225-235.

Knapp, A.G. and Schiller, P. (1984) The contribution of ON-bipolar cells to the electroretinogram of rabbits and monkeys, *Vision Res.* 24:1841-1846.

Nakajima, Y., Iwakabe, H., Akazawa, C., Nawa, H., Shigemoto, R., Mizuno, N. and Nakanishi, S. (1993) Molecular characterisation of a novel retinal metabotropic glutamate receptor mGluR6 with a high agonist selectivity for L-2-amino-4-phosphonobutyrate, *J. Biol. Chem.* 268:11868-11873.

Nawy, S. and Jahr, C. E. (1990) Suppression by glutamate of cGMP-activated conductance in retinal bipolar cells, *Nature* 346:269-271.

Newman, E.A. and Odette, L.L. (1984) Model of electroretinogram b-wave generation: a test of the K hypothesis, *J. Neurophysiol.* 51:164-182.

Schiller, P.H., Sandell, J.H. and Maunsell, J.H.R. (1986) Functions of the ON and OFF channels of the visual system, *Nature* 322:824-825

Shiells, R.A. and Falk, G. (1990) Glutamate receptors of rod bipolar cells are linked to a cyclic GMP cascade via a G-protein, *Proc. R. Soc. Lond. B* 242:91-94.

Shiells, R.A. and Falk, G. (1992a) The glutamate receptor linked cGMP cascade of retinal on-bipolar cells is pertussis and cholera toxin-sensitive, *Proc. R. Soc. Lond. B* 247:17-20.

Shiells, R.A. and Falk, G. (1992b) Retinal on-bipolar cells contain a nitric oxide-sensitive guanylate cyclase, *Neuroreport* 3:845-848.

Shiells, R. A. and Falk, G. (1992c) Properties of the cGMP-activated channel of retinal on-bipolar cells, *Proc. R. Soc. Lond. B* 247:21-25.

Shiells, R.A. and Falk, G. (1994) Responses of rod bipolar cells isolated from dogfish retinal slices to concentration-jumps of glutamate, *Visual Neurosci.* 11:1175-1183.

Shiells, R.A. and Falk, G. (1995) Signal transduction in retinal bipolar cells. *Prog. Retinal and Eye Res.* 14:223-247.

Shiells, R. A., Falk, G. and Naghshineh, S. (1981) Action of glutamate and aspartate analogues on rod horizontal and bipolar cells, *Nature* 294:592-594.

Slaughter, M.M. and Miller, R.F. (1981) 2-amino-4-phosphonobutyric acid: a new pharmacological tool for retina research, *Science* 211:182-185.

Stockton, M. and Slaughter, M.M. (1989) B-wave of the electroretinogram. A reflection of ON bipolar cell activity, *J. Gen. Physiol.* 93:101-122.

# EFFECTS OF GABA RECEPTOR AGONISTS ON CHOLINERGIC AMACRINE CELLS IN THE RABBIT RETINA

Michael J. Neal

Department of Pharmacology
St Thomas' Hospital (UMDS)
Lambeth Palace Road, London SE1 7EH, United Kingdom

## INTRODUCTION

The release of ACh from the rabbit retina in vivo has been used to indicate the activity of cholinergic amacrine cells. The response of these cells to drugs acting at GABA receptors has been studied in order to obtain information on the role of GABA receptor subtypes in modulating the activity of cholinergic amacrine cells. Drugs that stimulate $GABA_A$ receptors inhibited the light-evoked release of ACh (flickering light, 3Hz). In contrast, stimulation of $GABA_B$ receptors by baclofen enhanced the light-evoked release of ACh. In separate experiments on isolated rat retinas, it was found that baclofen inhibited the potassium-evoked release of glycine, but not GABA, suggesting that baclofen might enhance the light-evoked release of ACh from the rabbit retina by inhibiting the release of glycine from glycinergic amacrine cells. This idea was supported by further experiments on the rabbit retina in which it was found that strychnine enhanced the light-evoked release of ACh in a similar manner to baclofen. These results suggest that baclofen enhances the light-evoked release of ACh by reducing inhibition which originates from a glycinergic feedback loop onto the cholinergic amacrine cell. The effect of baclofen disappeared at high flicker frequency (10Hz), perhaps because the inhibitory presynaptic $GABA_B$ receptors were fully occupied by endogenously released GABA. If this explanation is correct, then GABA co-released with ACh might provide a mechanism of facilitating ACh release at higher rates of cell firing

The first report of ACh in the (cow) retina was that of Loewi & Hellaver (1938) and in the following decade ACh was found in the retinas of the pig, dog, ox and frog. In the following thirty or so years all the components of the retinal cholinergic system were studied (for refs. see review by Neal 1983) and indirect evidence, eg distribution of choline acetyltransferase, suggested that ACh was a transmitter in the inner plexiform layer. However, it was not until the late 1970's that the identity of the cholinergic neurones in the retina was established by using the technique of dry emulsion autoradiography to localise the sites of [$^3$H]-choline uptake in the chicken retina (Baughman & Bader, 1977) and rabbit retina (Masland & Mills, 1979). The latter authors found that when rabbit retinas were incubated

*Basic and Clinical Perspectives in Vision Research*, Edited by Jon Robbins et al.
Plenum Press, New York, 1995

under conditions that preserved [$^{3}$H]ACh (20mM Mg, 0.2mM Ca, eserine 30μM) a substantial proportion of the total radioactivity in the tissue was present as [$^{3}$H]ACh, although the major metabolites were phosphorylcholine and Ch-phospholipid in approximately equal amounts. Autoradiographical examination revealed that in addition to photoreceptors, two groups of neurones bordering the inner plexiform layer accumulated radioactivity, which was shown to be [$^{3}$H]ACh. The distal neurones occurred adjacent to the inner plexiform layer and were identified as amacrine cells. The second group of neurones occurred in the ganglion cell layer and were found to be displaced amacrine cells.

The cholinergic neurones in the rabbit retina are now often referred to as starburst amacrine cells because of their characteristic morphology (Famiglietti, 1983a). The cells are unistratified and those in the inner nuclear layer are OFF-cells ramifying in sublamina 2 of the inner plexiform layer, whilst the displaced cholinergic amacrines are ON-cells and ramify in sublamina 4. The cholinergic cells receive excitatory inputs from cone bipolar cells and inputs from unidentified amacrine cells (Famiglietti 1983b; Brandon, 1987) which are almost certainly inhibitory. The outputs of the cholinergic amacrine cells are mainly to complex ganglion cells (Masland & Ames, 1976; Ariel & Daw, 1982). Recently, the surprising discovery was made that GABA was co-localised with ACh in cholinergic amacrine cells (Vaney & Young, 1988; Brecha et al, 1988; Kosaka et al, 1988) and it seems likely that all the cholinergic amacrine cells contain GABA, although there is also a separate subpopulation of GABAergic amacrine cells which do not contain ACh. There is some evidence that ATP is co-released with ACh but it is not known whether the ATP is released from the cholinergic neurones themselves (Neal & Cunningham, 1994).

When the rabbit retina in vitro or in vivo is stimulated with flickering light ACh is released but changes in constant illumination (dark-light) have no effect, indicating that the cholinergic amacrine cells are a class of transiently responding cells (Masland & Livingstone, 1976; Massey & Neal, 1979).

Since the release of ACh reflects the activity of the cholinergic neurones, we and others, have used measurements of ACh release and pharmacological agents to obtain information on both excitatory inputs (Cunningham & Neal, 1985; Neal & Cunningham, 1989; Linn et al, 1991) and inhibitory inputs (Massey & Redburn, 1982; Cunningham & Neal, 1983; Cunningham, Dawson & Neal, 1983; Linn & Massey, 1992) to the cholinergic amacrine cells. In this paper, the pharmacology of cholinergic amacrine cells is discussed with emphasis on drugs acting on GABA and glycine receptors.

## MATERIALS AND METHODS

### Eye-Cup Preparation

The methods have been described in detail previously (Neal & Cunningham, 1982). Briefly, adult albino rabbits were anaesthetised with urethane (1.5g/kg I.P.) and a stainless-steel ring was sutured to the episclera to prevent collapse of the eye-cup. The cornea and iris were removed and this was followed by an intracapsular lens extraction and vitrectomy. The resulting eye-cup was filled for 30 min with Krebs Ringer bicarbonate containing [$^{3}$H]Ch (0.5μM) and then the retina was continuously irrigated for 60 min with fresh medium. A syringe mounted on a micromanipulator was used to place 0.5ml of medium in the eye-cup and this was replaced at 5 min intervals. The total radioactivity in the resulting samples was measured by liquid scintillation counting. In previous experiments [$^{3}$H]ACh and [$^{3}$H]Ch were separated by high voltage electrophoresis before counting. The spontaneously released radioactivity was mainly [$^{3}$H]Ch ([$^{3}$H]ACh, 20%; [$^{3}$H]Ch, 80%). However, it was found that [$^{3}$H]ACh represented 95-100% of the increase in total radioactivity evoked by light flashes.

Therefore, in the present experiments, the increase in total radioactivity produced by stimulation with light was assumed to be due to an increase in the release of [$^3$H]ACh.

The light source was a quartz halide bulb (100W) and the initially dark-adapted retina was stimulated for 5 min periods by flickering light at (usually) 3Hz (25% duty cycle; retinal illuminance, 650 lux).

The viability of the eye-cup preparation and the effects of cholinergic drugs on the gross physiological response of the retina were assessed by electroretinography (Neal & Cunningham, 1982). Since none of the drugs used in the present study significantly affected the b-wave of the e.r.g., the assumption has been made that they had no important effects on the distal part of the retina.

## Expression of Results

The effects of light stimulation on the release of total radioactivity from the retina are expressed as the ratio of the peak-evoked release of radioactivity to the average release of radioactivity in the two samples immediately preceding stimulation. The resulting ratio is hereafter referred to as the "light-evoked release of [$^3$H]ACh".

In a previous study (Cunningham & Neal, 1983) we established that when the retina was subjected to six periods of stimulation (3Hz, 25% duty cycle, 650 lux), the light-evoked releases of [$^3$H]ACh (as defined above) did not differ significantly in experiments lasting 2-5 hours. These control studies also revealed that during the course of a 2-5 hour release experiment, only 28% of the radioactivity initially present in the retina was lost, and that at the end of these experiments, 42% of the radioactivity remaining in the tissue was present as [$^3$H]ACh (Cunningham & Neal, 1983).

## Analysis of Results

Each drug was tested in four to six separate rabbit experiments and the light-evoked release of [$^3$H]ACh in the presence of the drug was expressed as a percentage of that obtained in the absence of drug (control light-evoked release). The results are expressed as the mean $\pm$ s.e. of mean and the effects of the drugs were tested for significance by the use of Student's t test.

## Isolated Retinas

In some experiments the release of endogenous GABA and glycine was measured using isolated rat retinas and, for comparison, slices of cerebral cortex and spinal cord.

## Preparation of Tissues

Male Wistar rats (150-200g) were killed by cervical dislocation. The eyes were enucleated and the retinas dissected in Krebs bicarbonate medium at room temperature. The brains were removed and slices of cerebral cortex (0.2mm thick) were prepared by means of a McIlwain tissue chopper. The spinal cord was removed following extensive laminectomy and transverse (0.3mm thick) slices were again prepared by a McIlwain chopper. In experiments on GABA release, tissues were taken from animals that had been treated with the irreversible GABA-T inhibitor γ-vinyl-GABA (GVG) (250mg/kg$^{-1}$ I.P.) 18 hours before death.

In order to measure the release of GABA and glycine from the tissues, slices of cortex (5-6 slices, weight 15mg), spinal cord (5-6 slices, weight 15mg), or two entire retinas (approx 20mg) were placed in a small perspex chamber (volume 0.5ml) containing Krebs-bicarbonate medium (250μl) at room temperature and were gently agitated by a stream of 5% $CO_2$ in oxygen.

At 10 min intervals the medium was drained off and replaced with fresh solution. The GABA and glycine in each of the resulting 10 min samples were measured by HPLC. Potassium evoked release of GABA and glycine was achieved by exposing the tissues for one 10 min period to KCl (50mM). The effects of drugs on the K-evoked release were usually examined by exposing the tissue to the drug in the collection period immediately preceding the addition of KCl and during exposure to KCl.

## Analysis of Samples by HPLC

GABA and glycine were analysed by reverse phase chromatography on a Varian 5000 HPLC using precolumn derivatisation with o-phthalaldehyde (Turnell & Cooper, 1981).

## Medium and Chemicals

Krebs bicarbonate Ringer of the following composition was used (mM): NaCl, 118.4; KCl, 4.84; $CaCl_2$,.2.4; $MgSO_4$, 1.8; $KH_2PO_4$, 1.2; $NaHCO_3$, 25; glucose 9.5; gassed with a mixture of 5% $CO_2$ in oxygen.

[Methyl-$^3$H]choline chloride (sp.act. 85 Ci mmol$^{-1}$) was obtained from Amersham International. Vigabatrin (γ-vinyl-GABA) was a gift from Merrell Dow, Strasbourg, France. CGP-35348 was a gift from Professor MM Slaughter. Other chemicals (Analar grade) were obtained from BDH or Sigma.

# RESULTS AND DISCUSSION

## Effects of $GABA_A$ and $GABA_B$ Receptor Agonists

The effect of exposing the retina to medium containing muscimol (1μM) is illustrated in Fig 1a and shows that the drug at this concentration almost abolished the light-evoked release of ACh. Similar results were obtained with GABA (Fig 1b) and 3-aminopropanesulphonic acid (3-APS). By examining the effects of different concentrations of these agonists, the concentrations that inhibited the light-evoked release of ACh by 50% ($IC_{50}$) were found to be: muscimol 0.3μM; 3-APS 5μM; GABA 900μM (Cunningham & Neal, 1983). The high potency of muscimol was confirmed by Massey & Redburn (1982) who reported an $IC_{50}$ of approximately 1μM, using similar techniques. In the retina, muscimol was approximately 3000 times more potent than GABA itself in inhibiting the light-evoked release of ACh. In other areas of the CNS the relative potency of muscimol and GABA is much lower, the highest value being found in cuneate slices where muscimol is 60 times more potent than GABA in producing depolarisation (Pickles & Simmonds, 1980). In the bovine retina, the affinity of muscimol for GABA receptors is not more than 10 times that for GABA (Osborne, 1980) and it seems unlikely that differences in affinity for the $GABA_A$ receptor can account for the remarkable potency of muscimol. The relatively high potency of muscimol might partly be due to the fact that muscimol is accumulated by cells at only 1/15th the rate of GABA (Neal & Cunningham, unpublished observations). However, another possibility is that GABA acts on additional receptors to muscimol and 3-APS. This possibility was suggested by experiments in which it was found that exposure of the retina to the $GABA_B$ receptor agonist, (-)baclofen, enhanced the light-evoked release of ACh (Cunningham & Neal, 1983). This was the first report of $GABA_B$ receptors in the retina and raised the possibility that

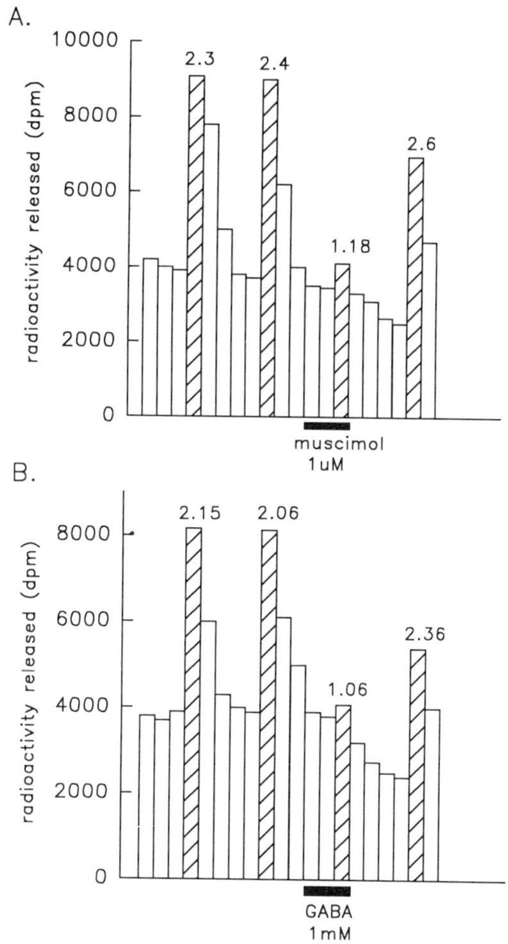


**Figure 1.** Typical experiments showing (A) the effect of muscimol and (B) the effect of GABA on the release of [$^3$H]ACh from the rabbit retina. Each histobar represents a 5 min collection period. Hatching indicates periods during which the retina was stimulated by flickering light (3Hz, 25% duty cycle, 650 lux). The numbers above hatched columns indicate the peak release/average release of radioactivity in the two samples preceding stimulation (light-evoked release of [$^3$H]ACh).

exogenously applied GABA might stimulate both $GABA_A$ and $GABA_B$ receptors in the retina with opposing actions on ACh release. This could at least partially explain the low potency of GABA compared with muscimol, because the latter drug has no affinity for the $GABA_B$ sites (Bowery et al, 1981). Friedman & Redburn (1990) using receptor binding techniques found that about a quarter of the GABA binding sites in the rabbit retina are of the $GABA_B$ subtype and they also reported that baclofen increased the release of ACh. Recently we have found that the effect of baclofen on the light-evoked release of ACh is blocked by the $GABA_B$ receptor antagonist CGP-35348 (Fig 2) (Neal, Cunningham & Slaughter, 1993; Neal & Cunningham, 1995).

A few years ago, the introduction of $GABA_B$ receptor antagonists stimulated examination of the physiological role of $GABA_B$ receptors in the retina using electrophysiological techniques (for review see Slaughter & Pan, 1992) and these have revealed that one effect of $GABA_B$ receptor activation is to enhance the light responses of phasic amacrine and ganglion cells (Bai & Slaughter, 1989; Ikeda et al, 1990). Although both electrohysiological experiments and ACh release studies indicate that baclofen enhances light responses in transiently firing cells, the mechanisms involved are far from clear.

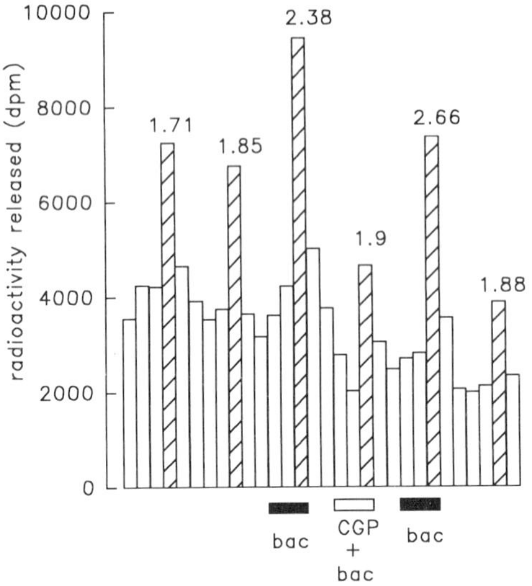


**Figure 2.** Typical experiment showing that baclofen (bac) (100μM) enhances the light-evoked release of [$^3$H]ACh and that the effect of baclofen is blocked by the $GABA_B$ antagonist CGP-35348 (500μM). Baclofen increased the light-evoked release of [$^3$H]ACh by 43 ± 2.1% (n=5, P<0.001).

## Mechanism of Action of Baclofen on ACh Release

Many $GABA_B$ receptors are located on presynaptic nerve terminals and their activation inhibits transmitter release. Bowery et al, (1979) demonstrated that baclofen and GABA inhibited the release of [$^3$H]-noradrenaline from rat isolated atria. Subsequently, inhibitory presynaptic $GABA_B$-receptors have been found in many peripheral tissues (see review by Bowery, 1993 for references) and in central neurones utilising several transmitters, including dopamine (Bowery et al, 1980), 5-hydroxytryptamine (5-HT, Gray & Green, 1987) and cholecystokinin (Conzelmann et al, 1986). GABAergic neurones have been shown to possess inhibitory presynaptic $GABA_B$ autoreceptors in several areas of the brain (Anderson & Mitchell, 1985; Limberger et al, 1986; Pittaluga et al, 1987). However, we have recently found that although baclofen inhibited the potassium-evoked release of GABA from slices of rat cerebral cortex and spinal cord, it had no effect of GABA release from the retina (Fig

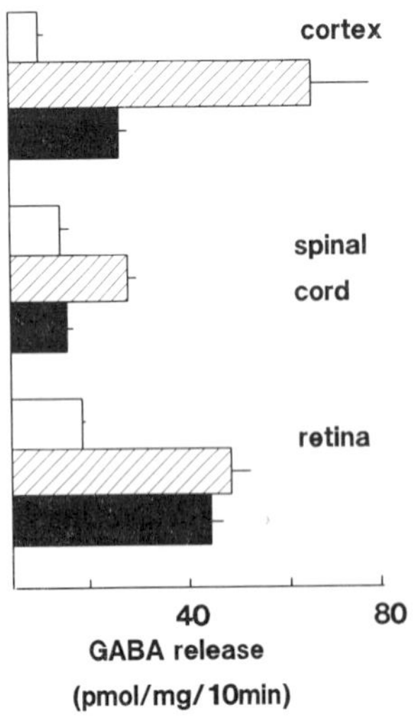


**Figure 3.** Effect of baclofen (500μM) on the K-evoked release of GABA from cortical slices, spinal cord slices, and isolated retinas prepared from the CNS of γ-vinyl GABA treated rats. Baclofen significantly inhibited the K-evoked release of GABA from cortical slices (P<0.01) and spinal cord slices (P<0.002) but did not affect release from the retina. Open columns (□) release preceding stimulation; cross-hatched columns (□) K-stimulation (50mM); solid columns (■) K-stimulation in presence of baclofen. Each result is the mean of 4-5 experiments, bars show the s.e. mean. (Data from Neal and Shah, 1989).

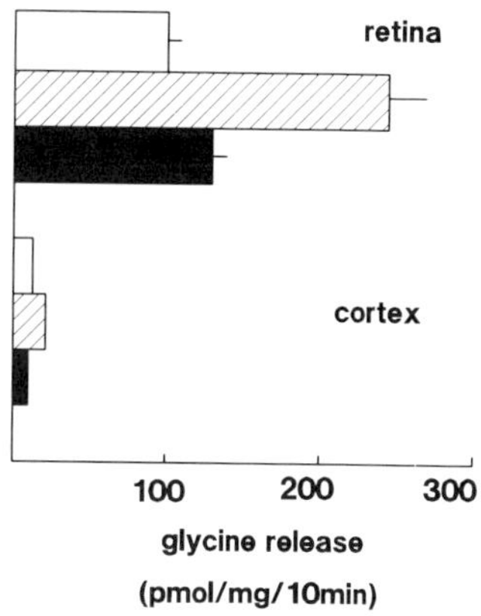


**Figure 4.** Effect of baclofen (500μM) on the K-evoked release of glycine from isolated retinas and cortical slices. In both tissues baclofen significantly inhibited the K-evoked release of glycine (P<0.01, n=5). Open columns (□) release preceding stimulation; cross-hatched columns (□) K-stimulation (KCl 50mM); solid columns (■) K-stimulation in presence of baclofen.

3) (Neal and Shah, 1989). Thus, it seems unlikely that baclofen enhances the light-evoked release of ACh by inhibiting GABA release from GABAergic amacrine cells.

Another important class of amacrine cells in the retina release the inhibitory transmitter glycine. We have previously shown that glycine inhibits the light-evoked release of ACh by acting at a strychnine sensitive receptor (Cunningham & Neal, 1983). Thus, the possibility arises that baclofen might enhance the release of ACh by inhibiting a glycinergic input. This idea was supported by experiments in which we found that the potassium-evoked release of glycine from isolated rat retinas was inhibited by baclofen (Fig 4). If baclofen enhances the light-evoked release of ACh by inhibiting glycine release, then it would be expected that strychnine would also enhance the light-evoked release of ACh, but we previously found that this was not the case (Cunningham, Dawson & Neal, 1983). However, we have recently re-examined the effects of strychnine on ACh release, and whilst the concentration of the drug used in the original experiments (20μM) had no effect on the light-evoked release of ACh, when the concentration was raised to 40μM, strychnine caused the predicted enhancement in ACh release (Fig. 5) (Neal & Cunningham, 1994b). These experiments are consistent with the idea

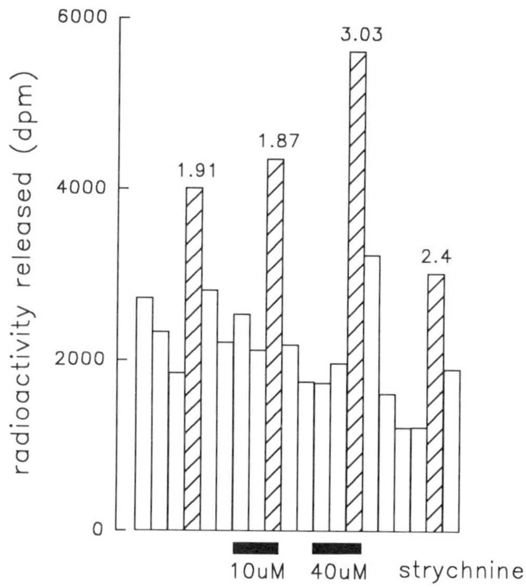


**Figure 5.** Effect of strychnine on [$^3$H]ACh release from the rabbit retina. Strychnine at a concentration of 40μM enhanced the light-evoked release of [$^3$H]ACh. This experiment was repeated three times with similar results. Reprinted from Journal of Physiology, with permission.

that baclofen activates inhibitory presynaptic $GABA_B$ receptors on glycinergic amacrine cells and hence enhances the light-evoked release of ACh.

## Glycinergic Neurones May Form an Inhibitory Feed-Back Loop

We have previously reported evidence for a cholinergic inhibitory feed-back mechanism in the retina (Cunningham, Dawson & Neal, 1983). In these experiments we examined the effects of muscarine and atropine on the light-evoked release of ACh from the rabbit retina. The results are summarised in Fig 6 and show that activation of muscarinic receptors by muscarine reduced the light-evoked release of ACh, whilst the antagonist atropine, enhanced the light-evoked release of ACh. The effect of these drugs on ACh release was not due to an action on inhibitory autoreceptors on the cholinergic amacrine cells themselves because the drug effects were abolished by strychnine (Fig 6).

The results of all the experiments described suggest that baclofen may act on inhibitory presynaptic $GABA_B$ receptors on glycinergic amacrine cells that form an inhibitory feed-back loop from the cholinergic amacrine cells (Fig. 7). Thus, ACh released from the cholinergic amacrine cell activates muscarinic receptors on the glycinergic amacrine cell causing depolarisation and the release of glycine onto inhibitory receptors on the cholinergic cell. Baclofen could enhance the light-evoked release of ACh by activating inhibitory presynaptic $GABA_B$ receptors on the glycinergic neurones. This scheme offers an explanation as to why, in contrast to bicuculline which causes a striking increase in the resting release of ACh, strychnine does not increase the resting release of ACh. Presumably glycine is only released when the glycinergic cell is stimulated by ACh released by flickering light. Support for this mechanism has been obtained in a preliminary experiment in which we have found that muscarine stimulates the release of glycine from the isolated rat retina. The inhibitory actions of muscimol and GABA on the light-evoked release of [$^3$H]ACh are not shown in Fig 7. It is possible that $GABA_A$ receptors occur on the cholinergic amacrine cells, although Linn & Massey (1992) have argued that GABA reduces ACh release by an inhibitory action on the bipolar cell input to the cholinergic amacrine cells.

## Function of $GABA_B$ Receptors

The question arises as to whether the putative inhibitory $GABA_B$ presynaptic receptors on glycinergic amacrine cells have a physiological function. In another preliminary experiment we have found that the effect of baclofen on ACh release disappears when the flicker rate is increased to 10Hz (Fig 8). One explanation for these results is that at low flicker

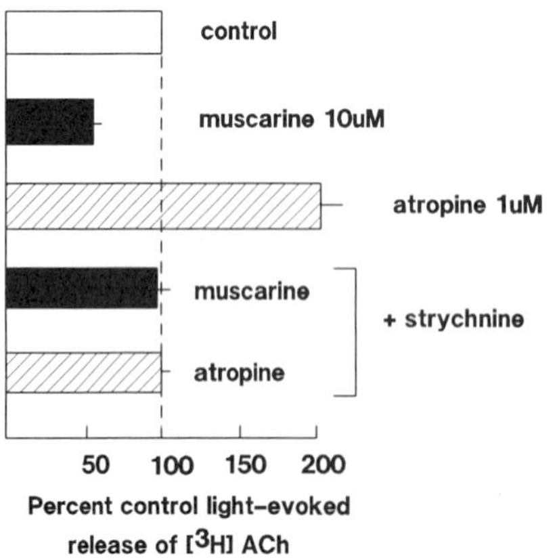


**Figure 6.** Summary of effects of muscarine and atropine on the light-evoked release of [$^3$H]ACh from the rabbit retina. Muscarine significantly reduced the light-evoked release of ACh (P<0.02) whilst atropine enhanced it (P<0.01). The effects of both drugs were blocked by strychnine (10μM), which on its own, had no effect on [$^3$H]ACh release. (Data from Cunningham, Dawson & Neal, 1983).

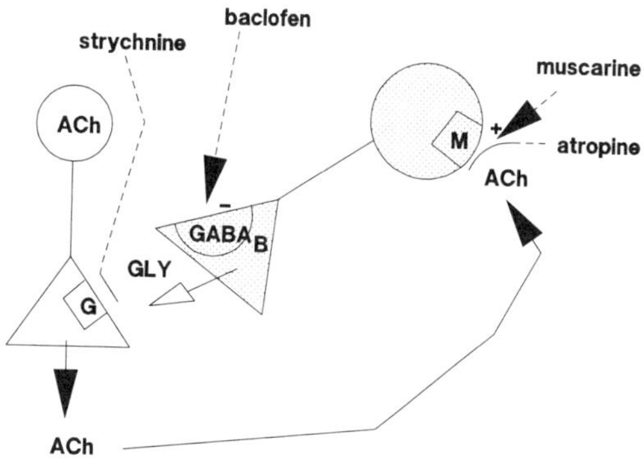


**Figure 7.** Summary of possible sites of action of some drugs that reduce and enhance the light-evoked release of [$^3$H]ACh from cholinergic amacrine cells (open cell). In the proposed scheme, flashes of light result in the release of ACh (solid arrow) which stimulates muscarinic receptors (M) on a glycinergic amacrine cell (shaded) which feeds back and releases glycine (open arrow) onto the cholinergic neurone. Activation of glycine receptors (G) results in a reduction in the light-evoked release of ACh. Muscarine decreases ACh release by stimulating the release of glycine via activation of muscarinic receptors (M). Atropine enhances ACh release by preventing stimulation of the glycinergic neurone, whilst baclofen enhances ACh release by activating presynaptic inhibitory receptors ($GABA_B$) and reducing glycine release. Strychnine enhances ACh release by blocking the inhibitory actions of glycine on the cholinergic neurone. $GABA_A$ receptors are not shown (see text for their location). Reproduced with permission of the Journal of Physiology.

frequencies (1 and 3Hz), the $GABA_B$ receptors are not activated by endogenously released GABA, but at 10Hz the receptors are fully occupied by GABA and therefore not susceptible to further activation by baclofen. A mechanism that decreases glycine inhibition of the cholinergic amacrines with increasing frequency might be important in maintaining ACh release at higher frequencies of cell firing. Since GABA is co-located with ACh in the cholinergic amacrine cells, and can be co-released (O'Malley, Sandell & Masland, 1992; Neal et al, 1992) it is possible that GABA released from the cholinergic cells themselves may facilitate ACh release at high rates of stimulation.

## ACKNOWLEDGMENTS

I am most grateful to Dr Jo Cunningham who carried out most of the experiments described in this paper. Parts of the work were supported by the Medical Research Council.

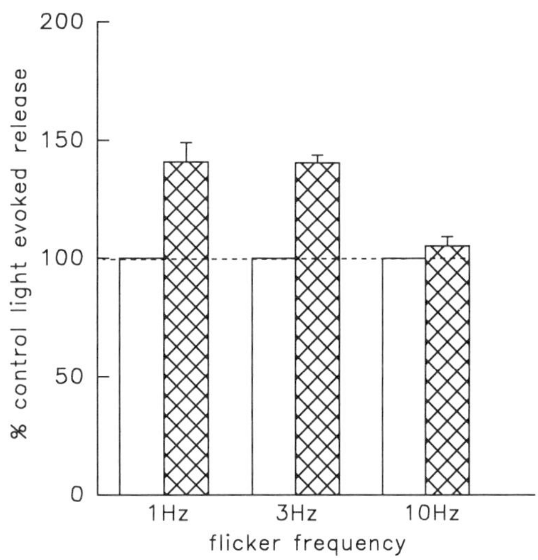


**Figure 8.** Summary of effects of baclofen (100 μM) on the release of [$^3$H]ACh stimulated by flickering light at different frequencies. The drug enhances release at 1Hz and 3Hz but release evoked by flashes at 10Hz was not affected. Open columns (□) control light-evoked release; cross-hatched columns (□) light-evoked release of ACh in the presence of baclofen. Each result is the mean of 3-5 experiments.

Finally, I would like to express my gratitude to Professor Hisako Ikeda who, since my arrival at St Thomas', has been a charming and most helpful colleague.

## REFERENCES


Anderson, R.A. and Mitchell, R. (1985). Evidence for $GABA_B$ autoreceptors in median eminence, *Eur. J. Pharmacol.* 118:355-358.

Ariel, M. and Daw, N.W. (1982). Effect of cholinergic drugs on receptive field properties of rabbit retinal ganglion cells, *J. Physiol.* 324:135-160.

Bai, S.H. and Slaughter, M.M. (1989). Effects of baclofen on transient neurones in the mudpuppy retina: Electrogenic and network actions, *J. Neurophysiol.* 61:382-390.

Baughman, R.W. and Bader, C.R. (1977). Biochemical characterisations and cellular localisation of the cholinergic system in the chicken retina, *Brain Res.* 138:469-485.

Bowery, N.G. (1993). $GABA_B$ receptor pharmacology, *Ann. Rev. Pharmacol.Toxicol.* 33:109-147.

Bowery, N.G., Doble, A., Hill, D.R., Hudson, A.L., Shah, J.S. and Turnbull, M.J. (1979). Baclofen: a selective agonist for a novel type of GABA receptor, *Brit. J. Pharmacol.* 67:444P.

Bowery, N.G., Hill, D.R., Hudson, A.L., Doble, A., Middlemiss, D.N., Shaw, J. and Turnbull, M.J. (1980). (-)Baclofen decreases neurotransmitter release in the mammalian CNS by an action at a novel GABA receptor, *Nature* 283:92-94.

Bowery, N.G., Doble, A., Hill, D.R., Hudson, A.L., Shaw, J.S., Turnbull, M.J. and Warrington, R. (1981). Bicuculline-insensitive GABA receptors on peripheral autonomic nerve terminals. British Journal of Pharmacology 71, 53-70.

Brandon, C. (1987). Cholinergic neurons in the rabbit retina: dendritic branching and ultrastructural connectivity, *Brain Res.* 426:119-130.

Brecha, N.C., Johnson, D., Peichl, L. and Wässle, H. (1988). Cholinergic amacrine cells of the rabbit retina contain glutamate decarboxylase and γ-aminobutyrate immunoreactivity, *Proc. Nat. Acad. Sci. USA* 85:6187-6191.

Conzelmann, U., Meyer, D.K. and Sperk, G. (1986). Stimulation of receptors of γ-aminobutyric acid modulates the release of cholecystokinin-like immunoreactivity from slices of rat neostriatum, *Brit. J. Pharmacol.* 89:845-852.

Cunningham, J.R., Dawson, C. and Neal, M.J. (1983). Evidence for a cholinergic inhibitory feed-back mechanism in the rabbit retina, *J. Physiol.* 340:455-468.

Cunningham, J.R. and Neal, M.J. (1983). Effect of γ-aminobutyric acid agonists, glycine, taurine and neuropeptides on acetylcholine release from the rabbit retina, *J. Physiol.* 336:563-577.

Cunningham, J.R. and Neal, M.J. (1985). Effect of excitatory amino acids and analogues on [$^3$H]acetylcholine release from amacrine cells of the rabbit retina., *J. Physiol.* 366:47-62.

Damsma, G., Westerink, B.H.C. Boer de, P., Vries de, J.B. and Horn, A.S. (1988). Basal acetylcholine release in freely moving rats detected by on-line trans-striatal dialysis: pharmacological aspects, *Life Sci.* 43:1161-1168.

Famiglietti, E.V. (1983a). ON and OFF pathways through amacrine cells in the mammalian retina: The synaptic connections of "starburst" amacrine cells, *Vision Res.* 23:1265-1279.

Famiglietti, E.V. (1983b). "Starburst" amacrine cells and cholinergic neurones: Mirror-symmetric ON and OFF amacrine cells of the rabbit retina, *Brain Res.* 261:138-144.

Friedman, D.L. and Redburn, D.A. (1990). Evidence for functionally distinct subclasses of γ-aminobutyric acid receptors in rabbit retina., *J. Neurochem.* 55:1189-1199.

Gray, J.A. and Green, A.R. (1987). $GABA_B$-receptor mediated inhibition of potassium-evoked release of endogenous 5-hydroxytryptamine from mouse frontal cortex., *Brit. J. Pharmacol.* 91:517-522.

Ikeda, H., Hankins, M.W. and Kay, C.D. (1990). Actions of baclofen and phaclofen upon ON- and OFF-ganglion cells in the cat retina., *Eur. J. Pharmacol.* 190:1-9.

Kosaka, T., Tauchi, M. and Dahl, J.L. (1988). Cholinergic neurons containing GABA-like and/or glutamic acid decarboxylase-like immunoreactivities in various brain regions of the rat., *Exp. Brain Res.* 70:605-617.

Limberger, N. Spath, L. and Starke, K. (1986). A search for receptors modulating the release of γ-[$^3$H]aminobutyric acid in rabbit caudate nucleus slices., *J. Neurochem.* 46:1109-1117.

Linn, D.M., Blazynski, C., Redburn, D.A. and Massey, S.C. (1991). Acetylcholine release from the rabbit retina mediated by kainate receptors, *J. Neurosci.* 11:111-122.

Linn, D.M. and Massey, S.C. (1992). GABA inhibits ACh release from the rabbit retina: A direct effect or feedback to bipolar cells, *Visual Neurosci.* 8:97-106.

Loewi, O. and Hellaver, H. (1938). Uber das acetylcholin im peripheren nerven, *Pflügers Arch.* 240:769-775.

Masland, R.H. and Ames, A. (1976). Responses to acetylcholine of ganglion cells in an isolated mammalian retina., *J. Neurophysiol.* 39:1220-1235.

Masland, R.H. and Livingstone, C.J. (1976). Effect of stimulation with light on synthesis and release of acetylcholine by an isolated mammalian retina, *J. Neurophysiol.* 39:1210-1219.

Masland, R.H. and Mills, J.W. (1979). Autoradiographic identification of acetylcholine in the rabbit retina., *J. Cell Biol.* 83:159-178.

Massey, S.C. and Neal, M.J. (1979). The light evoked release of acetylcholine from the rabbit retina in vivo and its inhibition by gamma-aminobutyric acid., *J. Neurochem.* 32:1327-1329.

Massey, S.C. and Redburn, D.A. (1982). A tonic γ-aminobutyric acid-mediated inhibition of cholinergic amacrine cells in rabbit retina., *J. Neurosci.* 2:1633-1643.

Neal, M.J. (1983). Cholinergic mechanisms in the vertebrate retina. In: Progress in Retinal Research 2, pp191-212 (ed.) N.N. Osborne and G.J. Chader. Pergamon Press, Oxford.

Neal, M.J. and Cunningham, J.R. (1982). The vertebrate retina in vivo: a preparation for studying acetylcholine release from central cholinergic neurons. In: Progress in Cholinergic Biology: Model Cholinergic Synapses pp249-270 (ed.) I Hanin and A Goldberg. Raven Press, New York.

Neal, M.J. and Cunningham, J.R. (1989). L-Homocysteic acid - a possible bipolar cell transmitter in the rabbit retina, *Neurosci. Lett.* 102:114-119.

Neal, M.J. and Cunningham, J.R. (1994). Endogenous ATP modulates the light-evoked release of ACh from retinal cholinergic neurones, *Brit. J. Pharmacol.* 113:1085-1087.

Neal, M.J. and Cunningham, J.R. (1995). Baclofen enhances acetylcholine release from amacrine cells in the rabbit retina by reducing glycinergic inhibition, *J. Physiol.* 482:363-372.

Neal, M.J., Cunningham, J.R., Hutson, P.H. and Semark, J.E. (1992). Calcium dependent release of acetylcholine and γ-aminobutyric acid from the rabbit retina., *Neurochem. Internat.* 20:43-53.

Neal, M.J., Cunningham, J.R. and Slaughter, M.M. (1993). Synergy of acetylcholine and gamma-aminobutyric acid in rabbit cholinergic amacrine cells., *Invest. Ophthalmol.* 34:1062.

Neal, M.J. and Massey, S.C. (1980). The release of acetylcholine and amino acids from the rabbit retina in vivo, *Neurochem.* 1:191-208.

Neal, M.J. and Shah, M.A. (1989). Baclofen and phaclofen modulate GABA release from slices of rat cerebral cortex and spinal cord but not from retina, Brit. J. Pharmacol. 98:105-112.

O'Malley, D.M., Sandell, J.H. and Masland, R.H. (1992). Co-release of acetylcholine and GABA by the starburst amacrine cells, J. Neurosci. 12:1394-1408.

Osborne, N.N. (1980). Binding of $^3$H-muscimol, a potent-γ-aminobutyric acid receptor agonist to membranes of the bovine retina, *Brit. J. Pharmacol.* 71:159-264.

Pickles, H. and Simmonds, M.A. (1980). Antagonism by penicillin of GABA depolarizations at presynaptic sites in rat olfactory cortex and cuneate nucleus in vitro, *Neuropharmacol.* 19:35-38.

Pittaluga, A., Asaro, D., Pellegrini, G. and Raiteri,M. (1987). Studies on $^3$H-GABA and endogenous GABA release in rat cerebral cortex suggest the presence of autoreceptors of the $GABA_B$ type, *Eur. J. Pharmacol.* 144:45-52.

Slaughter, M.M. and Pan, Z-H. (1992). The physiology of $GABA_B$ receptors in the vertebrate retina., *Prog. Brain Res.* 90:47-60.

Turnell, D.C. and Cooper, J.D.H. (1981). Rapid assay for amino acids in serum or urine by precolumn derivatization and reversal phase liquid chromatography., *Clin. Chem.* 28:101-118.

Vaney, D.I. and Young, H.M. (1988). GABA-like immunoreactivity in cholinergic amacrine cells of the rabbit retina., *Brain Res. 438:369-373.*

9

# RAMIFICATIONS OF GABA RECEPTOR SUBTYPES ON RETINAL INFORMATION PROCESSING

Malcolm Slaughter, Jian Zhang, and Ning Tian

Departments of Biophysical Sciences and Ophthalmology
School of Medicine
State University of New York
Buffalo, New York 14214

GABAergic inhibition shapes light responses in both plexiform layers of the vertebrate retina. GABA is contained in a population of horizontal cells and in about one-third of amacrine cells in the amphibian retina. Almost all second and third order neurons have GABA receptors. GABA's influence is not only diffuse, but also diverse. Pharmacological studies, coupled with amino acid sequence analysis, have identified several subtypes of the GABA receptor, including: the classical GABAa, the GABAb, GABAρ, and GABAc receptors. We have begun investigating the role of these different receptors in information processing in the proximal retina. Although some of our results are preliminary, and our conclusions tentative, it appears that: 1) the GABAa receptor produces a strong, shunting inhibition of amacrine and ganglion cell EPSPs; 2) the GABAb receptor suppresses sustained signals while augmenting transient responses; 3) GABAρ receptors preferentially suppress transient ON responses; and 4) GABAc receptors appear to suppress sustained activity, somewhat similar to the action of GABAb receptors. Overall, GABA can modulate the retinal network through a variety of mechanisms, utilizing different receptor systems, to selectively suppress particular information pathways.

Professor Ikeda and I have shared a mutual interest in excitatory and inhibitory amino acid neurotransmitters. In recent years, we have witnessed the explosion in receptor diversity associated with these transmitters and have attempted to relate this emerging amino acid pharmacology to retinal physiology. It has been a fruitful area of reserach, revealing and reemphasizing the robust diversity of retinal synaptology. It is with pleasure and regret that I participate in this symposium to celebrate the acheivements of Hisako Ikeda and mark her retirement.

The experiments on GABA, glycine, and glutamate represent only a portion of Ikeda's research accomplishments. Nevertheless, she has made major contributions to our understanding of these transmitter systems in the cat retina. Looking at retinal output by recording ganglion cell spike activity, she found that glutamate receptors mediate sustained responses, leaving transient response generation to acetylcholine (Ikeda and Sheardown,

*Basic and Clinical Perspectives in Vision Research*, Edited by Jon Robbins et al.
Plenum Press, New York, 1995

1982). Also, comparing the inhibitory amino acid transmitters, she found GABA inhibited the ON but not the OFF pathway (Ikeda and Sheardown, 1983). We have pursued similar experiments in the amphibian retina, using intracellular recordings in the superfused retina and retinal slice to evaluate the action of excitatory and inhibitory amino acids. One of the interesting and evolving stories is the GABAergic system.

The GABA receptor had undergone a pharmacological population explosion in recent years and retinal research has provided fertile ground for new discoveries. Until only a few years ago, there appeared to be only one GABA receptor, which could be differentially regulated by allosteric binding at benzodiazepine and barbituate recognition sites. Now, this GABA receptor is only one class, called the GABAa subtype. It is joined by GABAb and GABAc receptors. Even the GABAa receptor may have to share its moniker, since it is a hetero-oligomer with variation in the constituent polypeptides. One particularly interesting variant in the retina is the recently cloned ρ-subunit (Cutting et al, 1991).

With diversity comes uncertainty. The classical GABAa receptor is well defined, activated by the agonists muscimol and THIP and blocked by the conventional antagonists, bicuculline and picrotoxin. This receptor is directly linked to a chloride channel. Bowery et al (1980) described another GABA receptor, GABAb, that was unaffected by these agonists and antagonists but activated by baclofen. GABAb receptors activate G-protein systems which can regulate two channels, either opening a potassium channel (Newberry and Nicoll, 1984) or closing a voltage dependent calcium channel (Dolphin and Scott, 1986). It has been suggested, but still unresolved, whether there are two subtypes of GABAb receptor which mediate these two different channel mechanisms (Scherer et al, 1988; Dutar and Nicoll, 1988b).

Our understanding of GABA pharmacology becomes less clear as we look further. Heidelberger and Matthews (1991) have described a GABA receptor in isolated goldfish bipolar cells that is presumably not GABAa, based on insensitivity to bicuculline, and not GABAb, based on insensitivity to baclofen. GABA binding to this receptor shifts the activation curve of L-type voltage dependent calcium channels to the right, reducing the calcium current within the physiological range. This receptor is sensitive to very low micromolar concentrations of CACA (cis-amino crotonic acid), an analog that matches extended conformations of GABA. The regulation of a voltage dependent calcium conductance suggests that this CACA receptor acts through a second messenger system. This harkens back two decades to experiments by G. A. R. Johston and colleagues (1975), who found that CACA competed with GABA for a novel binding site which has been termed the GABAc receptor.

Two other laboratories have found retinal neurons that also express CACA sensitivity. Qian and Dowling (1993), experimenting on rod driven (H4) horizontal cells isolated from the white perch retina and Feigenspan et al (1993), investigating rod driven bipolar cells from the rat retina, have found that these neurons are CACA sensitive, but not baclofen nor bicuculline sensitive. Nor are these receptors modulated by barbituates or benzodiazepines. In both of these systems CACA regulates a chloride conductance. Qian and Dowling found the fish horizontal cell GABA receptor is sensitive to muscimol, which appears to act as a partial agonist. It is completely blocked by 500μM picrotoxin. Feigenspan et al. found that the rat ON bipolar GABA receptor was only weakly antagonized by 100μM picrotoxin. They also looked at single channel properties in outside-out patches and found that this CACA sensitive channel has a lower conductance (7 vs 27pS) but longer mean open time (150 vs 25ms) than the classical GABAa receptor. This provides further evidence that this is a distinct, ionotropic receptor.

All these retinal studies point to a bicuculline insensitive GABA receptor that can be activated by CACA. However, there are numerous differences between the three studies. The receptor described by Heidelberger and Matthews is clearly different. It is much more

**Table 1.** Proposed classification of retinal GABA receptors

| Receptor | Agonists | Antagonists | Channel conductance |
|---|---|---|---|
| GABAa | muscimol, THIP | bicuculline, picrotoxin | increase chloride |
| GABAa- ρ | CACA (weak) | picrotoxin | increase chloride |
| GABAb | baclofen | CGP-35348 | increase potassium or decrease calcium |
| GABAc | CACA (potent) | none | decrease calcium |

sensitive to CACA, does not regulate a chloride conductance, and is probably a metabotropic receptor. In many respects, it is similar to the GABAb receptor described by Maguire et al (1989b). The receptors described in fish horizontal cells and rat bipolars are very similar. They are chloride channels, probably ionotropic. CACA serves as a weak but selective agonist. The major difference is the strong blocking action of picrotoxin in horizontal cells and weaker effect in bipolars. Both authors suggest that their studies identify a GABAc receptor.

Overall, GABA receptor classification is in flux. CACA sensitivity identifies at least two different receptors. The low sensitivity CACA receptor, found in fish horizontal cells and rat bipolar cells, has a pharmacology that is very similar to the ρ-subunit homo-oligomer expressed in oocytes. This could be considered a distinct GABAc receptor, or a subset of the GABAa receptor. We favor the latter, based on chloride conductance and sequence homology to other subunits of the GABAa receptor. The GABA receptor found on fish bipolar cells may be considered so distinct that it may be a GABAc receptor. However, it may be a variant of the GABAb receptor. Functionally, it is very similar to the GABAb receptor, but pharmacologically it has no similarity. Therefore, as a working model, we suggest the classification in table 1.

In the amphibian retina, application of exogenous GABA hyperpolarizes the membrane potential and blocks the light response of almost all amacrine and ganglion cells. This strong inhibitory action is duplicated by muscimol (figure 1). This has been interpreted to demonstrate that the classical GABAa receptor is responsible for mediating the large shunting action produced by GABA. However, relatively high concentrations of muscimol are required, muscimol is also effective at the ρ-subunit, and the low sensitivity CACA receptor is also muscimol sensitive. Thus, muscimol activity may not unequivocably define GABAa receptor activation. Nevertheless, GABA application does produce a strong shunting inhibition of light responses in third order neurons and this effect is blocked by bicuculline, as well as picrotoxin.

In the presence of picrotoxin, GABA produces a hyperpolarization and a decrease in membrane noise in amacrine and ganglion cells. This effect can also be elicited by baclofen or 3-amino-propyl phosphinic acid, indicating activation of a GABAb receptor. Unlike the GABAa receptor, this receptor produces a suppression of only sustained responses. In most cases, the transient EPSPs (figure 1) are made larger in amplitude, although shorter in duration, by exogenous baclofen application (Slaughter and Bai, 1989; Bai and Slaughter, 1989). Interestingly, in the cat retina Ikeda et al (1990) found a similar shift, but only in the OFF pathway. Another example of baclofen's effect is in the acetylcholine releasing amacrine cells in the rabbit retina. These are known to be transient, one set releasing acetylcholine at light onset, the other at offset. In preliminary studies with Cunningham and Neal we have found that baclofen enhances acetylcholine release in rabbit retina (Neal et al, 1993).

This shift in weighting toward transient responses is emphasized by the effect of baclofen on many sustained cells, in which the loss of the sustained light response is combined with the appearance of transient responses. This effect of exogenous baclofen indicates what the ramifications would be if GABAb receptors were activated, but does not

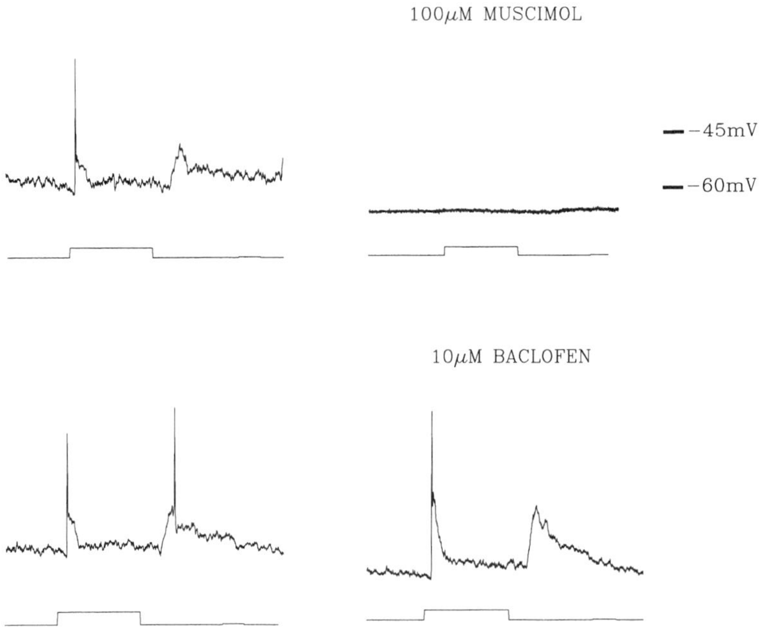


**Figure 1.** Figure 1 compares the effects of muscimol and baclofen on a transient ON-OFF third order neuron from the superfused tiger salamander retina. Muscimol caused a hyperpolarization of approximately 9mV, a decrease in dark noise, and a total suppression of the light response. Baclofen produced a very similar hyperpolarization, but an enhancement of the ON EPSP (from 10mV to 21mV) and OFF EPSP (from 12 mV to 16mV). In this cell, baclofen produced only a very slight decrease in dark noise. In addition, the EPSPs became more transient (approximately 30% decrease in time to half amplitude after baclofen). The square pulses beneath the voltage traces indicate the timing of diffuse light stimuli.

indicate the normal level of endogenous stimulation. We have had difficulty finding an effective GABAb antagonist. Ikeda et al (1990) found that phaclofen was an effective antagonist of iontophoresed baclofen. But neither this antagonist, nor saclofen, nor 2-hydroxy saclofen (Kerr et al, 1987, 1988, 1989) blocked the action of applied baclofen in the amphibian retina. Bittinger and colleagues (1990) developed an antagonist, CGP35348, that we found to be a weak but effective antagonist of applied baclofen (figure 2). This antagonist produced effects on the light responses of amacrine and ganglion cells that was just the opposite of applied baclofen. The resting membrane potentials showed slight depolarizations, the membrane noise increased, and light responses became more sustained. This indicates that GABAb receptors were activated in both tonic and phasic modes, and suppression of these receptors enhanced sustained signals. The conclusion from these studies is that, during GABAb receptor stimulation, the output signals leaving the retina become biased toward the transient information channel.

This shift suggests that the visual information content sent from the retina to the brain is modified by GABAb receptor stimulation. Sustained signals are ideal for relaying information on steady state conditions, such as the intensity or location of a light stimulus,

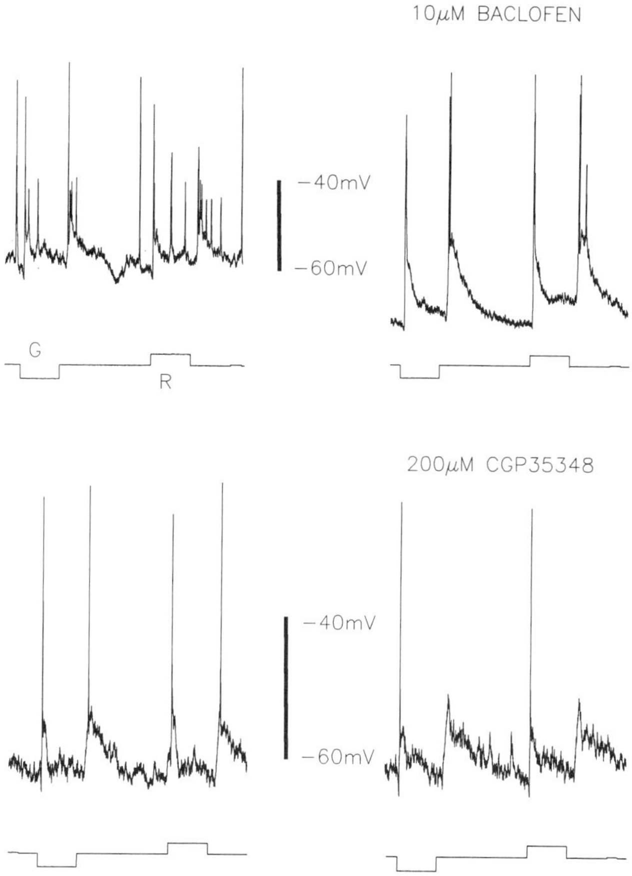


**Figure 2.** Figure 2. demonstrates that CGP35348 blocks the effects of applied baclofen. The top set of traces show the response of a transient ON-OFF cell before (upper left) and during the application of baclofen. Baclofen produced a 16mV hyperpolarization and an enhancement of the amplitudes of both the ON and OFF EPSPs. The lower set of traces shows control (bottom left) and CGP35348's effect on baclofen application in the same cell. CGP35348 prevented the effects seen with baclofen alone. In addition, CGP35348 increased the baseline noise and prolonged the duration of the EPSPs, particularly the ON responses. The square pulses beneath the voltage traces represent diffuse light stimuli, green is down and red up.

while transient signals are preferred for conveying information on change of state, such as contrast modulation or movement. We therefore looked at the effect of baclofen on motion detection, particularly directional selectivity (Pan and Slaughter, 1991). We found that baclofen enhanced directionally selective responses in third order cells. These experiments were performed in the tiger salamander retina, and under our experimental conditions only about 10% of third order cells demonstrated directionally selective responses under control conditions. In these cells, baclofen enhanced the directionality of the responses (measured as the ratio of peak amplitudes in the preferred compared to the null directions). Interestly, about a third of the cells expressed directional selectivity only after baclofen application. In most of these cases, APB (2-amino-4-phosphonobutyrate) (Slaughter and Miller, 1981) had to be applied in conjunction with baclofen to fully reveal this directionality. Experiments suggested that the ON and OFF pathways may not have been well aligned in many naturally non-directional cells, and the use of APB simplified the network by eliminating cross-talk between misaligned ON and OFF pathways. There is evidence from other studies that the ON and OFF pathways may handle directional information separately. These baclofen experiments indicate that signal processing in the retina can be regulated and that the pathways through which these signals pass may not be the discrete, hardwired conduits that we previously pictured. One way of interpreting the data is that information is distributed throughout the retinal network, but that differences in synaptic weighting give the appearance of disparate channels. And this weighting may be altered, not only to emphasis a particular channel, but to use multiple channels to relay the designated information.

Turning to other GABA receptors, we have looked at the effect of GABA while suppressing the GABAa receptor with bicuculline and preventing GABAb receptor activation by prior application of saturating concentrations of baclofen (figure 3). It would be preferable to block the GABAb receptor, but the only effective GABAb antagonist we have found is a weak, competitive agent and we could not be sure that effects of applied GABA were not due to chasing CGP35348 off the GABAb receptor site. Bicuculline is also a competitive antagonist, but it is reasonably potent and previous studies indicate that 100μM bicuculline should effectively block 500μM GABA. Under these conditions, exogenously applied GABA typically did not change the membrane potential, nor reduce the membrane noise. This suggests that GABA was not having a direct effect on third order cells. However, very frequently the ON response was significantly more suppressed than the OFF response. This data correlates with many previous studies, including the work of Ikeda and Sheardown (1983), indicating that GABA preferentially suppresses the ON pathway. In the amphibian retina this is not true for superfused GABA, but is evident when classical GABAa and GABAb receptors are pretreated. The action of GABA appears to be presynaptic to third order neurons, probably an effect on ON bipolar cells. This is reminescent of Heidelberger and Matthews (1991) studies in ON bipolars in fish retina, but 500μM CACA does not duplicate the GABA effect in our preparation (figure 3).

Which GABA receptor produces this effect is unclear. Presumably it is not the GABAa or GABAb receptor because these studies were carried out in the presence of bicuculline and baclofen. The other two identified receptors (see table 1) are both sensitive to CACA, while this effect was not produced by 500μM CACA. However, evaluation of CACA's effectiveness is based on studies performed on isolated cells. Pharmacological studies in the eyecup preparation are impaired by diffusion barriers and uptake systems which reduce the effective concentration of applied agonists. Nevertheless, it is unlikely to be the GABAc receptor of Matthews et al (1991), because their receptor was exquisitely sensitive to CACA and it is unlikely that the effective concentration in the eyecup will be reduced from 500μM to less than 1μM. The GABAρ receptor, although sensitive to CACA, requires high micromolar doses even in the isolated cell preparation. Both Feiganspan et al (1993) and Qian and Dowling (1993) routinely used 500μM to activate these channels. This may mean that the same concentration in the eyecup would be insufficient. Therefore, we tentatively propose that the GABAρ receptor mediates inhibition of this transient ON response.

In the presence of bicuculline and baclofen, 500μM CACA often produced a hyperpolarization and a reduction of spike activity. This effect is not linked to the selective reduction of ON transient responses (figure 3). It is tempting to relate this CACA effect to the GABAc receptor, but we have observed a similar effect applying muscimol in the presence of bicuculline and baclofen. Heidelberger and Matthews (1991) reported that, in isolated goldfish bipolar cells, 1μM CACA was effective while 5μM muscimol had no effect. But in our preparation, 500μM CACA had no greater effect than 100μM muscimol. This reversal in potency suggests these are not the same receptors. However, since this is the most distinct effect of CACA in our preparation, we have provisionally categorized this as a GABAc receptor effect.

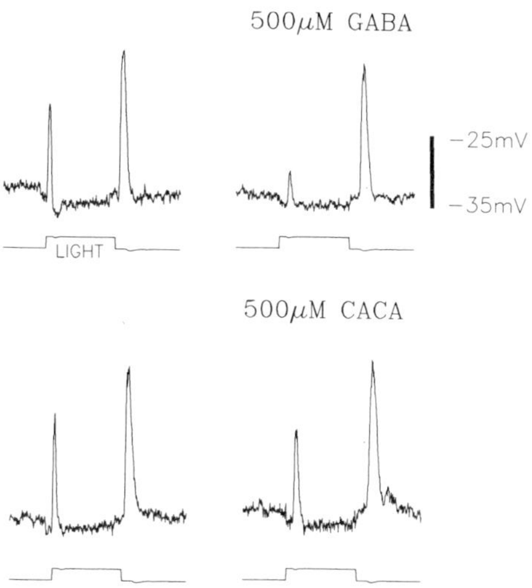


**Figure 3.** Figure 3. shows the effect of GABA and CACA on the ON and OFF light responses of a third order neuron after the GABAa receptors have been suppressed by 100μM bicuculline and the GABAb receptors have been saturated by 100μM baclofen. Under these conditions, 500μM GABA reduced the ON response to 35% of control while the OFF response was 92% of control. Application of 500μM CACA reduced the ON EPSP to 88% of control, and the OFF EPSP to 95% of control.

What are the functions of these sundry GABA receptors? The classical GABAa receptor seems to sculpt the shape of EPSPs. The diacritical attributes of rapid action and large conductance change make it well suited to limit the size and time course of excitatory signals (Frumkes et al, 1981). The GABAa receptor seems to share attributes, and probably function, with the glycine receptor. Both have very similar conductances and kinetics (Bormann et al 1987; Cohen et al, 1989). The differential aspects of GABA and glycine function are still debated. For example, Ikeda and Sheardown (1983) found that GABA suppressed ON while glycine suppressed OFF responses in cat ganglion cells. Caldwell et al (1978) found that GABAergic inhibition mediated several ganglion cell trigger features in rabbit retina. In the amphibian retina, McReynolds and colleagues (1984) found that glycine inhibited phasic signals while GABA suppressed tonic signals in ganglion cells. Many previous studies relied on picrotoxin, as well as bicuculline, in assessing the function of GABAa receptors. This could have led to inadequate differentiation of the role of GABAa receptors.

The GABAb receptors have a far less potent inhibitory action, tending to skew light response properties rather than suppress them completely. The metabotropic mechanism of action also means a slower onset, and perhaps more prolonged action. Maguire et al (1989b) have found that the GABAb receptors reduce an L-type, voltage dependent calcium channel in some bipolar cell terminals, which would reduce excitatory input to third order neurons. For a specific subset of bipolar to amacrine and ganglion cell connections, they propose that delayed GABA feedback serves to truncate bipolar cell transmitter release (Maguire et al, 1989a). They suggest that this could be the method of transient response generation. Werblin et al (1988) have incorporated this mechanism into a model of motion detection based on time delayed GABAb feedback that may form the substrate for Barlow and Levick's (1965) model of directional selectivity.

We have found that GABAb receptors suppress sustained responses while enhancing transient responses. In a feedback to bipolar cell circuit model, our observations would suggest that GABAb receptors were localized to bipolar cells that conveyed tonic information to the inner retina. In addition, GABAb receptors appear to produce a delayed opening of potassium channels on third order cells. This slow inhibition may be synchronized with rapid GABAa inhibition, to produce a fast, powerful suppression of EPSPs followed by a weaker inhibition. This pattern of fast and slow GABA inhibition has been found in the hippocampus by Dutar and Nicoll (1988a).

The GABA response produced in the presence of bicuculline and baclofen, which we tentatively have labelled as a GABAρ effect, suppresses transient ON responses. In part it is complementary to the GABAb effect, in that the former suppresses transient responses while the later suppresses sustained responses. But the GABAb system affects both the ON and OFF pathways about equally, while GABAρ preferentially affects the ON channel.

In summary, GABA acts through various receptors to inhibit neurons in the inner retina. GABA acts both tonically, to lower basal activity, and phasically to shape light responses. If we view the retina as processing visual input through multiple parallel channels, then the advantage provided by multiple GABA receptor subtypes is the capability of selective suppression of particular pathways. Whether this is accomplished by selective circuitry (where discrete GABAergic neurons synapse on only one type of GABA receptor) or based on differential binding affinities (where low concentrations of GABA activate one receptor, higher concentrations activate additional receptor subtypes) has yet to be determined. Perhaps both mechanisms work together to provide multifarious GABAergic regulation of visual signals.

## REFERENCES


Bai, S-H., and Slaughter, M.M. (1989) Effects of baclofen on transient neurons in the mudpuppy retina: Electrogenic and network actions, *J. Neurophysiol.* 61:382- 390.

Barlow, H.B.and Levick, W.R. (1965) The mechanism of directionally selective units in rabbit's retina, *J. Physiol.* 178:477-504.

Belgum, J.H., Dvorak, D.R.and McReynolds, J.S. (1984) Strychnine blocks transient but not sustained inhibition in mudpuppy retinal ganglion cells, *J Physiol.* 354:273-286.

Bittinger, H., Froestl, W., Hall, R., Karlsson, G., Klebs, K., Olpe, H-R., Pozza, M.F., Steinmann, M.W.and Van Riezen, H. (1990) Biochemistry, electrophysiology, and pharmacology of a new GABAb antagonist: CGP 35348. In: GABAb receptors in mammalian function (Bowery ,N.G., Bittinger, H.,and Olpe, H-R, eds) pp47-60. New York: Wiley.

Bormann, J., Hamill, O.P.and Sakman, B. (1987) Mechanisms of anion permeation through channels gated by glycine and gamma-aminobutyric acid in mouse cultured spinal neurones.*J. Physiol.* 385:243-286.

Bowery, N.G., Hill, D.R., Hudson, A.L., Doble, A., Middlemiss, D.N., Shaw, J.and Turnbull, M.J. (1980) (-)Baclofen decreases neurotransmitter release in the mammalian CNS by an action at a novel GABA receptor, *Nature* 283: 92-94.

Boycott, B.B.and Wassle, H .(1974) The morphological types of ganglion cells of the domestic cat's retina, *J. Physiol.* 240:397-419.

Caldwell, J.H., Daw, N.W.and Wyatt, H.J. (1978) Effects of picrotoxin and strychnine on rabbit retinal ganglion cells: Lateral interactions for cells with more complex receptive fields, *J. Physiol.* 276:277-298.

Cohen, B.N., Fain ,G.L.and Fain, M.J .(1989) GABA and glycine channels in isolated ganglion cells from goldfish retina, *J. Physiol.* 418:53-82.

Cunningham, J.R.and Neal, M.J. (1983) Effect of gamma-aminobutyric acid agonists, glycine, taurine, and neuropeptides on acetylcholine release from the rabbit retina, *J. Physiol.* 336: 563-577.

Cutting, G., Lu, L., O'Hara, B.F., Kasch, L.M., Montrose-Rafizdeh, C., Donovan, D.M., Shimada, S., Antononarakis. S,E,, Guggino, W.B., Uhl, G.R.and Kazazian, H.H. (1991) Cloning of the γ-aminobutyric acid (GABA) ρ1cDNA: a GABA receptor subunit highly expressed in the retina, *Proc. Natl. Acad. Sci. USA* 88:2673-2677.

Dolphin, A.C.and Scott, R.H. (1986) Inhibition of calcium currents in cultured rat dorsal root ganglion neurones by (-)baclofen, *Br. J. Pharmacol.* 88:213-220.

Dutar, P.and Nicoll, R.A.(1988a) A physiological role for GABAb receptors in the central nervous system, *Nature* 332:156-158.

Dutar, P.and Nicoll, R.A. (1988b) Pre- and postsynaptic GABAb receptors in the hippocampus have different pharmacological properties, *Neuron* 1:585-591.

Feigenspan, A., Wässle, H.and Bormann, J. (1993) Pharmacology of GABA receptor $Cl^-$ channels in rat retinal bipolar cells, *Nature* 361:159-162.

Frumkes, T.E., Miller, R.F., Slaughter, M.and Dacheux, R.F. (1981) Physiological and pharmacological basis of GABA and glycine action on neurons of mudpuppy retina III. Amacrine-mediated inhibitory influences on ganglion cell receptive-field organization: A model, *J. Neurophysiol.* 45:783-804.

Heidelberger, R.and Matthews, G. (1991) Inhibition of calcium influx and calcium current by γ-aminobutyric acid in single synaptic terminals, *Proc. Natl. Acad. Sci. USA* 88:7135-7139.

Ikeda, H., Hankins, M.W.and Kay, C.D. (1990) Actions of baclofen and phaclofen upon ON- and OFF-ganglion cells in the cat retina, *Eur. J. Pharmacol.* 190:1-9.

Ikeda, H.and Sheardown, M.J. (1982) Aspartate may be an excitatory transmitter mediating visual excitation of 'sustained' but not 'transient' cells in the cat retina: Iontophoretic studies , in *vivo, Neurosci.* 7:25-36.

Ikeda, H.and Sheardown, M.J. (1983) Transmitters mediating inhibition of ganglion cells in the cat retina: Iontophoretic studies in vitro, *Neurosci.* 8:837-853.

Johnston, G.A.R., Curtis, D.R., De Groat, W.C.and Duggan,A.W. (1968) Central actions of ibotenic acid and muscimol, *Biochem. Pharmacol.* 17:2488-2489.

Johnston, G.A.R., Curtis, D.R., Beart, P.M., Game, C.J.A., McCulloch, R.M.and Twitchin, B. (1975) Cis and trans-4-aminocrotonic acid as GABA agonists of restricted conformation, *J. Neurochem.* 24:157-160.

Kerr ,D.I., Ong, J., Johnston, G.A.R., Abbenante, J.and Prager, R.H. (1988) 2-Hydroxy-saclofen: An improved antagonist at central and peripheral GABAb receptors, *Neurosci. Lett.* 92:92-96.

Kerr, D.I., Ong, J., Johnston ,G.A.R., Abbenante, J.and Prager, R.H. (1989) Antagonism at GABAb receptors by saclofen and related sulphonic analogues of baclofen and GABA, *Neurosci. Lett.* 107:239-244.

Kerr, D.I., Ong, J., Prager, R.H., Gynther, B.D. and Curtis, D.R. (1987) Phaclofen: A peripheral and central baclofen antagonist, *Brain Res.* 405:150-154.

Krnjevic, K.and Schwartz, S. (1967) The action of gamma-amino-butyric acid on cortical neurones, *Exp. Brain Res.* 3:320-336.

Maguire, G., Lukasiewicz, P.and Werblin, F. (1989a) Amacrine cell interactions underlying the response to change in the tiger salamander retina, *J. Neurosci.* 9:726-735.

Maguire, G., Maple, B., Lukasiewicz, P.and Werblin, F. (1989b) Gamma-aminobutyrate type B receptor modulation of L-type calcium channel current at bipolar cell terminals in the retina of the tiger salamander, *Proc. Natl. Acad. Sci. USA* 86:10144-10147.

Massey SC Redburn DA (1987) Transmitter circuits in the vertebrate retina. Prog Neurobiol 28:55-96.

Matthews, G., Ayoub, G.and Heidelberger, R. (1991) Inhibition of presynaptic calcium currents via GABAc receptors, *Soc. Neurosci. Abstr.* 17:900.

Neal, M.J., Cunningham, J.R.and Slaughter, M.M. (1993) Synergy of acetylcholine and gamma-aminobutyric acid in rabbit cholinergic amacrine cells, *Invest. Ophthalmol. Vis. Sci. (suppl.)* 34:1062.

Newberry, N.R.and Nicoll, R.A. (1984) Baclofen directly hyperpolarizes hippocampal cells, *Nature* 308:450-452.

Pan, Z-H. and Slaughter, M.M. (1991) Control of retinal information coding by GABAb receptors, *J. Neurosci.* 11:1810-1821.

Qian, H.and Dowling, J.E. (1993) Novel GABA responses from rod-driven retinal horizontal cells, *Nature* 361:162-164.

Redburn, D.A., Friedman, D.L.and Massey, S.C. (1989) The function of multiple subclasses of GABA receptors in rabbit retina. In: Neurobiology of the Inner Retina (Weiler, R.and Osborne, N.N. eds)pp 41-52. Berlin: Springer-Verlag.

Scherer, R.W., Ferkany, J.W.and Enna, S.J. (1988) Evidence for pharmacologically distinct subsets of GABAb receptors, *Brain Res. Bull.* 21:439-443.

Slaughter, M.M.and Bai, S-H. (1989) Differential effects of baclofen on sustained and transient cells in the mudpuppy retina, *J. Neurophysiol.* 61:374-381.

Slaughter, M.M.and Miller, R.F. (1981) 2-Amino-4-phosphonobutyric acid: A new pharmacological tool for retina research, *Science* 211:182-185.

Werblin, F., Maguire, G., Lukasiewicz, P., Eliasof, S.and Wu, S.M. (1988) Neural interactions mediating the detetion of motion in the retina of the tiger salamander, *Visual Neurosci.* 1:317-329.

10

# POSTNATAL DEVELOPMENT OF NEUROTRANSMITTER SYSTEMS IN THE MAMMALIAN RETINA

Jon Robbins[1] and Hisako Ikeda[2]

[1] Pharmacology Group
Biomedical Sciences Division, King's College
Manresa Road, London, SW3 6LX, United Kingdom
[2] Vision Research Unit
UMDS, Rayne Institute, St Thomas's Hospital
London SE1 7EH, United Kingdom

## INTRODUCTION

The mammalian retina is not fully developed at birth and many mammals eyes do not open for many days post partum. Much of the gross morphological development is complete before the eyes open. For example in the cat retina most of the morphological development is complete by postnatal day (PND) 40 (Fig. 1.), whilst spatial resolution of the cells in the lateral geniculate nucleus (Ikeda & Tremain , 1978), visual acuity measured by visual evoked potentials (Freeman & Marg, 1975) or behaviourally (Mitchell et al, 1976) have not reached adult levels until PND 100. Likewise in humans, the time to reach adult like visual acuity is 3-5 years (Teller & Movshon, 1986).

Once the eyes are open, the visual system is exposed to modification by the visual environment as described in the classical work of Hubel & Wiesel (1970). Furthermore even at a late stage of postnatal life electrophysiological responses of the visual system are still immature. Indeed the retinal ganglion cells in young cats have larger receptive fields and weaker surround inhibition (Hamasaki & Flynn, 1977: Rusoff & Dubin, 1977; Hamasaki & Sujita, 1979; Ikeda & Tremain, 1979; Ikeda & Robbins, 1985a). These differences in receptive field characteristics cannot be explained by optical or anatomical factors (Rusoff & Dubin: 1977; Bonds & Freeman, 1978; Ikeda, 1979; Rusoff, 1979). As the cells in the lateral geniculate nucleus of 3-10 week old cats have lower spatial contrast sensitivity than 12 week old cats (Ikeda & Tremain, 1979), it may be expected then that in young animals the retina displays a reduced spatial contrast resolution. However studies in cat retina have not yet demonstrated this unequivocally (Tootle & Friedlander, 1989).

Our hypothesis was then, that the functional immaturity of the retina might be due to immaturity in the neurotransmitter systems controlling retinal signalling. In order to test

*Basic and Clinical Perspectives in Vision Research*, Edited by Jon Robbins et al.
Plenum Press, New York, 1995

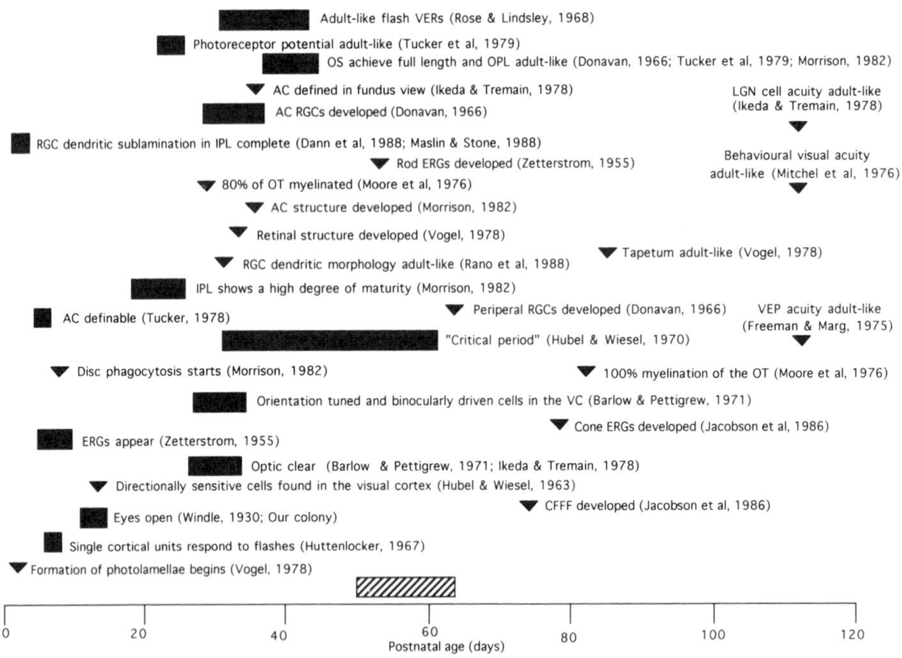


**Figure 1.** Development events in the feline visual system. hatched bar indicates age of cats used in our electrophysiological studies. AC - area centralis, RGC - retinal ganglion cell, VER visually evoked potential, LGN - lateral geniculate nucleus, ERG - electroretinogram, CFFF - critical flicker fusion frequency, OS - outer segments, OPL - outer plexiform layer, IPL - inner plexiform layer, OT - optic tract, VC - visual cortex.

this hypothesis, we investigated the postnatal development the neurotransmitter systems in the cat retina.

## METHODS OF MEASURING NEUROTRANSMITTER DEVELOPMENT

A number of individual or a combination of approaches can be used to monitor changes in the neurotransmitter systems during development of the retina.

### Neurotransmitters

The measurement of the neurotransmitter itself either using traditional biochemistry or by using antibodies directed against the neurotransmitter in question. However low levels of neurotransmitter and the availability of specific antibodies may limit these studies. Incubating retina in a radiolabelled neurotransmitter or precursor can be performed. This is, however, limited to those substances with a specific uptake system.

### Metabolic Enzymes

The study of the synthetic and or degradative enzymes involved in the metabolism of the neurotransmitter is usually performed with fluorescent antibodies. This is a very indirect monitor of neurotransmitter development.

## Receptors

Characterisation of receptors can be achieved by radioligand binding or antibody labelling. The radioligand binding has the advantage in that quantitative data can be obtained of the number of receptors present as well as the affinity for the ligand used. Incorporating autoradiography can identify the location of the receptors. Antibody labelling can only demonstrate spatial location of the receptors and the cell types expressing them. More recent molecular biological techniques allows the demonstration of mRNA for the neurotransmitter receptor (*In situ* hybridisation). Although these procedures demonstrate the presence of the receptors, they do not indicate whether they are functionally important in visual signalling.

## Second Messengers and their Metabolic Enzymes

An important aspect of a family of neurotransmitter receptors is that they act through one or more second messenger systems. These receptors activate or inhibit enzymes controlling synthesis or degradation of second messenger molecules. These receptor types are termed metabotrophic and are generally coupled to G-proteins. Therefore studies on the changes during development of these second messenger or the enzymes controlling them are of importance. Here specific antibodies to enzymes or *in situ* hybridisation can be used. A more functional assay is to measure the generation of second messengers by stimulating retinal tissue with neurotransmitter receptor agonists.

## Functional Studies

These may take a number of forms including *in vitro* methods of measuring light stimulated neurotransmitter release to co-culturing retinal neurones with skeletal muscle to monitor functional cholinergic synapse formation. A question arises with the *in vitro* studies in that, once the tissues are removed from the animal either as a whole retina or as cell cultures, do they then behave as they would *in vivo*?

*In vivo* studies are generally limited to electrophysiological recordings of extracellular action potential activity (restricted to retinal ganglion cells) and the application of neurotransmitter agonists and antagonists either systematically or locally by iontophoresis.

None of the above techniques in isolation can give us all the information of the changes to neurotransmitter systems during development. A combination of them, however can give us insights into the complexity of the changes occurring during the maturation of the central nervous system.

# POSTNATAL DEVELOPMENTAL CHANGES IN RETINAL NEUROTRANSMITTERS

In the following section, we have grouped neurotransmitters into their chemical types in the traditional manner. We realise that the recent studies on the different receptor mechanisms that these neurotransmitters utilise, suggests that this method of classification may be questionable.

## Amino Acids

Receptor binding studies show an increase in receptor number for GABA, benzodiazepines and glycine in early postnatal life (Redburn & Mitchell, 1981: Guaneri et al,

1982: Mates & Redburn, 1982: Parkinson & Rando 1984: Schliebs et al, 1986). Uptake systems for GABA and glycine appear very early on in development and reaches maturity before synthesis or release (Lam et al, 1980; Fung et al, 1982). Early in postnatal life, the GABA synthetic enzyme, glutamic acid decarboxlyase, and GABA immunoreactivity is present in a large number of cell types compared to adult retinae (Schnitzer & Rusoff, 1984; Osborne et al, 1986; Fry et al, 1991; Messersmith & Redburn, 1992).

Our electrophysiological studies of sustained (X) retinal ganglion cells in the area centralis of the kitten (7-9 weeks old) and cat (18-22 weeks old) *in vivo* (Ikeda & Robbins, 1984; Ikeda & Robbins, 1985a, b; Ikeda, 1985) demonstrated the developmental modification in the sensitivity and selectivity of iontophoretically applied GABA and glycine agonists and antagonists. Visually evoked inhibitions in kitten ON- and OFF-type cells were more readily blocked by the GABA antagonist, bicuculline and the glycine antagonist, strychnine respectively, suggesting that the inhibitory input is weaker at this age. Inversely, the kitten cells were less sensitive to the receptor agonists, GABA and glycine than the adult cells, suggesting that the sensitivity of these receptors still had not reached adult levels. In kitten retina, furthermore, both, ON- and OFF cells responded to GABA and Glycine, whereas in the adult ON-cells where inhibited predominantly by GABA and OFF-cells, predominantly by glycine. This suggests the selectivity of the inhibitory transmitters in the kitten cells had not been fully established at this age. Parallel changes in retinal ganglion cell selectivity were seen in a similar study using benzodiazepine receptor agonists and antagonists (Robbins & Ikeda 1989). Furthermore, cells in the adult peripheral retina showed a degree of sensitivity and selectivity to the inhibitory transmitters somewhere between those in the area centralis (the most stimulated area of the retina) of the adult and those in the kitten (Ikeda, 1985; Priest et al, 1985). These studies then provide evidence that the selectivity and sensitivity of inhibitory neurotransmitters develop postnataly and that such development parallels the receptive field properties of retinal ganglion cells.

Glutamate immunoreactivity is located in four cell types in the immature rabbit retina, whilst it is found only in photoreceptors and ganglion cells in the mature animal. Uptake systems for radiolabelled glutamate are present at birth but metabolism of glutamate to glutamine does not reach adult levels until after PND 12 (Redburn et al, 1992). Glutamate is now established as the excitatory neurotransmitter released by ON- and OFF-bipolar cells onto non-NMDA receptors of On- and OFF-retinal ganglion cells (Ikeda et al, 1989). The sensitivity of cat retinal ganglion cells to excitatory amino acid receptor agonists have been demonstrated to alter during postnatal development (Ikeda et al, 1990).

## Catecholamines

Dopamine content of the retina increases from birth to reach adult levels by PND 30 (Parkinson & Rando, 1984), similarly in a range of mammalian species the dopamine synthetic enzyme, tyrosine hydroxylase, shows changes in content and distribution over the first 30 days of postnatal life (Mitrofanis et al, 1988; 1989; Martin-Martinelli et al, 1989; Wulle & Schnitzer, 1989; Mitrofanis & Finlay, 1990; Wang et al, 1990; Casini & Brecha, 1992). Retinal dopamine receptors show a decrease in affinity during postnatal life (Parkinson & Rando, 1984). A number of studies have further suggested the need for adequate light stimulus for the normal development of the retinal dopaminergic system (Kato et al, 1980; Cohen & Neff, 1982; Morgan & Kamp, 1982; Parkinson et al, 1985; Spira & Parkinson, 1991).

Our iontophoretic studies have shown that despite the fact that all classes of cat retinal ganglion cells in inhibited by dopamine acting on a D2, neither D1 nor D2 receptor antagonists had any action indicating that there was no endogenous release of dopamine onto any of the retinal ganglion cells in the adult (Ikeda et al, 1986). This is consistent with anatomical data from the adult cat retina showing that dopaminergic amacrine cells do not synapse with retinal ganglion cells (Pourcho 1982). In the kitten retina however, all types of retinal ganglion cells were found to be sensitive to dopamine which acted on both D1 and D2 receptor subtypes (Ikeda et al, 1987). Iontophoretic studies with dopamine antagonists further showed evidence for endogenous release of dopamine onto retinal ganglions cells not demonstrable in the adult cat retina (Ikeda et al, 1987).

The enzyme that synthesises adrenaline, phenylethanolamine-N-methytransferase increases from 10% of adult levels at birth to adult levels by PND 40 (Cohen, 1987). However our iontophoretic studies of adrenoreceptor anagonists failed to demonstrate the existence of adrenergic receptors on retinal ganglion cells in the adult cat. The action of noradrenaline on retinal ganglion cells seemed to be mediated by dopamine (D2) receptors (Robbins et al, 1988), although we cannot rule out the possibility that adrenergic receptors are expressed during early postnatal life.

## Indolamines

Serotonin (5HT) is taken up by a wider range of cells in young rabbit retina compared with just a subclass of amacrine cells in the adult (Osborne, 1985) however, the stimulation of significant amounts of inositol phosphate formation by 5HT only occurs in adult retina (Cutcliffe & Osborne, 1987).

## Acetylcholine

In rat retina acetylcholine (ACh) synthesis and degradation starts at around birth but does not reach adult levels until after PND 14 (Puro et al, 1982). Numbers of cholinergic amacrine cells in feline retina increase dramatically after birth (Dann, 1989) and the same cells in the rabbit lose their immature dendritic morphology (dendritic spines) by PND 21 (Wong & Collin, 1989). Nicotinic ACh receptors increase in heterogeneity and number after birth (Hoover & Goldman, 1992).

Co-cultures of retinal cholinergic neurons from embryonic rats and skeletal myotubes demonstrate that spontaneous ACh release occurs at very early stages of development however evoked ACh release did not occur in the majority of synapses, unless retinal cells were taken from animals near birth age (Puro et al, 1982; Yeh et al, 1983).

## Peptides

Embryonic and postnatal changes in substance P (Yew et al, 1991; Zhang & Yeh, 1992), somatostatin (Ferriero & Sagar, 1987; Mitrofanis et al, 1989; Ferriero et al, 1990; Bodenant et al, 1991; Ferriero et al, 1992; Ferriero, 1992; White & Chalupa, 1992), corticotrophin releasing factor (Zhang & Yeh, 1990; 1991a; Zhang et al, 1990) neuropeptide Y (Ferriero et al, 1989) and their respective receptors has been reported in a number of mammalian retinae.

# DEVELOPMENTAL CHANGES IN SECOND MESSENGERS ENZYMES AND ION CHANNELS

Changes in Protein kinase A and C have been documented in mammalian retina during development (Gentleman et al, 1989; Zhang & Yeh, 1991b). These are two of the many enzymes that are modulated by some G-protein coupled receptors and exert the actions by the phosphorylation of receptors and ion channels among other proteins. It has been further suggested that the changes in muscarinic stimulation of inositol phosphate in rat retina during development are consistent with an increase in coupling efficiency of the receptors to G-proteins (Osborne, 1988). A recent study (Mitrofanis et al, 1992) has demonstrated changes in amacrine cells that were labelled histochemically with nicotinamide adenine phosphate diaphorase, a substance that is restricted to cells that use nitric oxide as a second messenger/neurotransmitter.

Developmental changes in ion channels have also been shown. Low levels of mRNA for the α subunit of the sodium channel can be detected before birth in rat retina, although the ß subunit mRNA is not detectable until after birth. Adult level of mRNA and protein of both subunits do not occur until after PND 45 (Wollner et al, 1988).

In summary, all neurotransmitters so far studied show changes in both quantity and distribution during postnatal development. Changes also occur in synthetic and degradative enzyme, uptake systems and receptor type and distribution. Our electrophysiological studies on some neurotransmitters further demonstrate that the physiological functions of synapses undergo developmental changes to reach adult-like specificity and sensitivity during early postnatal life. These results suggest complex dynamic changes occur throughout early postnatal life and may explain the long period needed for many mammalian visual systems to acquire adult levels of visual function.

## DISCUSSION

With only a small insight into the complexity of the events involved in the maturation of the retina and in particularly the neurotransmitter systems, it is difficult to make uncontroversial statements in coming to any conclusions. However the following are our opinions of what are the important concepts behind retinal neurotransmitter development.

Early in development there is a multiplicity of neurotransmitters and synaptic connections impinging on cells expressing many receptor types and subtypes. The reason for this promiscuity maybe that at this developmental stage the neurotransmitters are not functioning as they do in the adult to relay and encode information of the visual world but are involved in the process of development itself. It is conceivable that the neurotransmitters and their receptors are acting as growth factors and guidance molecules involved the basic modelling of the interconnections in the retina. Indeed Redburn and coworkers (Redburn & Madtes, 1986; Redburn, 1992) has put forward a strong case for GABA having a role in these respects. Other substances that are neurotransmitters but have been suggested to act as growth regulators in the developing retina and central nervous system include dopamine (Lankford et al, 1988) and somatostatin (Ferriro, 1992).

Once the basic frame work of connections is established, then the fine tuning takes place whereby the neurotransmitter systems convert to their role to signal visual information. In this process, there is an elimination of some inputs and a strengthening of others. Consistent with this idea, many morphological studies monitoring the development of neurotransmitter containing cells show that the number of cells is higher at birth but which declines to adult levels in the weeks after (e.g. Ferriero & Sagar 1989; Mitrofanis et al, 1992). Similarly, the morphological studies of cat retinal ganglion cells show that developmental changes in their dendritic trees are characterised by a reduction of branches and the loss of short spine like processes (Dann et al, 1988: Ramoa et al, 1988: Wässle, 1988). This evidence would support the notion of postnatal elimination of some synaptic (neurotransmitter) inputs. Furthermore the stratification of ON- and OFF-type retinal ganglion cell dendrites in the inner plexiform layer is achieved by the "pruning" of dendrites in the "wrong" sublamina (Dann et al, 1988; Maslin & Stone, 1988). This stratification process has recently been shown to be disrupted by intravitreal injections of 2-amino-4-phosphonbutyric acid, a glutamate analogue which prevents the release of glutamate from ON-bipolar cells (Bodnarenko & Chalupa, 1993). Cholinergic amacrines also show the loss of spines during postnatal development (Wong & Collin, 1989).

The increased selectivity and sensitivity of the postsynaptic membrane could be achieved by changes in the expression of neurotransmitter receptor subunit composition. It has been established that ligand gated ion channels are composed of at least five subunits (e.g. $GABA_A$ receptor, nicotinic ACh receptor). Some of these subunits have a number of variants there are, for example at least 7 types of α and three types of ß subunit for the neuronal nicotinic ACh receptor. If there is a change in the composition of the receptor during development as occurs in the muscle nACh receptor, from the embryonic form $((\alpha 1)_2\ \beta 1\ \delta\ \gamma)$ to the mature form $((\alpha 1)_2\ \beta 1\ \delta\ \varepsilon)$, there could be changes in agonist sensitivity as well as function. Indeed recent *in situ* hybridisation studies have shown that in rat retina, there are increases in the number and type of nicotinic ACh receptor subunit genes expressed (Hoover & Goldman, 1992).

Interestingly, the α and β subunits of the $GABA_A$ receptor confer the GABA binding site and the ion channel, whilst the gamma subunit is required for modulation by benzodiazepines (Pritchett et al, 1989). This heterogeneity of $GABA_A$ receptors, for example, may explain the lack of benzodiazepine effects on adult peripheral OFF cells which are sensitive to GABA (muscimol) and bicuculline (Robbins & Ikeda, 1989).

It is tempting to suggest that these changes in sensitivity and selectivity of the postsynaptic membrane during development are driven by an activity dependent system that reflects visual environment and experience. An example which supports this idea may be found in the differences between the neurotransmitter selectivity which we found between the retinal ganglion cells in the area centralis and those in the peripheral retina (Fig. 2.). Is it that retinal ganglion cells located where the visual image on the retina is the sharpest that these cells are driven to show the strongest dichotomy in their neurotransmitter inputs? Indeed this would be an ideal way to segregate the ON- and OFF pathways which respond to opposite contrast as we have previously suggested (Ikeda & Robbins, 1989) It is also tempting to suggest that the highest selectivity to neurotransmitters is expressed by the cells in the area centralis, where the cell density is highest and may be susceptible to unwanted visual cross-talk.

These questions and ideas have not been tested yet, however, when they are, it will give a greater understanding of events in the normal developing retina. Maybe then we can ask questions about what happens when it goes wrong?

On a more general note, recent studies (Zhou et al, 1995; Thomas et al, 1995) have demonstrated that inactivation of the genes that code for the enzymes that synthesise catecholamines leads to death of these mutant mice *in utero*. Furthermore, administration of the products of these enzymes can rescue the mutant mice. It is suggested from these studies that catecholamines are essential for foetal development and postnatal survival in rodents. These studies further support the idea that neurotransmitters may have an important role in mammalian development.

## ACKNOWLEDGMENTS

This work was supported by project grants from the Medical Research Council and from The Special Trustees of St Thomas's Hospital to H.I., during the years 1982-1988, when the majority of the experiments described in this paper were performed at the Vision Research Unit, The Rayne Institute. We thank Mr Peter deVillers for technical assistance.

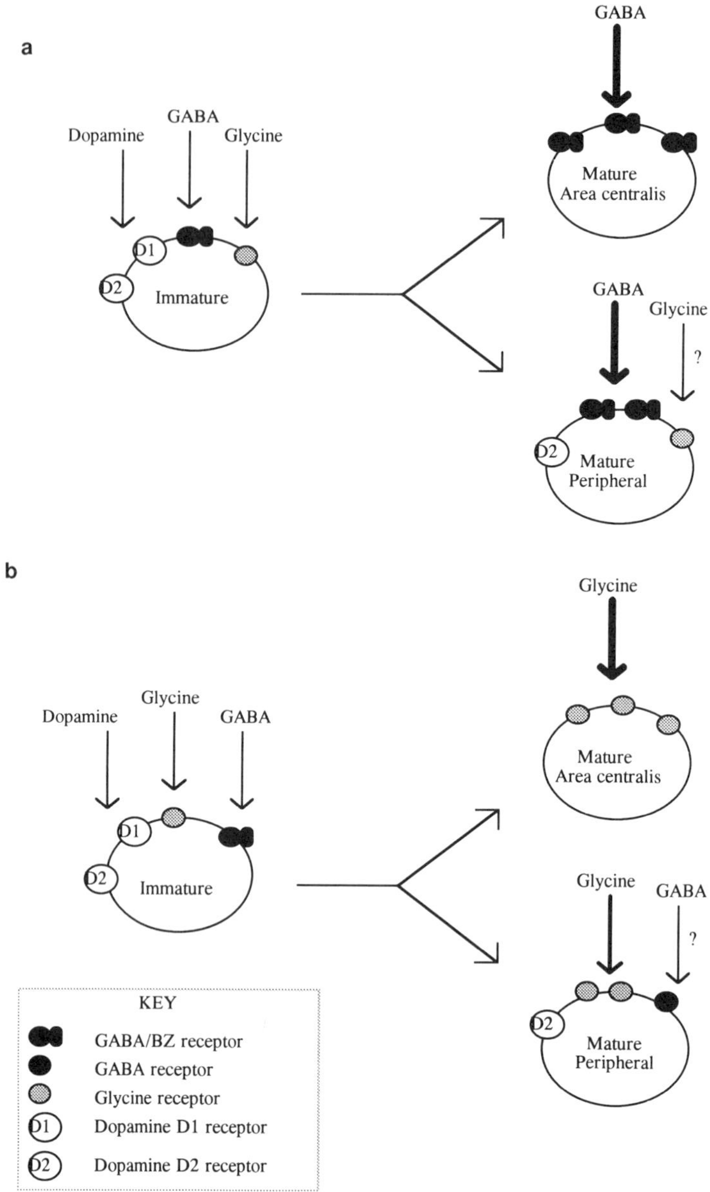


**Figure 2.** A schematic explanation of the observed changes in inhibitory neurotransmitter sensitivity and selectivity in cat retinal ganglion cells. Shown are the changes from the immature ON-cells (A) and OFF-cells (B) into mature cells in the area centralis and peripheral retina. The maturation of ON-cells involves an increase in GABAegic input as well as increased sensitivity to GABA. Cells in the area centralis lose sensitivity to dopamine and glycine whereas peripheral cells retain D2 dopamine receptors and a weak gylcinergic input. Conversely OFF-cells increase their glycinergic input. OFF-cells in the area centralis lose both GABAergic and dopaminegic inputs and sensitivity. Peripheral cells retain D2 receptors and a $GABA_A$ receptor that is not sensitive to benzodiazepines. The basis of this figure is a combination of data previously published (Ikeda & Robbins, 1985a; 1985b; Priest et al, 1985; Ikeda et al, 1987; Robbins et & Ikeda, 1989).

## REFERENCES

Barlow, H.B. and Pettigrew, J.D. (1971) Lack of specificity in neurones in the visual cortex of young kittens, *J. Physiol.* 218:739-744.

Bodenant, C., Leroux, P., Gonzalez, B.J. and Vaudry, H. (1991) Transient expression of somatostatin receptors in the rat visual system during development, *Neurosci.* 41:595-606.

Bodnarenko, S.R. and Chalupa, L.M. (1993) Stratification of ON and OFF ganglion cell dendrites depends on glutamate-mediated afferent activity in the developing retina, *Nature* 364:144-146.

Bonds, A.B. and Freeman, R.D. (1978) Development of optical quality in the kitten eye, *Vision Res.* 18:391-398.

Casini, G. and Brecha, N.C. (1992) Postnatal development of tyrosine hydroxylase immunoreactive amacrine cells in the rabbit retina: I morphological characterisation, J. *Comp. Neurol.* 326:283-301.

Cohen, J. (1987) Postnatal development of phenylethanolamine-N-methyltransferase activity of rat retina, *Neurosci. Lett.* 83:138-142.

Cohen, J. and Neff, N.N. (1982) Retinal amacrine cell system tyrosine hydroxylase: the development of responsiveness to light and neuroleptic drugs, *Develop. Brain Res.* 3:160-163.

Cutcliffe, N. and Osborne, N.N. (1987) Serotonergic and cholinergic stimulation of inositol phosphate formation in the rabbit retina. Evidence for the presence of serotonin and muscarinic receptors, *Brain Res.* 421:95-104.

Dann, J.F. (1989) Cholinergic amacrine cells in the developing cat retina, J. *Comp. Neurol.* 289:143-155.

Dann, J.F., Buhl, E.H. and Peichl, L. (1988) Postnatal dendritic maturation of alpha and beta ganglion cells in cat retina, *J. Neurosci.* 8:1485-1499.

Donovan, A. (1966) The postnatal development of the cat retina, *Exp. Eye Res.* 5:249-254.

Ferriero, D.M. (1992) Developmental expression of somatostatin receptors in the rat retina, *Develop. Brain Res.* 67:309-315.

Ferriero, D.M., Head, V.A., Edwards, R.H. and Sagar, S.M. (1990) Somatostatin mRNA and molecular forms during development of the rat retina, *Develop. Brain Res.* 57:15-19.

Ferriero, D.M. and Sagar, S.M. (1987) Development of somatostatin immunoreactive neurons in rat retina, *Develop. Brain Res.* 34:207-214.

Ferriero, D.M. and Sagar, S.M. (1989) Development of neuropeptide Y-immunoreactive neurons in the rat retina, *Develop. Brain Res.* 48:19-26

Ferriero, D.M., Sheldon, R.A. and Domingo, J. (1992) Somatostatin is altered in developing retina from ethanol-exposed rats, *Neurosci. Lett.* 147:29-32.

Freeman, D.N. and Marg, E. (1975) Visual acuity development coincides with the sensitive period in kittens, *Nature* 254:614-615.

Fry, K.R., Chen, N-X., Glazebrook, P.A. and Lam, D.M-K. (1991) Postnatal development of ganglion cells in the rabbit retina: characterisations with AB5 and GABA antibodies, *Develop. Brain Res.* 61:45-53.

Fung, S-K., Kong, Y-C. and Lam, D.M-K. (1982) Prenatal development of GABAergic, glycinergic and dopaminergic neurons in the rabbit retina, *J. Neurosci.* 2: 1623-1632.

Gentleman, S., Hemmings, B.A., Russell, P. and Chader, G.J. (1989) Developmental expression of the RI subunit of cyclic AMP-dependent protein kinase in retina, *Exp. Eye Res.* 48:717-731.

Guarneri, P., Corda, M.G., Concas, A., Salis, M., Caldeini, G., Toffano, G. and Biggio, G. (1982) Age related changes of benzodiazepine and GABA binding sites in the rat retina, *Neurobiol. Aging* 3:227-231.

Hamasaki, D.I. and Flynn, J.T. (1977) Physiological properties of retinal ganglion cells of 3-week-old kittens, *Vision Res.* 17:275-284.

Hamasaki, D.I. and Sutija, V.G. (1979) Development of X- and Y-cells in kittens, *Exp. Brain Res.* 35:9-23.

Hoover, F. and Goldman, D. (1992) Temporally correlated expression of nAChR genes during development of mammalian retina, *Exp. Eye Res.* 54:561-565.

Hubel, D.H., and Wiesel, T.H. (1963) Receptive fields of cells in the striate of very young, visually inexperienced kittens, *J. Neurophysiol.* 26:994-1002.

Hubel, D.H. and Wiesel, T.N. (1970) The period of susceptibility to the physiological effects of unilateral eye closure in kittens, *J. Physiol.* 206:419-436.

Huttenlocker, P.R. (1967) Development of cortical neuronal activity in the neonatal cat, *Exp. Neurol.* 17:347-415.

Ikeda, H. (1979) Physiological basis of visual acuity and its development in kitten, *Child Care Health Develop.* 5:375-383.

Ikeda, H. (1985) Transmitter actions at the cat retinal ganglion cells, *Prog. Retinal Res.* 4:1-32.

Ikeda, H., Kay, C.D., and Robbins, J. (1989) Properties of excitatory amino acid receptors on sustained retinal ganglion cells in the cat retina, *Neurosci.* 32:27-38.

Ikeda, H., Priest, T.D., Robbins, J. and Wakakuwa K. (1986) Silent dopaminergic synapse at feline retinal ganglion cells, *Clin. Vision Sci.* 1:25-38.

Ikeda, H. and Robbins, J. (1984) Development of inhibitory transmission at the retinal ganglion cells in cats. In Development of visual pathways in mammals, Eds Stone, J. Dreher, B. and Rapaport D.H., Alan R. Liss Inc. New York, pp 115-124.

Ikeda, H. and Robbins, J. (1985a) Postnatal development of GABA- and glycine-mediated inhibition of feline retinal ganglion cells in the area centralis, *Develop. Brain Res.* 23:1-17.

Ikeda, H. and Robbins, J. (1985b) Postnatal development of GABA and glycine actions on the surround inhibition of cat retinal ganglion cells in the area centralis. In. Neurocircuitry of the retina: a Cajal memorial. Eds Gallego, A. & Gouras, P. Elsevier, New York, pp. 257-264.

Ikeda, H. and Robbins, J. (1989) Development of neurochemical segregation of ON and OFF retinal channels which subserve contrast vision. In: Seeing contour and colour, Eds Kulikowski, J.J., Dickinson, C.M. and Murray, I.J. Pergamon Press, Oxford, pp 164-166.

Ikeda, H., Robbins, J. and Kay, C. (1990) Excitatory amino acid receptors on sustained retinal ganglion cell in the kitten during the critical period of development, *Develop. Brain Res.* 51:85-91.

Ikeda, H., Robbins, J. and Wakakuwa, K. (1987) Evidence for dopaminergic innervation on kitten retinal ganglion cells, *Develop. Brain Res.* 35:83-89.

Ikeda, H. and Tremain, K.E. (1978) The development of spatial resolving power of lateral geniculate neurones in kittens, *Exp. Brain Res.* 31: 193-206.

Ikeda, H. and Tremain, K.E. (1979) Amblyopia occurs in retinal ganglion cells in cats reared with convergent squint without alternating fixation, *Exp. Brain Res.* 35:559-582.

Jacobson, S.G., Ikeda, H. and Ruddock, K. (1987) Cone mediated retinal function in cats during development, *Docum. Ophthalmol.* 65:7-14.

Kato, S., Nakamura, T. and Negishi, K. (1980) Postnatal development of dopaminergic cells in the rat retina, *J. Comp. Neurol.* 191:227-236.

Lam, D.M-K., Fung, S-C. and Kong, Y-C. (1980) Postnatal development of GABA-ergic neurons in the rabbit retina, *J. Comp. Neurol.* 193:89-102.

Lankford, K.L., DeMello, F.G. and Klein, W.L. (1988) $D_1$-type dopamine receptors inhibit growth cone motility in cultured retina neurons: evidence that neurotransmitters act as morphogenic growth regulators in the developing central nervous system, *Proc. Natl. Acad Sci. USA* 85:4567-4571.

Madtes, Jr, P. and Redburn, D.A. (1982) [$^3$H]GABA binding in developing rabbit retina, *Neurochem. Res.* 7:495-503.

Martin-Martinelli, E., Simon, A., Vigny, A. and Nguyen-Legros, J. (1989) Postnatal development of Tyrosine-Hydroxylase-Immunoreactive cells in the rat retina, *Develop. Neurosci.* 11:11-25.

Maslim, J. and Stone, J. (1988) Time course of stratification of the dendritic fields of ganglion cells in the retina of the cat, *Develop. Brain Res.* 44:87-93.

Messersmith, E.K. and Redburn, D.A. (1992) $\gamma$-Aminobutyric acid immunoreactivity in multiple cell types of the developing rabbit retina, *Visual Neurosci.* 8:201-211.

Mitchell, D.E., Giffin, F., Wilkinson, F., Anderson, P. and Smith, M.L. (1976) Visual resolution in young kittens, *Vision Res.* 16:363-366.

Mitrofanis, J. and Finlay, B.L. (1990) Developmental changes in the distribution of retinal catecholaminergic neurones in hamsters and gerbils, *J. Comp. Neurol.* 292:480-494.

Mitrofanis, J., Maslim, J. and Stone, J. (1988) Catecholaminergic and cholinergic neurons in the developing retina of the rat, *J. Comp. Neurol.* 276:343-359.

Mitrofanis, J., Maslim, J. and Stone, J. (1989) Ontogeny of catecholaminergic and cholinergic cell distributions in the cat's retina, *J. Comp. Neurol.* 289:228-246.

Mitrofanis, J., Robinson, S.R. and Ashwell, K. (1992) Development of catecholaminergic, Indoleamine-accumulating and NADPH-diaphorase amacrine cells in rabbit retinae, *J. Comp. Neurol.* 319:560-585.

Mitrofanis, J., Robinson, S.R. and Provis, J.M. (1989) Somatostatinergic neurones of the developing human and cat retinae, *Neurosci. Lett.* 104:209-216.

Moore, C.L., Kalil, R. and Richards, W. (1976) Development of myelination in optic tract of the cat, *J. Comp. Neurol.* 165:125-136.

Morgan, W.W. and Kamp. C.W. (1982) Postnatal development of the light response of the dopaminergic neurons in the rat retina, *J. Neurochem.* 39:283-285.

Morrison, J.D. (1982) Postnatal development of the area centralis of the kitten retina: an electron microscopic study, *J. Anat.* 135:255-271.

Osborne, N.N. (1985) Interplexiform, horizontal and bipolar-like cells of the rabbit retina take up exogenous serotonin during early developmental stages, *Int. J. Develop. Neurosci.* 3:643-646.

Osborne, N.N. (1988) Muscarinic stimulation of inositol phosphate formation in rat retina: developmental changes, *Vision Res.* 28:875-881.

Osborne, N.N., Patel, S. Beaton, D.W. and Neuhoff, V. (1986) GABA neurones in retinas of different species and their postnatal development in situ and in culture in the rabbit retina, *Cell Tissue Res.* 243:117-123.

Parkinson, D. and Rando, R.R. (1984) Ontogenesis of dopaminergic neurons in the post-natal rabbit retina: pre- and post-synaptic elements, *Develop. Brain Res.* 13:207-217.

Parkinson, D., Spira, A., Wyse, J.P. and Patten, M. (1985) The ontogenesis of the dopaminergic cell in the pre- and postnatal guinea pig retina, *Int. J. Develop. Neurosci.* 3:157-167.

Pourcho, R.G. (1982) Dopaminergic amacrine cells in the cat retina. *Vision Res.* 252: 101-109.

Priest, T.D., Robbins, J. and Ikeda, H. (1985) The action of inhibitory neurotransmitters γ-aminobutyric acid and glycine may distinguish between the area centralis and the peripheral retina in cats, *Vision Res.* 25:1761-1770.

Pritchett, D.B., Sontheimer, H., Shivers, B.D., Ymer, S., Kettenmann, H., Schofield, P.R. and Seeburg, P.H. (1989) Importance of a novel $GABA_A$ receptor subunit for benzodiazepine pharmacology, *Nature* 338:582-585.

Puro, D.G., Battelle, B-A. and Hansmann, K.E. (1982) Development of cholinergic neurons of the rat retina, *Develop. Biol.* 91:138-148.

Ramoa, A.S., Campbell, G. and Shatz C.J. (1988) Dendritic growth and remodelling of cat retinal ganglion cells during fetal and postnatal development, *J. Neurosci.* 8: 4239-4261.

Redburn, D.A. (1992) Development of GABAergic neurons in the mammalian retina, *Prog. Brain Res.* 90:133-147.

Redburn, D.A., Agarwal, S.H. Messersmith, E.K. and Mitchell, C.K. (1992) Development of the glutamate system in rabbit retina, *Neurochem. Res.* 17:61-66.

Redburn, D.A. and Madtes Jr, P. (1986) GABA - Its role and development in retina, *Prog. Retinal Res.* 6:69-84.

Redburn, D.A. and Mitchell, C.K. (1981) 3H-muscimol binding in synaptosomal fractions from bovine and developing rabbit retinas, *J. Neurosci. Res.* 6:487-495.

Robbins, J. and Ikeda, H. (1989) Benzodiazepines and the mammalian retina: III postnatal development, *Brain Res.* 479:334-338.

Robbins, J., Wakakuwa, K. and Ikeda, H. (1988) Noradrenaline action on cat retinal ganglion cells is mediated by dopamine ($D_2$) receptors, *Brain Res.* 438:52-60.

Rose, G.H. and Lindsley, D.B. (1968) Development of visually evoked potentials in kittens: specific and non-specific responses, *J. Neurophysiol.* 31:607-623.

Rusoff, A.C. (1979) Development of ganglion cells in the retina of the cat. In Developmental neurobiology of vision, Ed. Freeman, R.D. Plenum Press. New York, pp 19-30.

Rusoff, A.C. and Dubin, M.W. (1977) Development of receptive-field properties of retinal ganglion cells in kittens, *J. Neurophysiol.* 40:188-1198.

Schnitzer, J. and Rusoff, A.C. (1984) Horizontal cells of the mouse retina contain glutamic acid decarboxylase-like immunoreactivity during early developmental stages, *J. Neurosci.* 4:2948-2955.

Schliebs, R., Rothe, T. and Bigl, V. (1986) Dark-rearing affects the development of benzodiazepine receptors in the central visual structures of rat brain, *Develop. Brain Res.* 24:179-185.

Spira, A.W. and Parkinson, D. (1991) Effects of dark-rearing on the retinal dopaminergic system in the neonatal and postnatal guinea pig, *Develop. Brain Res.* 62:142-145.

Teller, D.Y. and Movshon, J.A. (1986) Visual development, *Vision Res.* 26:1483-1506.

Thomas, S.A., Matsumoto, A.M. and Palmiter, R.D. (1995) Noradrenaline is essential for mouse fetal development, *Nature* 374:643-646.

Tootle, J.S. and Friedlander, M.J. (1989) Postnatal development of the spatial contrast sensitivity of X- and Y-cells in the kitten retinogeniculate pathway, *J. Neurosci.* 9:1325-1340.

Tucker, G.S. (1978) Light microscopic analysis of the kitten retina: postnatal development in the area centralis, *J. Comp. Neurol.* 180:489-500.

Tucker, G.S., Hamasaki, D.I., Labbie, A. and Muroff, J. (1979) Anatomic and Physiologic development of the photoreceptor of the kitten, *Exp Brain Res.* 37:459-474.

Vogel, M. (1978) Postnatal development of the cat's retina, *Adv. Anat. Embryol. Cell Biol.* 54:1-66.

Wang, H-H., Cuenca, N. and Kolb, H. (1990) Development of morphological types and distribution patterns of amacrine cells immunoreactive to tyrosine hydroxylase in the cat retina, *Visual Neurosci.* 4:159-175.

Wassle, H. (1988) Dendritic maturation of retinal ganglion cells, *Trends Neurosci.* 11:87-89.

White, C.A. and Chalupa, L.M. (1992) Ontogeny of somatostatin immunoreactivity in the cat retina, *J. Comp. Neurol.* 317:129-144.

Windle, W.F. (1930) Normal behavioural reactions of kittens correlated with the postnatal development of nerve-fibre density in spinal grey matter, *J. Comp. Neurol.* 50:479-503.

Wollner, D.A., Scheinman, R. and Catterall, W.A. (1988) Sodium channel expression and assembly during development of retinal ganglion cells, *Neuron* 1:727-737.

Wong, R.O.L. and Collin, S.P (1989) Dendritic maturation of displaced putative cholinergic amacrine cells in the rabbit retina, *J. Comp. Biol.* 287:164-178.

Wulle, I. and Schnitzer, J. (1989) Distribution and morphology of tyrosine hydroxylase-immunoreactive neurons in the developing mouse retina, *Develop. Brain Res.* 48:59-72.

Yeh, H.H., Battelle, B-B. and Puro, D.G. (1983) Maturation of Neurotransmission at cholinergic synapses formed in culture by rat retinal neurones: regulation by cyclic AMP, *Develop. Brain Res.* 10:63-72.

Yew, D.T., Luo, C.B., Zheng, D.R. Guan, Y.L. Tsang, D. and Stadlin, A. (1991) Immunohistochemical localisation of substance P, enkephalin and serotonin in the developing human retina, *J. Hirnforsh.* 32:61-71.

Zetterstrom, B. (1955) The effect of light on the appearance and development of the electroretinogram in newborn kittens, *Acta Physiol. Scand.* 35:272-279.

Zhang, D., Gallagher, M., Sladek, C.D. and Yeh, H.H. (1990) Postnatal development of corticotropin releasing factor-like immunoreactive amacrine cells in the rat retina, *Develop. Brain Res.* 51:185-194.

Zhang, D. and Yeh, H.H. (1990) Histogenesis of corticotrophin releasing factor-like immunoreactive amacrine cells in the rat retina, *Develop. Brain Res.* 53:194-199.

Zhang, D. and Yeh, H.H. (1991a) Corticotrophin releasing factor-like immunoreactivity (CRF-LI) in horizontal cells of the developing rat retina, *Visual Neurosci.* 6:383-391.

Zhang, D. and Yeh, H.H. (1991b) Protein kinase C-like immunoreactivity in rod bipolar cells of the rat retina: a developmental study, *Visual Neurosci.* 6:429-437.

Zhang, D. and Yeh, H.H. (1992) Substance-P-like immunoreactive amacrine cells in the adult and the developing rat retina, *Develop. Brain Res.* 68:55-65.

Zhou, Q-Y., Quaife, C.J. and Palmiter, R.D. (1995) Targeted disruption of the tyrosine hydroxylase gene reveals that catecholamines are required for mouse fetal development, *Nature* 374:640-643.

# 11

# IMPAIRED DYNAMICS OF RETINAL PROCESSING IN DOPAMINERGIC DEFICIENCY STATES IN MAN AND MONKEY

Ivan Bodis-Wollner

Department of Neurology
State University of New York – Health Sciences Center
450 Clarkson Avenue, Box 1213
Brooklyn, NY 11203

## INTRODUCTION

In 1976 we reported and then published in 1978 in greater detail results showing that visual evoked potentials (VEPs), are abnormal in many patients affected by Parkinson's disease (Bodis-Wollner and Yahr, 1978). This conclusion was somewhat unbelievable, since the disease was considered as an exclusively *motor system* disorder. Furthermore it was held, since the original description by Halliday and his colleagues (1972) that a 'delayed' VEP represents abnormal conduction in the visual pathway due to demyelination, usually caused by multiple sclerosis. Thus, finding that VEPs are delayed in Parkinson's disease both challenged prior knowledge concerning the *extent* and *nature* of the disease, and broadened the interpretation concerning the *causes* and significance of delayed visual responses. The VEP results in PD suggested that VEP abnormalities may relate to primary neuronal dysfunction in addition to demyelination. Although our results were surprising, there was prior information concerning neuronal dopaminergic mechanisms in visual processing.

The original, albeit rarely quoted, description of the effect of systemic catecholamine depletion on vision stems from Malmfors (1963). He observed that reserpine-treated rats have a remarkable aversion to light: they turn away from illumination and close their eyes. This defensive behaviour could not be explained by an enlarged pupil, since in fact the animals' pupils were small as a result of adrenergic 'symapathectomy' caused by the reserpine. Hence the light-aversive behaviour of the reserpinised rats could not be attributed to an increased amount of light reaching the eye. Malmfors made the seminal suggestion that "the cause of this increased sensitivity to light might be that an aminergic inhibitory mechanism has got out of order". He proceeded to show the presence of cells with specific catecholamine histofluorescence in the normal rat retina, and that these cells were normal even after section of the optic nerve and sympathetic nerves. Thus, he concluded that there are "intrararetinal adrenergic neurons with intraretinal synaptic terminals". Subsequently, Haggendal and Malmfors (1965) identified these intraretinal catecholaminergic neurons as

*Basic and Clinical Perspectives in Vision Research*, Edited by Jon Robbins et al.
Plenum Press, New York, 1995

dopaminergic. In the following decade a number of studies have demonstrated the presence of retinal dopamine (DA) neurones in all examined species. Since then it has become apparent, however, that while dopaminergic neurons exist in all vertebrate retinas (see Negishi, 1990) there are certain interspecies similarities. When animals are exposed to light there is activation of tyrosine hydroxylase and increased synthesis and release of DA (Kramer, 1971; Iuvone et al , 1978) and increased formation of DA metabolites (Cohen & Dowling, 1983). It is also apparent, however, that there are substantial species differences in the morphology of DA neurons (Ikeda et al, 1986; Oyster et al, 1988; Witkowsky & Schutte, 1991) and in the functional role of DA in the retina. In the monkey, two types of retinal DA neurons were localised in the inner plexiform layer (Mariani & Hokoc, 1988) using tyrosine hydroxylase (TH) antibody labelling. Morphological techniques also revealed the presence of two types of DA neurons in the human retina, regular amacrines and also interplexiform cells (Frederick et al 1982, and Savy et al, 1991). The functional role of DA in the primate retina is not yet clear. Do retinal DA neurons control light sensitivity in mammalians as suggested by Malmfors? If so, at which retinal layer do they affect signal transmission?

In lower species and in some lower mammalians (Hankins and Ikeda, 1991) a role for dopamine has been suggested in a preganglionic circuit which regulates centre-surround interaction of ganglion cells (for a summary of relevant studies see Bodis-Wollner and Piccolino 1988). This suggested physiological role is in agreement with the results of studies on the visual effect of retinal and systemic dopaminergic deficiency in patients with Parkinson's disease (PD) and the Parkinsonian monkey model. In this chapter I shall summarise visual, electrophysiological and psychophysical effects of DA deficiency that we and others have observed in man and monkey. Research in humans affected by Parkinson's disease (PD) and monkeys experimentally treated with MPTP (methyl-phenyl-tetrahydropyridine), which causes a parkinsonian-like syndrome and retinal DA deficiency, will be described in relationship to the visual effects of $D_2$ receptor blockade. Emphasis will be on electroretinographic results. Studies will be summarised which demonstrate that $D_2$ receptor blockade in humans and monkeys cause stimulus specific deficits in human visual processing at the retinal level, similar to the defects seen in PD patients. The role of retinal $D_2$ receptor-linked dopaminergic processes raises new questions concerning the *dynamics* of DA-mediated neuromodulatory circuits in the human CNS.

## VISUAL EVOKED POTENTIALS (VEPs)

### Spatial Factors

A significant number of PD patients have delayed transient (Bodis-Wollner & Yahr 1978) or abnormal steady-state VEPs. The abnormalities are asymmetrical for the two eyes. An abnormality is mostly or perhaps only evident if the size of the pattern elements (spatial frequency) of the visual stimulus is above 2 cpd (Tartaglione et al 1985, Onofrj et al 1986). This size selectivity was also shown in VEP studies using dopamine receptor blockers in normal or schizophrenic patients (Bodis-Wollner et al 1982; Onofrj et al 1986). This spatial frequency dependent action of dopamine is consistent with the results of psychophysical studies (Bodis-Wollner et al 1987) and with the results of electrophysiological recordings in monkeys rendered systematically DA deficient by MPTP (Ghilardi et al 1988a) or by intraocular injection of 6-hydroxydopamine (Ghilardi et al 1989). In both species the hypodopaminergic state due to either presynaptic cell destruction or to postsynaptic blockade predominantly attenuate mid-spatial frequency responses (Tagliati et al. 1994, Tagliati et al. 1995).

## Temporal Factors

While the dynamics of retinal DA circuits are unknown, circumstantial evidence suggests that slower than synaptic processes regulate DA's distal retinal action. This may take the form of paracrine release from proximal to distal retinal layers. Therefore, the possible role of DA in a feedback-type of regulatory circuit doe not demand that a discrete group of DA IPC cells exist in the human retina, although some studies (Frederick et al 1982, Savy et al 1991) have suggested that such cells do exist. It is possible, as Witkovsky & Schuette (1991) point out, that "the functional distinction between synaptic and paracrine modes of communication is more like a continuum than two very different entities". From this viewpoint, DA neurons of the inner retina could affect distal elements without having ordinary transmitter release. We have studied VEP in PD patients using three temporal frequencies of input and comparing the responses at input and output frequencies (Marx et al 1986). This VEP study (Marx et al 1986) suggests that abnormalities in PD are indeed more evident for 4 than for 8 and for 1Hz stimuli (Fig 1). Using non-linear systems analysis allowed us to infer that the temporal tuning abnormality most likely results from an input abnormality prior to a stage of non-linear transform, and hence is likely to represent retinal dysfunction (Fig 2). It appears that abnormal visual processing in PD can be explained by the insufficiency of retinal neuronal circuits with relatively long time constants compared to synaptic events.

Some, but not all psychophysical studies (Bodis-Wollner et al 1987) suggest that contrast sensitivity deficits are enhanced when the temporal modulation rate is near 4-6 Hz. This temporal frequency dependent deficit is consistent with contrast sensitivity changes in "on-off" PD patients, linked to a waxing and waning efficiency of DA at the postsynaptic receptor. Fig 3 Taken together the psychophysical and electrophysiological results suggest an abnormality of circuit "tuned" to 4 Hz. This "slow" circuit deficit is more consistent with a defect in paracrine DA release than with a conventional synaptic dysfunction in PD. ERG studies using $D_2$ blockers provide converging evidence to these results.

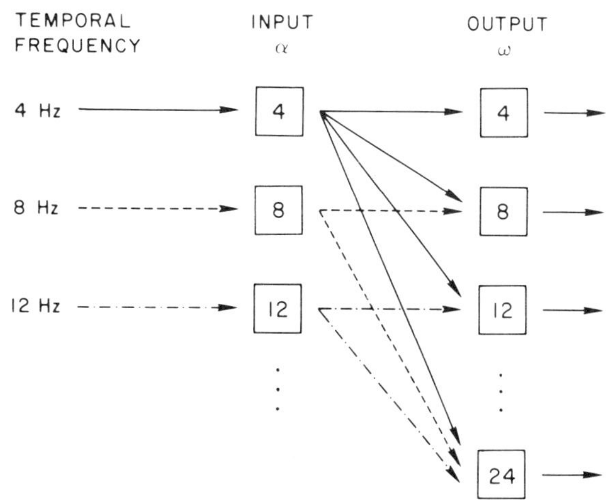


**Figure 1.** Schematic diagram showing the hypothetical effect of stimulation with discrete temporal frequencies at the early (a) and later (w) stages of processing in the visual system. Consistent with the known facts, the response to any input frequency consists of a response at the fundamental frequency and at the harmonic frequencies of a response at the fundamental frequency and at the harmonic frequencies of the input frequency, responses due to nonlinearities in the visual system. If a temporal frequency abnormality occurs at the α (input) stage for a specific input (say, 4 Hz), then the response at that frequency and at all its harmonic frequencies (i.e., 4, 8, 12 and 16 Hz) are affected. If a temporal frequency abnormality occurs at the w (output), then only the temporal frequency abnormality occurs at the w (output) stage, then only the response or output frequency will be affected, regardless of the input temporal frequency (for example, the response at Hz may appear abnormal whether the input frequency is 4 or 8 Hz).

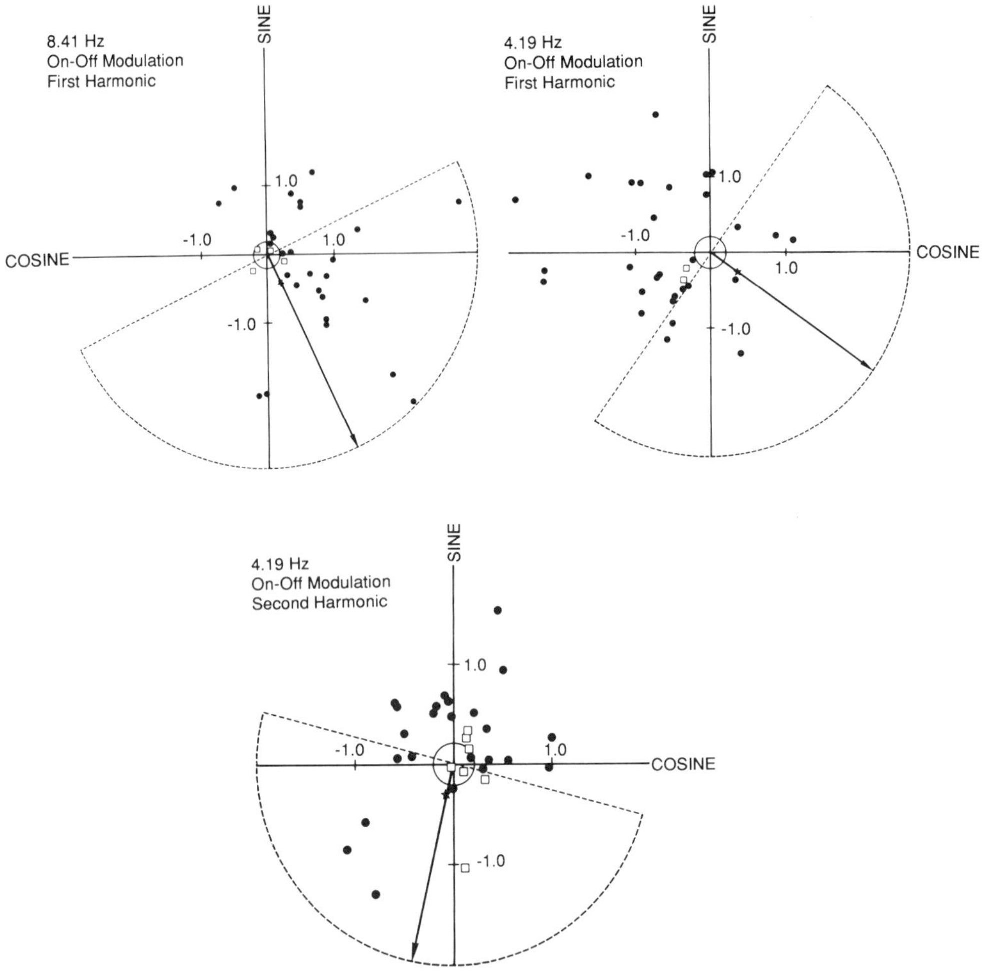


**Figure 2.** (a) This figure represents the VEP response of Parkinsonian patients and normals, presented in the form of a "vectogram". VEP amplitude is represented by the length of a vector. The mean vector of normal observers is shown with the heavy arrow and it reaches an amplitude of 2 microvolts. The phase of the response is shown by the angle of the arrow. The spread of the VEP of normals is indicated by interrupted lines. Patients' data are shown by heavy dots. Notice that 22 of 29 data points are outside normal limits using 4.19 Hz stimulation frequency and measuring the response at 4.19 Hz output. (b) The simultaneously obtained response of the second harmonic (8.32 Hz) of 4.19 Hz input frequency shows that 18 of 23 data points are abnormal in the same set of patients. (c) Using 8.41 Hz input frequency the response at the first harmonic shows an abnormality for 8 of the 27 responses. Taken together Figures 2a, b and c represent that in DA deficiency in PD a circuit which responds to near 4 Hz input suffers worst.

# PATTERN ERGs

## Spatial Factors

A number of PERG studies provided somewhat conflicting conclusions concerning the PERG effects of DA deficiency in PD, although by now morphological/histochemical studies provide evidence for retinal DA deficit in this disease (Nguyen-Legros et al 1993).

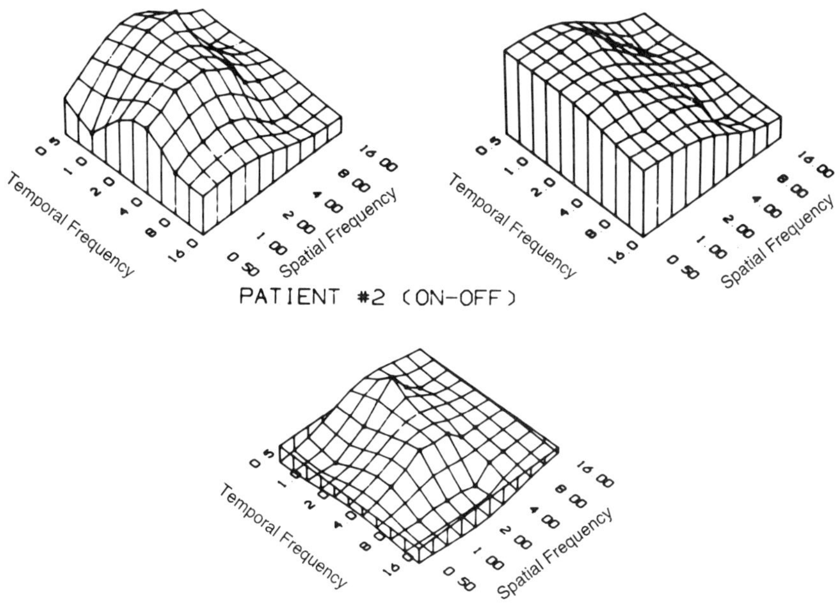


**Figure 3.** Spatiotemporal CS to a 9″ field is shown for six spatial frequencies and six temporal frequencies in a three-dimensional surface plot for a 46-year-old PD patient who shows the on-off syndrome. Snellen visual acuity was 20/20 and 20/30. (a) Representation of the three-dimensional surface measured when this patient was in the on phase. (b) The surface measured when this patient was in the off phase. Since this patient serves as his own control by providing data on both on and off periods, (c) shows the three-dimensional surface resulting when the off-phase responses are subtracted from the on-phase responses. The shape of this surface suggests that DA enhances sensitivity at the peak of the patiotemporal surface and attenuates sensitivity at the lower spatial frequencies (see text: Bodis-Wollner et al, 1987).

Using rather coarse checkerboard stimuli in PD patients, decreased transient PERG amplitudes were reported by Nightingale and co-workers (1986), who failed to show any abnormality for check sizes of 25′ of arc. On the other hand, Gottlob and co-workers (1987) reported decreased transient PERG amplitudes in PD patients using checks of 25′ of arc. In early parkinsonians, Calzetti and co-workers (1990) did not report any amplitude abnormality, showing only a mean latency delay for the finer stimulus used (17.5′ of arc checks).

A delayed PERG latency was reported in studies performed with medium spatial frequency grating stimuli (2 cpd), in the absence of significant amplitude abnormalities (Pierelli et al 1989). The effect of dopaminergic blockers on PERGs was reported to be a delayed latency both to 28′ checks (Bartel et al 1990a and b) and to 4 cpd gratings (Stanzione et al 1991). L-DOPA administration to healthy volunteers caused a decreased PERG latency to different sized stimuli (Gottlob et al 1987; Bartel et al 1990) or had no effect on grating stimuli (Stanzione et al 1990b). It should be noted that a true comparison is difficult because there is little consistency concerning the spatial frequency of the stimuli used in different studies, although spatial frequency has been shown to be crucial for VEP diagnosis in PD (Tartaglione et al 1984; Bodis-Wollner 1985; Onofrj et al 1986). One problem with checkerboard stimuli is that they provide a number of higher spatial frequencies besides the nominal size of the pattern elements, thus eliciting a response less specifically related to the fundamental dimension of the stimulus (Bodis-Wollner et al 1990b).

There is more consistency across those PERG studies which used stimuli with a spatial frequency above 1-2 cpd, near the peak (5 cpd) of the primate spatial contrast sensitivity function. Additionally, several studies (Nightingale et al 1986; Gottlob et al 1987; Stanzione et al 1989), demonstrate significant losses in PD and improvement under levodopa. The human results are comparable to data obtained in monkeys. PERG studies in old-world monkeys rendered systemically DA deficient (Ghilardi et al 1988a) show a loss of retinal dopamine content (Ghilardi et al 1989). The PERG effects of DA depletion caused by MPTP are similar to those induced by the direct intravitreal use of 6-OHDA (Ghilardi 1988b). In both, the processing of peak spatial frequency patterns suffers most.

## DIVERSE DOPAMINE RECEPTORS

Diverse ERG results may be due to several causes. First, flash and complex checkerboard stimuli may not adequately address, or may differently stimulate the retinal networks involved in dopaminergic transmission, compared to patterns. This may be due to the presence of two anatomically and functionally different DA-linked processes in the retina. In the monkey and human retina, two different types of DA cells (Mariani and Hokoc 1988; Frederick et al 1982) and two different DA receptors have been described (Zarbin et al 1986; McGonigle et al 1988; Denis et al 1989). Diverse dopaminergic agents may act differently on DA-linked functions of the retina and this may depend on drug concentration. Our current studies reveal possible reasons why ERG studies which do not consider the importance of spatial and temporal parameters of stimuli in humans may not provide an optimal strategy to evaluate the dynamics of retinal dopaminergic circuits in humans.

We studied $D_2$ receptor blocker effects in retinal activity in humans and monkeys. The effect of the $D_2$ blocker was tested for three spatial frequencies to evaluate the responses of different populations of retinal neurons (Hess & Baker, 1984). Stimuli were grating patterns of either 0.25, 1, or 4 cpd, modulated at either 1 or 7.5 Hz. In humans PERGs were performed in 24 healthy subjects before and after the i.m. administration of L-sulpiride. Furthermore, the effect of low (0.07 mgm/kg) and high (0.35 mgm/kg) doses of sulpiride were compared. L-sulpiride is a neuramide derivative with almost selective $D_2$-blocker properties (Spano et al 1979). The human ERG was recorded using a corneal silver wire electrode referenced to the ipsilateral temple. ERGs obtained at a modulation rate of 7.5 Hz and Fourier analysed for amplitude and phase. To compare results form pre- and post-sulpiride and across spatial frequencies, analysis of variance was performed using ANOVA. Low dose sulpiride ("high" dose) significantly decreased the amplitude and delayed the phase of the ERG for 4 cpd. Using 1 Hz stimulation the latencies of P50 of the PERG before l-sulpiride administration were between 53.0 and 56.6 msec. Consistent with previous studies, we found that P50 latency did not depend on spatial frequency.

Conversely, after L-sulpiride administration, 6 out of 8 groups in the "low" dose group and 7 out of 10 subjects in the "high" dose group showed a delay of P50 at 4 cpd. At 0.25 cpd, only 4 subjects showed a delayed P50 after either low or high dose, while at 1 cpd 3 were abnormal for the low dose and 7 subjects had delays ERGs after the high dose. These results analysed with the Wilcoxon Signed-Rank Test revealed only borderline significant values, albeit there was a trend consistent with the highly significant results of the steady-state data. Comparing the drug effects obtained at 0.25 and 4 cpd using 7.5 Hz stimulation rate we found a highly significant difference (p is less than 0.005).

In monkeys both doses of L-sulpiride reduced the amplitude of the PERG to the peak SF and didn't change much the response to 0.5 cpd. The higher dose, L-sulpiride was more consistently effective. These results are consistent with the human data. Both human and monkey results suggest that neurons with different receptive field sizes may be differently sensitive to the drug and, $D_2$ blockers at low temporal and low spatial frequencies of stimulation are less effective.

## SUMMARY

The spatio-temporal stimulus dependent effects of DA deficiency obtained are similar to the effects of D2 receptor blockade in humans and monkeys. Apparently, retinal $D_2$ receptors (Stormann et al 1989) are important for spatial and temporal frequency tuning of the pattern ERG. The relatively low temporal frequency dependent deficit suggests that pathology may disrupt a pracrine mode of dopaminergic retinal transmission. The similarity of the current data obtained in normals with a $D_2$ blocker substance compared to data obtained in presynaptic DA deficient primates raises the possibility that postsynaptic retinal $D_2$ receptors may also be dysfunctional in PD. At this point the relevance of $D_1$ receptors to specific aspects of visual processing in the primate is unknown and should be further explored by appropriate studies using selective compounds acting at the $D_1$ site. With appropriate attention to stimulus details, PERGs may provide a valuable *in vivo* tool to assess the dynamics of DA mechanisms and the functional state of DA receptors in the retina.

## REFERENCES

Bartel, P., Blom, M., Robinson, E., Van der Meyden, C., Sommers, D. K. and Becker, P. (1990a) Effects of chlorpromazine on pattern and flash ERGs and VEPs compared to oxazepam and to placaebo in normal subjects, *Electroenceph.Clin. Neurophysiol.* 77:330-339

Bartel, P., Blom, M., Robinson, E., Van de Meyden, C., Sommers, D.K. and Becker, P. (1990b) The effect of levodopa and haloperidol on flash and pattern ERGs and VEPs in normal humans, *Doc. Ophthalmol.* 76:55-64

Bodis-Wollner, I. (1985) Pattern evoked potential changes in Parkinson's disease are stimulus-dependent, *Neurology* 35:1675-1676.

Bodis-Wollner. I., Brannan, J., Ghilardi, M.F., and Mylin, L. (1990) The importance of phypsiology to visual evoked potentials. In : Visual Evoked Potentials, JE Desmedt (ed), Amsterdam : Elsevier, pp 1-24.

Bodis-Wollner, I., Marx, M.S., Mitra, S., Bobak, P., M:ylin, L. and Yahr, M. (1987) Visual dysfunction in Parkinson's disease, *Brain* 110:1675-1698.

Bodis-Wollner, I. and Regan, D. (1991) Spatiotemporal contrast vision in Parkinson's disease and MPTP-treated monkeys. Regan (ed) Spatial vision. Vision and Visual Dysfunction, Vol 10. Macmillan Press, Scientific & Medical.

Bodis-Wollner, I., Yahr, M., Mylin, L. and Thornton, J. (1982) Dopaminergic deficiency and delayed visual evoked potentials in humans, *Ann. Neurol.* 11:478-483.

Bodis Wollner, I. and Yahr, M. (1978) Measurement of visual evoked potentials in Parkinson's disease, *Brain* 101:661-671.

Bodis-Wollner, I. and Piccolino, M., eds, (1989) Dopaminergic Mechanisms in Vision. Alan R Liss Inc, New York City, NY.

Calzetti, S., Franchi, A.,Taratufolo, G.,Groppi, E. (1990) Simultaneous VEP and PERG investigations in early Parkinson's disease, *J. Neurol. Neurosurg. Psychiat.* 53:114-117.

Cohen, J.L. and Dowling, J.E. (1983) The role of the retinal interplexiform cell: effects of 6-hydroxydopamine on spatial properties of carp horizontal cells, *Brain Res.* 261:307-310.

Cracco, R.Q. and Bodis-Wollner, I., eds, (1986). Evoked Potentials. Frontiers of Clinical Neuroscience: Vol 3. Alan R Liss Inc, New York City, N Y.

Denis, P., Nordmann, J.P., Elena, P.P., Laroche, L., Lapalus, P. (1989) Autoradiographic localisation of retinal dopaminergic receptors in man, *Bull. Soc. Ophthalmol. Franc.* 89:807-810.

Frederick, J.M., Rayborn, M.E., Laties, A.M., Lam, D.M.K. and Hollyfield, J.G. (1982) Dopaminergic neurons in human retina, *J. Comp. Neurol.* 210:65-79.

Ghilardi, M.F.,Bodis-Wollner, I., Onofrj, M., Marx, M.S. and Glover, A. (1988a) Spatial frequency dependent abnormalities of the pattern electroretinogram and visual evoked potentials in a parkinsonian monkey model, *Brain* 111:131-149.

Ghilardi, M.F., Chung, E., Bodis-Wollner, I., Dvorzniak, M., Glover, A. and Onofrj, M. (1988b) Systemic 1-methyl,4-phenyl,1-2-3-6-tetrahydropyridine (MPTP) administration decreases retinal dopamine content in primates, *Life Sci.* 43:255-262.

Ghilardi, M.F., Marx, M.S., Bodis-Wollner, I., Camras, C. and Glover, A. (1989) The effect of intraocular 6-OH-dopamine on retinal processing in primates, *Ann. Neurol.* 25:357-364.

Gottlob, I., Schneider, E,. Heider, W. and Skrandies, W. (1987) Alteration of visual evoked potentials and electroretinograms in Parkinson's disease, *Electroencephalog. Clin. Neurophysiol.* 66:349-357.

Gottlob, I., Weghaupt, H. and Vass, C. (1990) Effect of levodopa on the human luminance elctroretinogram, *Invest. Ophthalmol. Visual Sci.* 31: 1252-1258.

Gutierrez, O. and Spiguel, R.D. (1973) Oscillatory potentials of the cat retina : effects of adrenergic drugs, *Life Sci.* 13:991-999.

Haggendal, J. and Malmfors, T. (1965) Identification and cellular localisation of the catecholamines in the retina and the choroid of the rabbit, *Acta Physiol. Scand.* 64: 58-66.

Halliday, A.M., McDonald, W.I., Mushin, J. (1972) Delayed visual evoked responses in optic neuritis, *Lancet* 1:982-985.

Hankins, M.W. and Ikeda, H. (1991) The role of dopaminergic pathways at the outer plexiform layer of the mammalian retina, *Clin. Vision Sci.* 6:87-93.

Hess, R.F. and Baker, C.L. Jr (1984) Human pattern-evoked electroretinogram, *J. Neurophysiol.* 51:939-951.

Ikeda, H., Priest, T.D., Robbins, J. and Wakakuwa, K. (1986) Silent dopaminergic synapse at feline retinal cells, *Clin. Vision. Sci.*1:25-38.

Iuvone, P.M., Galli, C.C., Garrison-Gund, C.K. and Neff, N.H, (1978) Light stimulates tyrosine hydroxylase activity and dopamine synthesis in retinal amacrine neurons, *Science* 202:901-902.

Kramer, S.G. (1971) Dopamine: A retinal neurotransmitter. I. Retinal uptake, storage and light-stimulated release of $^{3}$H-dopamine in vivo, *Invest. Ophthalmol.* 10:438-52.

Malmfors, T. (1963) Evidence of adrenergic neurons with synaptic terminals in the retina of rats demonstrated with fluorescence and electron microscopy, *Acta. Physiol. Scand.* 58:99-100.

Mariani, A.P. (1989) Synaptic organisation of type 2 catecholaminergic amacrine cells in the rhesus monkey retina, *Invest. Ophthalmol. Visual Sci.* 30:119.

Mariani, A.P. and Hokoc, J.N. (1988) Two types of tyrosine hydroxylase-immunoreactive amacarine cells in the rhesus monkey retina, *J. Comp. Neurol.* 276:81-91.

Marx, M., Bodis-Wollner, I., Bobak, P., Harnois, C., Mylin, L., Yahr, M. (1986) Temporal frequency-dependent VEP changes in Parkinson's disease, *Vision Res.* 26:185-193.

McGonigle, P., Wax, M.B. and Molinoff, P.B. (1988) Characterisation of binding sites for $^{3}$H-spiroperidol in human retina, *Invest. Ophthalmol. Visual Sci.* 29: 687-694.

Nguyen-Legros, J., Harnois, C,. DiPaolo, T. and Simon, A. (1993) The retinal dopamine system in Parkinson's disease, *Clin. Vision. Sci.* 8:1-12.

Negishi, K., Teranishi, T. and Kato, S. (1990) The dopamine system of the teleost fish retina, *Prog. Retinal Res.* 9:2-48.

Nightingale, S., Mitchell, K.W., and Howe, J.W. (1986) Visual evoked potentials and pattern electroretinograms in Parkinson's disease and control subjects, *J. Neurol. Neurosurg. Psychia.* 49:1280-1287.

Onofrj, M.,Ghilardi, M.F., Basciani, M., Gambi, D. (1986). Visual evoked potentials in Parkinson's disease and dopamine blockade reveal a stimulus-dependent dopamine function in humans, *J. Neurol. Neurosurg. Psychiat.* 49:1150-1159.

Oyster, C.W., Takahashi, E.S. and Brecha, N.C. (1988) Morphology of retinal dopaminergic neurons. In : Dopaminergic Mechanisms inVision, Bodis-Wollner and Piccolino M (eds), Alan Liss, NewYork, pp 19-30.

Pierelli, F., Stanzione, P., Peppe, A., Stefano ,E., Rizzo, P.A., Bernardi, G. and Morocutti, C. (1989) Electrophysiological (PERG, VEP)abnormalities in Parkinson's disease are reversed by L-DOPA. In: Dopaminergic Mechanisms in Vision, I Bodis-Wollner and M Piccolino (eds), NewYork : Alan R Liss, pp 253-265.

Savy, C., Simon, A. and Nguyen-Legros, J. (1991) Spatial geometry of the dopamine innervation in the avascular area of the human fovea, *Visual Neurosci.* 7:487-498.

Spano, P.F., Stefanini, E., Trabucchi, M. and Fresia, P. (1979) Stereospecific interaction of sulpiride on striatal and non-striatal dopamine receptors. In : Sulpiride and other Benzamides. P F Spano, M Trabucchi, G U Corsini, G L Gessa (Eds) Milano : Italian Brain Research Foundation Press, pp 1132.

Stanzione, P., Pierelli, F., Peppe, A., Rizzo, P.A. and Morocutti, C. (1989) Pattern visual evoked potentials and electroretinogram abnormalities in Parkinson's disease: effects of L-dopa therapy, *Clin. Vision Sci.* 4:115-127.

Stanzione, P., Tagliati, M., Silvestrini. M,, Porcu, S. and Bernardi, G. (1991) Haloperidol delays pattern electroretinograms more thanVEPs in normal humans, *Clin. Vision Sci.* 6:137-147.

Stormann, T., Gdula, D., Weiner, D. and Brann, M. (1990) Molecular cloning and expression of a dopamine $D_2$ receptor from human retina, *Mol. Pharmacol.* 37:1-6.

Tagliati, M., Bodis-Wollner, I., Kovanecz, I., and Stanzione, P. (1994). Spatial frequency tuning of the monkey pattern ERG depends on D2 receptor-linked action of dopamine. *Vision Res.* 34:2051-2057.

Tagliati, M., Bodis-Wollner, I. Yahr, M. (1995). The pattern electrorentingogram in Parkinson's disease reveals lack of retinal spatial tuning. *Electroenceph and Clin Neurophysiol.* (in press).

Tartaglione, A., Pizio, N., Bo, I., Spadavecchia, L., Favale, E. (1985) Spatial properties of pattern as determinants of visual evoked potential changes in Parkinson's syndrome. In: Evoked Potentials: Neurophysiological and Clinical Aspects, C Morocutti and PA Rizzo (eds), Elsevier,Amsterdam, pp 321-327.

Witkovsky, P. and Schuette, M. (1991) The organisation of dopaminergic neurons in vertebrate retinas, *Visual Neurosci.* 7:113-123.

Zarbin, M.A., Wamsley, J.K., Palacios, J.M. and Kuhar, M.J. (1986) Autoradiographic localisation of high affinity GABA, benzodiazepine, dopaminergic, adrenergic and muscarinic cholinergic receptors in the rat, monkey and human retina, *Brain Res.* 374:75-92.

# SECTION III

## Higher Visual Centers: Structure, Function and Disease

12

# THE ROLE OF THE PRETECTUM IN THE PUPILLARY LIGHT REFLEX

R. J. Clarke and P. D. R. Gamlin

Vision Science Research Center
University of Alabama at Birmingham
Birmingham, Alabama 35294

## INTRODUCTION

The pupillary light reflex (PLR) is the constriction muscle of the iris that occurs with increases in retinal illumination. The direct PLR, present in virtually all vertebrates, is the constriction of the pupil in the same eye as that stimulated with light. The consensual PLR is the constriction of the pupil in the eye opposite to the eye stimulated with light. In mammals with laterally placed eyes, such as the rat and rabbit, the direct PLR is more pronounced than the consensual PLR (Lowenstein & Loewenfeld, 1969; Clarke & Ikeda, 1985; Trejo & Cicerone, 1989). However, in those mammalian species with frontally placed eyes such as humans & monkeys, the direct and consensual PLR are essentially equal (Lowenstein & Loewenfeld, 1969).

The primary function of the pupil is to ensure that retinal illumination is optimal for visual perception. In dim light, a dilated pupil maximizes the number of photons reaching the retina. Under bright light, the pupil can reduce retinal illumination by up to 1.5 log units. While this reduction in retinal illumination cannot account for the retina's sensitivity over a 12 log unit range of illuminance (Rushton, 1965), it clearly aids in the initial phase of light adaptation. It has also been suggested that a small pupil may prepare the retina for speedier dark adaptation following prolonged exposure to bright light (Woodhouse & Campbell, 1975). Furthermore, a partially constricted pupil can contribute to improving image quality at the retina by reducing spherical and chromatic aberration as well as by increasing depth of focus (Campbell, 1957; Campbell & Green, 1965; Campbell & Gregory, 1960; Westheimer, 1964). However, constriction beyond a certain diameter results in image degradation due to increased diffraction (Charman et al., 1978).

The PLR has proven particularly useful to ophthalmologists and neurologists in the diagnosis of diseases of the central and peripheral visual and oculomotor systems. One example is the 'swinging light' test which is a simple, but objective means of detecting optic nerve neuropathies as well as other afferent defects (Thompson et al., 1982; Thompson, 1987). A second example is subjective perimetry in conjunction with objective pupil perimetry which is used to plot visual field scotomas (Johnson et al., 1988; Kardon et al.,

*Basic and Clinical Perspectives in Vision Research*, Edited by Jon Robbins et al.
Plenum Press, New York, 1995

1991). The advantage of the PLR in this latter application is that it does not rely on patient compliance to map out visual field abnormalities. The most powerful clinical uses of the PLR may be yet to come in the development of tests to predict the onset of glaucoma (Brown et al., 1987) and Alzheimer's disease (Scinto et al., 1994).

The pupil receives innervation from both the parasympathetic and sympathetic components of the autonomic nervous system. The parasympathetic pathway innervates the sphincter iris muscle that constricts the pupil. Inputs influencing the parasympathetic pathway include those related to the light reflex and to the near reflex, which is the pupilloconstriction that occurs in conjunction with the increases in convergence and ocular accommodation required for near viewing. In addition, other factors such as painful stimuli, emotions, arousal, and alertness can act through central pathways to influence the parasympathetic pathway (Lowenstein et al., 1953; Sillito & Zbrozyna, 1970a,b; Loewy et al., 1973). The first accurate descriptions of the central course of the parasympathetic component of the PLR pathway were made by Magoun, Ranson and colleagues in the 1930s (Ranson & Magoun, 1933; Magoun et al., 1935). Their pioneering studies showed that this pathway ran from the retinae, through the optic nerves and tracts and synapsed in the pretectum, a midbrain area rostral to the superior colliculus. From here they traced the course of the pathway both ipsilaterally, and contralaterally via the posterior commissure, around and in the periaqueductal grey to the oculomotor complex. They reported that the pathway decussates again ventral to the central aqueduct before synapsing with the pupilloconstrictor neurons of the Edinger-Westphal nucleus (EW) on both sides of the midline. Preganglionic fibers arising from EW neurons leave the brain in the ipsilateral oculomotor nerve and innervate postganglionic neurons in the ciliary ganglion. These cholinergic, postganglionic neurons give rise to the postganglionic fibers that innervate the sphincter pupillae muscle of the iris.

The sympathetic pathway innervates the radial iris muscle which dilates the pupil. The pupillodilator sympathetic pathway runs from the ciliospinal centers of Budge and Waller to the superior cervical ganglion in the neck, which in turn innervates the radial iris muscle via noradrenergic postganglionic fibers (Thompson, 1987; Loewenfeld, 1993). Normally, the sympathetic pathway provides a tonic drive to the dilator iris muscle. The loss of this tonic drive can occur in certain types of spinal injuries and results in a characteristic pupil constriction or Horner's pupil (Thompson, 1987; Loewenfeld, 1993). A number of other factors such as onset of darkness (dark reflex), painful stimuli, emotions, arousal, and alertness dilate the pupil by increasing the drive to the peripheral sympathetic pathway (Lowenstein & Loewenfeld, 1950; Nisida et al., 1960; Okada et al., 1960).

## ELECTROPHYSIOLOGICAL STUDIES IN THE PRETECTUM

We have carried out a series of studies in rats and primates on the role of the pretectum in the parasympathetic PLR pathway. Single cell recording in this region of the brain was used to search for neurons with activity that correlated with changes in pupil size. For studies in rats, animals were anaesthetized and the head placed in a stereotaxic headholder. The pupils were dilated with atropine and a craniotomy performed through which a glass microelectrode was lowered into the pretectum where single unit activity was recorded. A half ping-pong ball arrangement which acted as a diffuser for a light source provided a Ganzfeld stimulus for the eye contralateral to the recorded side. Luminance was changed by interspersing neutral density filters between the light source and diffuser. Since the size of the pupil is dependent on the constant level of retinal illumination, only tonically active pretectal neurons were considered suitable candidates for pupillomotor control. Such neurons were encountered in the rat pretectum and the

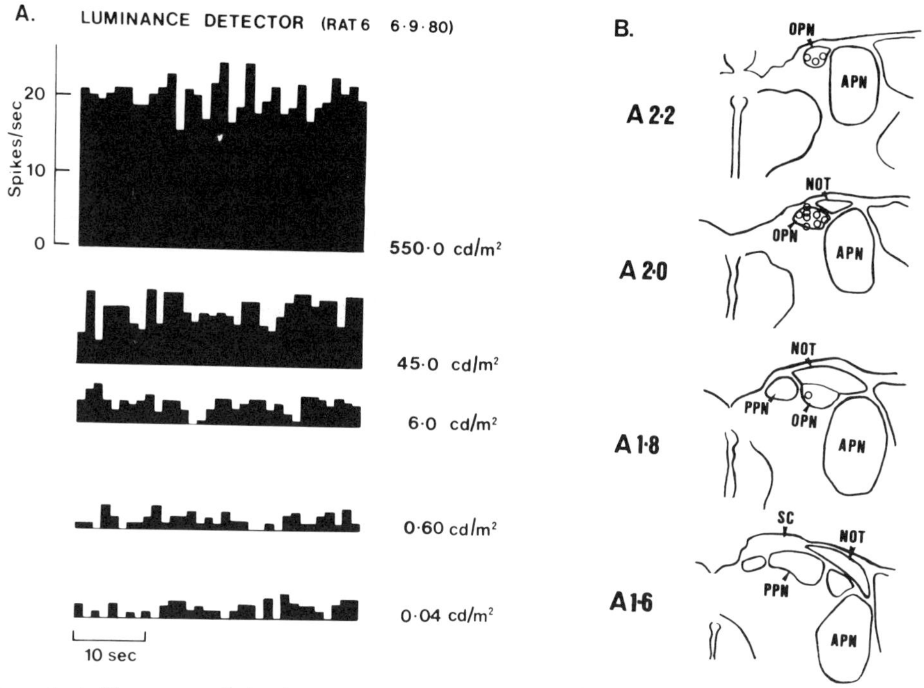


**Figure 1.** A. Histograms of a luminance detector's response to 5 different steady luminances over a range of 4 log units in intensity. Firing rate was sampled over 1.25 second intervals. B. Schematic diagrams of four coronal sections of the pretectum about 200 μm apart (based on nomenclature of Scalia, 1970, 1979; Campbell & Lieberman, 1985). The positions of the sections correspond approximately to the stereotaxic co-ordinates shown about each diagram. Open circles denote positions of luminance detector recording sites obtained from pontamine blue marks (modified from Clarke & Ikeda, 1985a).

histograms in figure 1A show the response characteristics of one such cell recorded in an anaesthetized pigmented rat which we classified as a 'luminance detector' (Clarke & Ikeda, 1981, 1985a,b). For the duration of the presentation of each steady luminance, the firing rate of the neuron remained approximately constant. The mean firing rate of the neuron during this period was related to the logarithm of luminance and increased from approximately 1 spike per second at the lowest luminance level to about 20 spikes per second at the highest luminance level used.

Pretectal neurons with these characteristics have also been referred to as 'tonic-on' cells in the literature (Smith et al., 1968, 1970; Harutiunian-Kosak et al., 1968; Siminoff et al., 1967; Cavaggioni et al.,1968; Trejo & Cicerone, 1984; Distler & Hoffmann, 1989a,b). Despite some earlier claims that these luminance neurons were located in the nucleus of the optic tract (NOT) (Cavaggionni et al., 1968), figure 1B shows that this is not the case. The recording sites of eight luminance detector neurons were marked with pontamine blue. All recording sites marked in this way were recovered by histology after the experiment, and all were located in the pretectal olivary nucleus (PON) (Figure 1B). Consistent with this, electrical stimulation of the PON in rats induces pupilloconstriction (Trejo & Cicerone, 1984), and lesions of the PON in rats produces deficits in pupillomotor function (Young & Lund, 1994). The NOT, on the other hand, is now considered part of the pathway for horizontal optokinetic nystagmus (Hoffmann & Schoppmann, 1975; Hoffmann & Distler, 1989; Mustari & Fuchs, 1990).

We have extended the study of the PLR from rats to primates (Clarke et al., 1993; Gamlin et al., 1995). The primate visual system more closely resembles the human visual system than it does the rat's. Furthermore, recording in alert primates allows the activity of single cells to be correlated with changes in pupil diameter without the compromising effects of anaesthesia. For studies of the primate PLR, monkeys were trained to binocularly fixate a small, dim, laser target back-projected centrally on a tangent screen for a juice reward. The accuracy of fixation was verified using the scleral search coil technique (Fuchs & Robinson, 1966). A second stimulus, comprised of a white, circular stimulus back-projected onto the tangent screen, probed the visual field. This probe stimulus was under computer control for position, size and intensity. The monkey was required to maintain fixation of the laser target for the duration of the stimulation period. Single units were recorded extracellularly in the pretectum using platinum-plated tungsten microelectrodes lowered through a cylinder implanted over a craniotomy in the skull. Pupil diameters were measured in both eyes under infra-red illumination with ISCAN RK406 pupillometers. Target parameters, eye positions, pupil diameters, and time of occurrence of action potentials were recorded on computer disk for subsequent off-line analysis.

Figure 2 shows pupillary responses and interspike interval plots of the response of a pretectal neuron to binocular presentation of three steady luminance levels of 100, 10, and 1 cd/m$^2$ respectively. During the initial fixation period, the pupil remained dilated and the

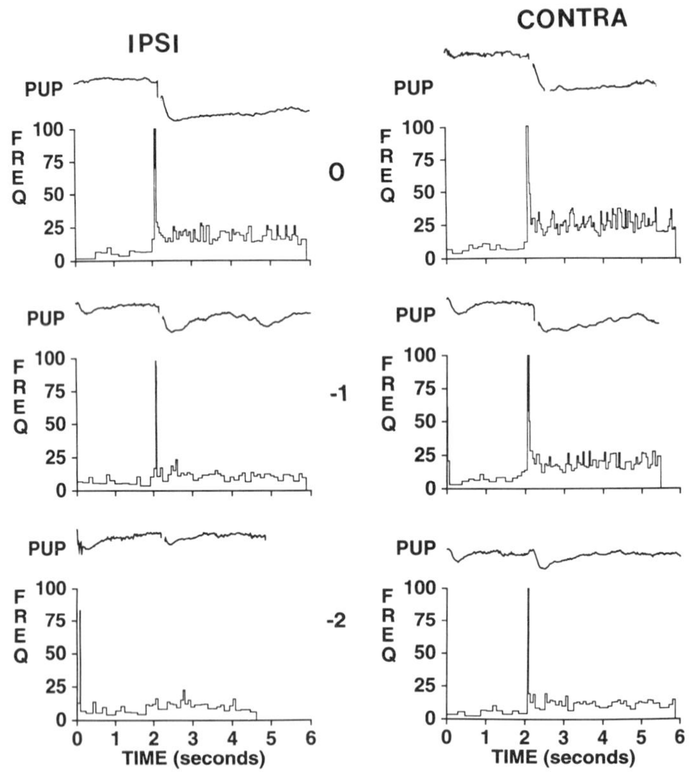


**Figure 2.** Histograms of a pretectal neuron's responses to three luminance levels of a 5 ° spot placed 30 ° away from the fovea contralateral (right column) and ipsilateral (left column) to the side of the recorded neuron. The highest luminance used was 100 cd/m$^2$ (0 log units). The monkey was viewing binocularly and fixated the laser target for 2 seconds before the probe stimulus appeared. Above each histogram, pupillary responses to the same stimuli, are displayed. The blanked portions of the pupillary responses denote blinks.

firing rate of the neuron remained low. After 2 seconds, the probe stimulus appeared 30° from the fixation spot either in the hemifield contralateral (right histograms) or ipsilateral (left histograms) to the side of the recorded cell. Both the pupil and the neuron responded to all three luminance levels irrespective of the location of the stimulus. With stimulus onset, the neuron responded with an initial burst followed by constant, elevated firing rate proportional to the luminance of the stimulus. The pupil constricted rapidly to a peak before redilating to a new constricted level for the duration of the stimulus light on. The amplitude of pupilloconstriction and the neuronal response increased with increase in stimulus strength. Clearly this pretectal neuron has the appropriate response characteristics necessary for subserving a pupilloconstrictor role. We had expected that the receptive fields of these pretectal neurons would be restricted to the contralateral hemifield; however, this and many other pretectal neurons had bilateral receptive fields.

If, as the electrophysiological data suggest, the PON is in the pathway for the PLR in the primate, then electrical stimulation in the vicinity of luminance sensitive neurons would be expected to produce pupilloconstriction. In figure 3 we show that pupilloconstriction is indeed elicited by microstimulation at the recording site of a luminance neuron. No other eye movements occurred during stimulation indicating that current had not spread to adjacent nuclei known to have other oculomotor functions, such as the nucleus of the optic tract, posterior pretectal nucleus and superior colliculus.

The recording sites of a few luminance neurons were marked with electrolytic lesions made by passing 20-40 μA anodal current for 20 seconds. The lesions were recovered after deeply anaesthetizing and perfusing the animal with a suitable fixative and then processing the brain for histology. The lesion site depicted in figure 4 was identified in a Nissl stained coronal section and shows the location of a recording site of a luminance pretectal neuron. The lesion was clearly located in the PON. When the position of other luminance neurons was calculated relative to the lesion sites, it was clear that in the primate, as in the rat, these luminance neurons are located within the confines of the PON. No luminance neurons were encountered in the nucleus of the posterior commissure (NPC), which projects to the EW (Gamlin et al., 1995).

## ANATOMICAL STUDIES OF MIDBRAIN PUPILLOMOTOR PATHWAYS

We have studied the anatomical substrate for the PLR in the primate using horseradish peroxidase conjugated with wheatgerm agglutinin (WGA-HRP) as a sensitive retrograde and anterograde anatomical tracer (Gamlin & Clarke, 1995).

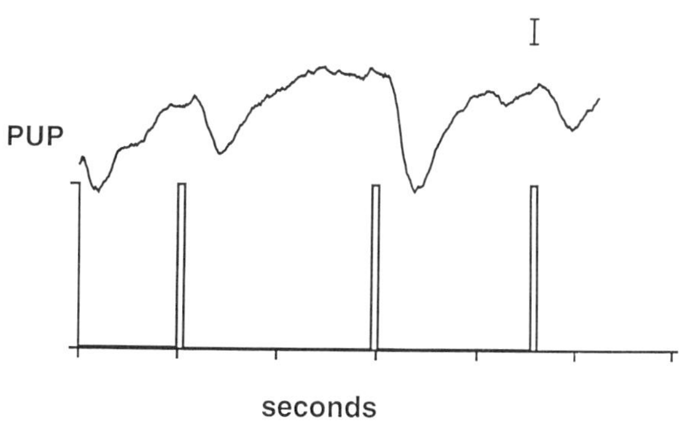


**Figure 3.** Pupillary constriction elicited at the recording site of a pretectal luminance cell. Stimulation current used was 60 μA at 300 Hz for 100 ms. Calibration bar is 0.1 mm.

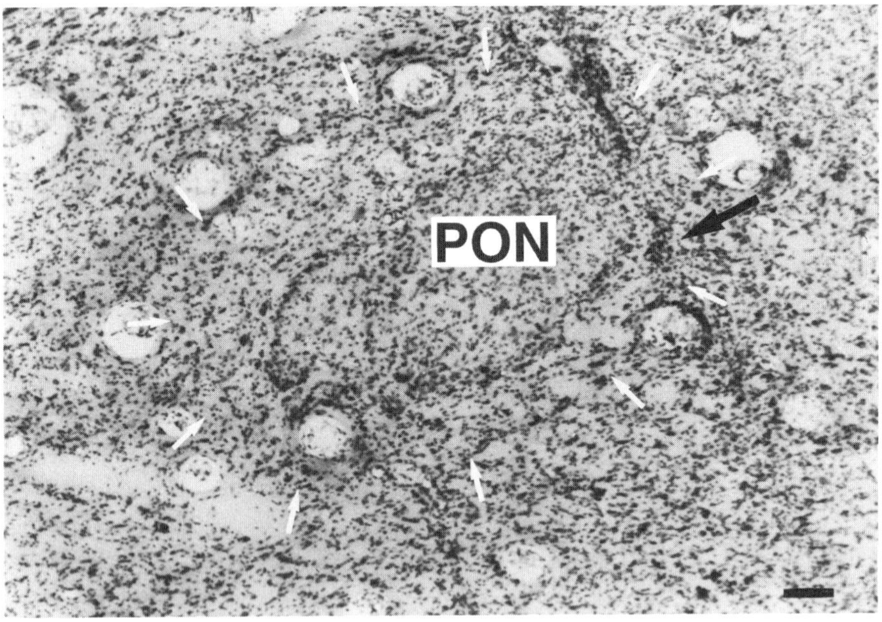


**Figure 4.** Photomicrograph of a coronal section taken through the pretectal olivary nucleus (PON) stained with cresyl violet. Small white arrows indicate the outer limits of the shell which surrounds a central neuropil. The black arrow indicates a small lesion produced by passing current through the electrode tip at the recording site of a luminance neuron. Calibration bar is 0.1 mm.

Figure 5A shows retrogradely labeled neurons in the PON following an injection of WGA-HRP under physiological guidance into the region immediately dorsal and lateral to the EW in a monkey. This injection spread to the ipsilateral EW, but the location of the injection site was chosen so that no tracer crossed the midline to the contralateral EW. This injection not only labeled the EW ipsilaterally, but it also labeled fibers en route from the pretectum to the EW. Numerous labeled cells were present contralaterally in the PON throughout its rostrocaudal extent. They were mostly located in the periphery of the nucleus in the shell surrounding the central neuropil, but were also found within the neuropil itself. A very few cells were found in the PON ipsilateral to the injection site. Although labeled cells were present bilaterally in the NPC, none were present in any retinorecipient pretectal nucleus other than the PON. Since the pretectal nuclei are poorly delineated cytoarchitecturally in primates, it was important to confirm that cells that were labeled by the injection described above were in the retinorecipient portion of the PON. We therefore placed an intravitreal injection of WGA-HRP in a second monkey to study the retinal terminal fields. The contralateral PON is clearly labeled throughout its extent (Fig. 5B). Slightly weaker anterograde labeling is present in the ipsilateral PON. These results clearly indicate that neurons projecting to the EW are located in the retinorecipient region of the PON.

## CONCLUDING REMARKS

The results from our electrophysiological and anatomical studies satisfy a number of criteria for supporting a role for the PON, and no other retinorecipient pretectal nucleus, in mediating the PLR in both primates and rats. Our results indicate that the neuronal discharge

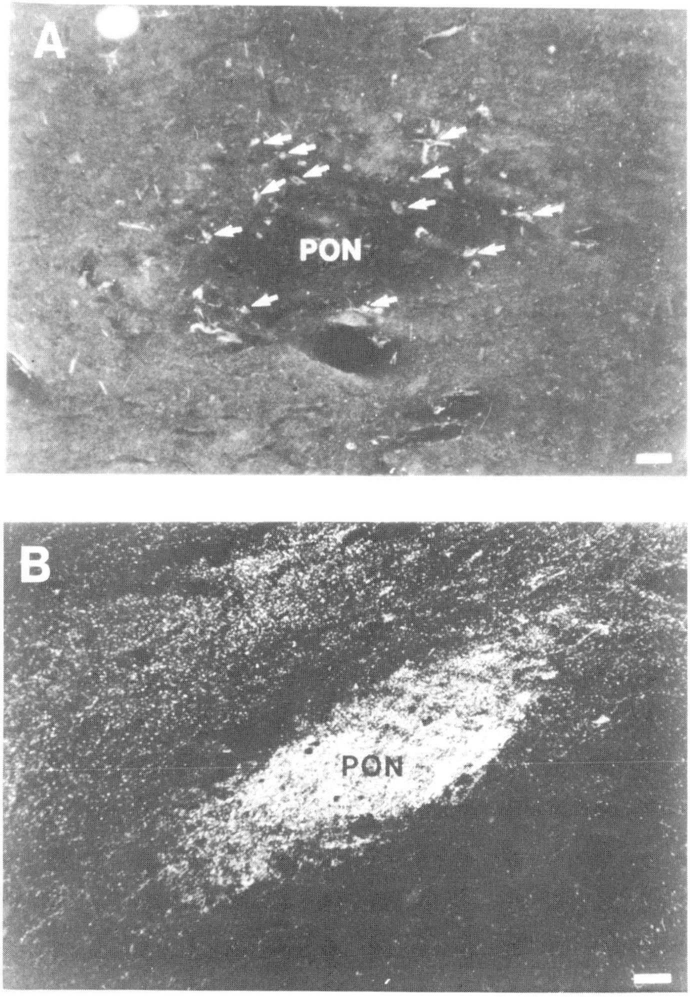


**Figure 5.** A. WGA-HRP retrogradely labeled neurons indicated with arrows in the PON contralateral to the injection site. Most neurons are encountered in the shell surrounding the central neuropil. B. WGA-HRP anterograde labeling of retinal afferent terminals in the PON after intraocular injections of the tracer on the contralateral side. Calibration bars = 50 μm (modified from Gamlin & Clarke, 1995).

characteristics of neurons in the PON correlate closely with changes in pupil size and that electrical microstimulation in the PON produces pupilloconstriction. Also, we have shown that the PON receives a direct retinal projection and projects directly to the EW nucleus.

Despite our progress using modern electrophysiological and anatomical techniques to study the PLR, there are many unresolved questions concerning the role of the pretectum in this reflex. Indeed, our results strongly suggest that the organization of the PLR pathway at the level of the pretectum is more complex than was originally proposed by Magoun and colleagues.

For example, many neurons encountered in the monkey PON, unexpectedly, possessed extensive bilateral, binocular receptive fields. A certain degree of convergence of retinal afferents onto PON neurons has been described in the rat (Trejo & Cicerone, 1984) and cat (Distler & Hoffmann, 1989). However, to our knowledge, no previous study has reported receptive fields in the primate retinorecipient nuclei that extend so far into the ipsilateral hemifield. Convergence of retinal inputs onto PON neurons can account for the large contralateral receptive fields of pretectal neurons it cannot account for the large ipsilateral visual field component. Since the primate pretectum receives input from ganglion cells in the nasal retina of the contralateral eye and the temporal retina of the ipsilateral eye (Perry & Cowey, 1984; Kondo et al., 1992), it is to be expected that, as in most other primate retinorecipient nuclei (Rodieck, 1979), there is a representation of only the contralateral hemifield in the pretectal nuclei. Receptive fields that include the ipsilateral hemifield must be constructed by the convergence of both retinal and non-retinal visual inputs to the PON. To our knowledge, the source of this input is not currently known. It is probable that it arises either from the contralateral PON or the extrastriate visual cortex. In support of this latter possibility, Hoffmann et al. (1992) have reported a cortical contribution to the ipsilateral receptive fields of pretectal neurons in the NOT. While this input is not as extensive as is reported here for the PON receptive fields, they nevertheless may be comparable pathways. Furthermore, studies in rhesus monkeys (Dineen & Hendrickson, 1983; Leichnetz, 1990) and cats (Distler & Hoffmann, 1989b) have shown a well-defined input to the PON from extrastriate visual cortex. In addition, studies in rabbits (Inoue & Kiribuchi, 1985) and cats (Bando, 1985) have reported extrastriate cortical pupillomotor areas. Overall, these results suggest that more integration of the signals for the PLR occurs at the level of the primate pretectum than had previously been suspected.

Some previous studies have suggested that the PLR operates in a push-pull fashion with some pretectal neurons increasing their activity with increases in luminance and other neurons decreasing their activity with increases in luminance, so-called "darkness neurons" (Clarke & Ikeda, 1985). However, in primates, no neurons were encountered with the characteristics of darkness detectors that were not also modulated by other oculomotor activity. Thus, in the alert Rhesus monkey, "darkness neurons" may be more related to fixation or saccades than to luminance. Consistent with this possibility, "darkness neurons" were usually encountered caudal to the PON in a region that could correspond to the fixation region of the anterior superior colliculus (Munoz & Wurtz, 1993). Furthermore, an anatomical pathway whereby "darkness neurons" could modulate pupilloconstrictory function has never been adequately demonstrated, and lesion and stimulation data is currently lacking.

Some anatomical studies had previously reported in cats (Graybiel & Hartweig, 1974; Berman, 1977), tree-shrews (Weber & Harting, 1980) and rats (Nicholson & Severin, 1981) that there was no direct projection from any retinorecipient pretectal nucleus to the EW. On this basis, these authors have suggested that the pupillary light reflex pathway involves an additional synapse in the NPC (Berman, 1977; Loewy, 1979; Weber & Harting, 1980). However, luminance sensitive neurons have not been encountered in the NPC of either rats (Clarke & Ikeda, 1985) or monkeys (Gamlin et al., 1995). Furthermore, neurons have been identified in the NPC with firing rates that are closely related to changes in vergence and accommodation (Straschill & Hoffmann, 1969; Judge & Cummings, 1986; Mays et al., 1986). It thus appears that the projection from the NPC to EW is consistent with a role in accommodation or the near reflex pathway, and is probably not related to a role in the PLR pathway.

## ACKNOWLEDGMENTS

The inspiration for the studies on the pupillary light reflex in rats belongs to Professor Ikeda and this chapter is dedicated to her on her retirement. The studies on the pupillary light reflex in the primate were carried out at the University of Alabama at Birmingham and were supported by NEI Grant R01 EY-09380 to P.D.R. Gamlin and by NEI CORE grant P30 EY-03039. We wish to thank Kelley Mitchell for technical assistance, and Sam Hayley and Chandrasekaran Venkatapathy for computer programming.

## REFERENCES

Bando, T. (1985) Pupillary constriction evoked from the posterior medial lateral suprasylvian (PMLS) area in cats, *Neurosci. Res.* 2:472-485.

Berman, N. (1977) Connections of the pretectum in the cat, *J. Comp. Neurol.* 174:227-254.

Brown, R.H., Zilis, J.D., Lynch, M.G. and Sandborn, G.E. (1987) The afferent pupillary defect in asymmetric glaucoma, *Arch. Ophthalmol.* 105:1540-1543.

Campbell, G. and Lieberman, A. (1985) The olivary pretectal nucleus: experimental anatomical studies in the rat, *Phil. Trans. R. Soc. Lond B.* 301:573-609.

Campbell, F.W. (1957) The depth of field of the human eye, *Optica Acta* 4:157-164.

Campbell, F.W. and Gregory, A.H. (1960) Effect of size of the pupil on visual acuity, *Nature* 187:1121-1123.

Campbell, F.W. and Green, D.G. (1965) Optical and retinal factors affecting visual resolution, *J. Physiol.* 181:576-593.

Cavaggionni, A., Madarasz, I. and Zampollo, A. (1968) Photic reflex and pretectal region, *Arch. Ital. Biol.* 106:227-242.

Charman, W.N., Jenning, J.A.M. and Whitefoot, H. (1978) The refraction of the eye in relation to spherical aberration and pupil size, *Vision Res.*17:737-738.

Clarke, R.J. and Ikeda, H. (1981) Pupillary response and luminance and darkness detector neurones in the pretectum of the rat, *Doc. Ophthalmol.* (The Hague). Series 30:53-61.

Clarke, R.J. and Ikeda, H. (1985a) Luminance and darkness detectors in the olivary and posterior pretectal nuclei and their relationship to the pupillary light reflex in the rat. I Studies with steady luminance levels, *Exp. Brain Res.* 57:224-232.

Clarke, R.J. and Ikeda, H. (1985b) Luminance detectors in the olivary pretectal nuclei and their relationship to the pupillary light reflex in the rat. II Studies with sinusoidal light, *Exp. Brain Res.* 59:83-90.

Clarke, R.J., Zhang, H.Y. and Gamlin, P.D.R. (1993) Visual response characteristics of olivary pretectal nucleus neurons in the alert monkey, *Soc. Neurosci. Abstr.* 19:330.

Dineen, J.T. and Hendrickson, A. (1983) Overlap of retinal and prestriate cortical pathways in the primate pretectum, *Brain Res.* 278:250-254.

Distler C. and Hoffmann, K-P. (1989a) The pupillary light reflex in normal and innate microstrabismic cats, I: Behavior and receptive-field analysis in the nucleus praetectalis olivaris, *Vis. Neurosci.* 3:127-138.

Distler C. and Hoffmann, K-P. (1989b) The pupillary light reflex in normal and innate microstrabismic cats, II: Retinal and cortical input to the nucleus praetectalis olivaris, *Vis. Neurosci.* 3:139-153.

Fuchs A.F and Robinson D.A. (1966) A method for measuring horizontal and vertical eye movement chronically in the monkey, *J. Applied Physiol.* 21:1068-1070.

Gamlin, P.D.R., Zhang, H. and Clarke, R.J. (1995) Luminance neurons in the pretectal olivary nucleus mediate the pupillary light reflex in the rhesus monkey, *Exp. Brain Res.* 106:177-180.

Gamlin, P.D.R. and Clarke, R.J. (1995) The pupillary light reflex of the primate, *J. Am. Opt. Assoc.* 66:415-418.

Graybiel, A.M. and Hartweig, E.A. (1974) Some afferent connections of the oculomotor complex in the cat: An experimental study with tracer techniques, *Brain Res.* 81:543-551.

Harutiunian-Kosak, B., Kosak, W. and Dec, K. (1968) Single units in the pretectal region of the cat, *Acta Biol. Exp. (Warsaw).* 28:333-343.

Hoffmann, K-P. and Schoppmann, A. (1975) Retinal input to direction selective cells in the nucleus tractus opticus of the cat, *Brain Res.* 99:359-366.

Hoffmann, K-P. and Distler, C. (1989) Quantitative analysis of visual receptive fields of neurons in nucleus of the optic tract and dorsal terminal nucleus of the accessory optic tract in macaque monkey, *J. Neurophysiol.* 62:416-428.

Hoffmann, K-P., Distler, C. and Ilg, U. (1992) Callosal and superior temporal sulcus contributions to receptive field properties in the macaque monkey's nucleus of the optic tract and dorsal terminal nucleus of the accessory optic tract, *J. Comp. Neurol.* 321:150-162.

Inoue, T. and Kiribuchi, T. (1985) Cortical and subcortical pathways for pupillary reactions in rabbits, *Jpn. J. Ophthal.* 29:63-70.

Johnson, L.N., Hill, R.A. and Bartholomew, M.J. (1988) Correlation of afferent pupillary defects with visual field loss on automated perimetry, *Ophthalmol.* 95:1649-1655.

Judge, S.J. and Cumming, B.G. (1986) Neurons in the monkey midbrain with activity related to vergence eye movement and accommodation, *J. Neurophysiol.* 55:915-930.

Kardon, R.H., Kirkali, P.A. and Thompson , H.S. (1991) Automated pupil perimetry. Pupil field mapping in patients and normal subjects, *Ophthalmology* 98:485-496.

Kondo, Y., Takada, M., Kayahara, T., Yasui, Y., Nakano, K. and Mizuno, N. (1992) Single retinal ganglion cells sending axon collaterals to the bilateral superior colliculi: a fluorescent double-labeling study in the japanese monkey (Macaca fuscata), *Brain Res.* 597:155-161.

Leichnetz, G. R. (1990) Preoccipital cortex receives a differential input from the frontal eye field and projects to the pretectal olivary nucleus and other visuomotor-related structures in the rhesus monkey, *Vis. Neurosci.* 5:123-133.

Lowenstein, O. and Loewenfeld, I.E. (1950) Role of sympathetic and parasympathetic systems in reflex dilation of the pupil, *Arch. Neurol. Psychiat.*, 64:313-377.

Lowenstein, O. and Loewenfeld, I.E. (1969) The pupil, In: The Eye, (Ed. H. Davson), New York-London, Academic Press. 3:255-337.

Lowenstein, O. Murphy, S.B. and Loewenfeld, I.E. (1953) Functional evaluation of the pupillary light reflex pathways, *Arch. Ophthalmol.* 49:657-670.

Loewenfeld, I.E. (1993) The pupil. Anatomy, physiology, and clinical applications, Vol.1, Iowa State University Press / Ames. Wayne State university Press/ Detroit.

Loewy, A.D., Aroujo, J.C. and Kerr, F.W.C. (1973) Pupillodilator pathways in the brainstem of the cat: anatomical and electrophysiological identification of a central autonomic pathway, *Brain Res.* 60:65-91.

Loewy, A.D. (1979) Neural regulation of the pupil, In: C.M. Brooks, K. Koizumi, and A. Sato (Eds.): Integrative Functions of the Nervous System. Amsterdam: Elsevier/North Holland. pp.131-141.

Magoun, H.W., Atlas, D., Hare, W.K. and Ranson, S.W. (1935) The afferent pathways of the pupillary light reflex in the monkey, *Brain* 59:234-249.

Mays, L.E., Porter, J.D., Gamlin, P.D.R. and Tello, C.A. (1986) Neural control of vergence eye movements: neurons encoding vergence velocity, *J. Neurophysiol.* 56:1007-1021.

Munoz, D.P. and Wurtz, R.H. (1993) Fixation cells in monkey superior colliculus. I. Characteristics of cell discharge, *J. Neurophysiol.* 70:559-575.

Mustari, M.J. and Fuchs, A.F. (1990) Discharge patterns of neurons in the pretectal nucleus of the optic tract (NOT) in the behaving primate, *J. Neurophysiol.* 64:77-90.

Nicholson, J. and Severin, C.M. (1981) Afferent projections of the Edinger-Westphal nucleus in the rat, *Acta Anat.* 199:183.

Nisida, I., Okada, H. and Nakano, O. (1960) The activity of the cilio-spinal centers and their inhibition in the pupillary light reflex, *Jap. J. Physiol.* 10:73-84.

Okada, H., Nakano, O., Okamoto, K., Hakayama, K, and Nishida, I. (1960) The central path of the light reflex via the sympathetic nerve in the cat, *Jap. J. Physiol.* 10:646-658.

Perry, V.H. and Cowey, A. (1984) Retinal ganglion cells that project to the superior colliculus and pretectum in the macaque monkey, *Neurosci.* 12:1125-1137.

Ranson, S.W. and Magoun, H.W. (1933) The central path of the pupilloconstrictor reflex in response to light, *Arch. Neurol. Psychiat.* 30:1193-1202.

Rodieck, R.W. (1979) Visual pathways, *Ann. Rev. Neurosci.* 2:193-225.

Rushton, W.A.H. (1965) (The Ferrier lecture). Visual adaptation, *Proc. Roy. Soc. B.* 162:20-46.

Scalia, F. (1970) Retinal projections of the olivary pretectal nucleus in the tree-shrew and comparison with the rat, *Brain, Behav. Evol.* 6:237-252.

Scalia, F. and Arango, V. (1979) Topographic organization of the projections of the pretectal region in the rat, *J. Comp. Neurol.* 186:271-292.

Scinto, L.F.M., Daffner, K.R., Dressler, D., Ransil, B.I., Rentz, D., Weintraub, S., Mesulam, M. and Potter, H. (1994) A potential non-invasive neurobiological test for Alzheimer's disease, *Science* 266:1051-1054.

Sillito, A.M. and Zbrozyna, A.W. (1970) The localization of pupilloconstrictor function within the midbrain of the cat, *J. Physiol.*. 211:461-477.

Sillito, A.M. and Zbrozyna, A.W. (1970) The activity characteristics of preganglionic pupilloconstrictor neurones, *J. Physiol.* 211:767-779.

Siminoff, R., Schwassmann, H.O. and Kruger, L. (1967) Unit analysis of the pretectal nuclear group in the rat, *J. Comp. Neurol.* 130:329-342.

Smith, J.D., Ichinose, C.Y., Masek, G.A., Watanabe, T. and Stark, L. (1968) Midbrain single units correlating with pupil response to light, *Science* 162:1302-1303.

Smith, J.D., Masek, G.A., Ichinose, C.Y., Watanabe, T. and Stark, L. (1970) Single neuron activity in the pupillary system, *Brain Res.* 24:219-234.

Straschill, M. and Hoffmann, K.P. (1969) Response characteristics of movement detecting neurones in the cat's pretectal region, *Exp. Neurol.* 25:165-176.

Thompson, H.S. (1987) The pupil, In: Adler's Physiology of the Eye. Clinical Application, 8th edition. (Eds. Robert A. Moses and William M. Hart). The C.V.Mosby Company, St Louis, Washington D.C., Toronto. pp 311-338.

Thompson, H.S., Montague, P., Cox, T.A. and Corbett, J.J. (1982) The relationship between visual acuity, pupillary defect and visual field loss, *Amer. J. Ophthalmol.* 93:681-688.

Trejo, L.J. and Cicerone, C.M. (1984) Cells in the pretectal olivary nucleus are in the pathway for the direct light reflex of the pupil in the rat, *Brain Res.* 300:49-62.

Trejo, L.J., Rand, M.N. and Cicerone, C.M. (1989) Consensual pupillary light reflex in the pigmented rat, *Vision Res.* 29:303-307.

Weber, J.T. and Harting, J.K. (1980) The efferent projections of the pretectal complex: an autoradiographic study and horseradish peroxidase analysis, *Brain Res.* 194:1-28.

Westheimer, G. (1964) Pupil size and visual resolution, *Vision Res.* 4: 39-45.

Woodhouse, J.M. and Campbell, F.W. (1975) The role of the pupil light reflex in aiding adaptation to the dark, *Vision Res.* 15:649-653.

Young, M.J. and Lund, R.D. (1994) The anatomical substrates subserving the pupillary light reflex in rats: Origin of the consensual pupillary response, *Neurosci.* 62:481-496.

13

# FROM SPATIOTEMPORAL TUNING TO VELOCITY ANALYSIS

M. J. Wright and K. N. Gurney

Department of Human Sciences
Brunel University
Uxbridge, Middlesex UB8 3PH, United Kingdom

## INTRODUCTION

One of the authors (MW) had the privilege of working with Hisako Ikeda between 1970 and 1975, and carried out extensive single cell recording studies in the retina, LGN and visual cortex of cats. Ikeda and Wright (1976) also conducted some of the earliest studies on animal models of amblyopia, in which they looked at the effects of surgically induced squint in kittens on the responsiveness of single cells in the LGN. This work was interpreted within a framework of parallel processing streams in the visual pathways. All these issues are very much still current concerns of the Vision Research community. One of these topics, parallel processing and spatiotemporal tuning of cortical neurons is the subject of current research at Brunel University. In this work, we try to develop models and explanations of human vision, for instance, human motion perception, which are firmly rooted in physiology, and make use of fundamental neurophysiological findings in deriving explanations of human visual performance.

## SPATIOTEMPORAL TUNING OF STRIATE CORTICAL NEURONS IN CAT

Until the pioneering studies of Hubel and Wiesel (1962; 1977) little was known about the kinds of analysis carried out by single neurons in primary visual cortex. One of their major contributions was to characterise the receptive fields of V1 cells, and show that the requirements for stimulating a cell in visual cortex were very specialised: a bar or edge with a particular orientation was required to make the cell fire, and diffuse light was generally ineffective. The cells were thus orientation selective: different cells had different preferred orientations. Some cells had separate regions in their receptive fields responding to the onset of light (or else to dark) bars or spots. These were called simple cells. Other cells responded both to onset and offset of light (or dark) bars, with no subdivisions selective for contrast polarity. These were called complex cells. The discovery that cells in V1 were very

*Basic and Clinical Perspectives in Vision Research*, Edited by Jon Robbins et al.
Plenum Press, New York, 1995

specialised in their requirements for stimulation led to the realisation that the visual image was not simply transmitted to the visual cortex, but was broken down into a special code. This was consistent with the prevailing view that recognition of an object took place by feature analysis: that a complex image could be recognised by identifying a list of simple components or features which it contained. These features would be relatively constant over a range of viewpoints and lighting conditions, thus solving the problem of generalisation.

However an important aspect of the response of V1 cells was omitted in these early studies. The visual input does not consist of a single retinal image, but is dynamic: continually changing over space and time. Accordingly, Ikeda and Wright (1975) set out to define the spatio-temporal characteristics of receptive fields of visual cortex neurons using essentially the same techniques as Hubel and Wiesel (1962; 1977), but with a more quantitative approach, using grating stimuli in addition to flashing bars.

The first stage was to plot the receptive fields in a conventional way, and confirmed the classical finding that receptive fields in V1 of the cat are orientation selective. Cells whose receptive fields are clearly separable into subregions which responded either to a localised light bar or a localised dark bar were classified as simple. Cells which responded to both types of bar at every location in the receptive field were classified as complex. The response of these cells to drifting gratings was then measured. Taking a square-wave grating, the first step was to define the optimum orientation for every cell. Then at the optimum orientation, the cell was stimulated with sinusoidal gratings of varying spatial frequency, and a spatial frequency tuning curve was plotted. Having found the optimum spatial frequency, the drift rate of the grating was then varied; this yielded a temporal frequency tuning curve. Data was obtained on the spatial response to bars (the receptive field), the temporal response to bars (the PSTH to onset and offset of both bright and dark bars at different locations in the field); the spatial frequency tuning curve at optimum orientation, and the temporal frequency tuning curve at optimum orientation and spatial frequency. Data from typical cells are shown in Figure 1.

On the basis of the tuning curves, cells fell into two broad classes; those with bandpass spatial tuning and lowpass temporal tuning, and those with lowpass spatial tuning and bandpass temporal tuning. Those with bandpass temporal tuning gave "transient" responses to the onset of a bar, and those with lowpass temporal tuning gave "sustained" responses to the onset of a bar. The tunings of simple and complex cells to spatial and temporal frequency were similar: the spatiotemporal tuning is relevant to the "sustained/transient" dichotomy, not the "simple/complex" dichotomy. The spatial and temporal tunings for the four groups of cells are shown in Figure 2 a,b. There are cells tuned to a whole range of spatial frequencies and a whole range of temporal frequencies, though the scatter of spatial frequencies is greater than that of temporal frequencies. This makes sense, because the natural world contains objects with a wide range of spatial scales.

## PARALLEL PROCESSING STREAMS AND MOTION ANALYSIS

One controversial aspect of this work concerned the classification of cell responses. At the time, there were three main dichotomous classifications; "sustained" cells versus "transient" cells (Ikeda & Wright, 1972), "X-cells" versus "Y-cells" (Enroth-Cugell & Robson, 1966) and "simple" cells versus "complex" cells (Hubel & Wiesel, 1962). The question arose whether these classifications were truly independent, or whether the different distinctions were synonymous.

Ikeda and Wright (1972, 1974) had already shown, in the retina and LGN, that most "transient" cells were Y-cells according to the criteria of Enroth-Cugell and Robson (1966); that is, they had non-linear spatial summation. They also gave transient responses to onset

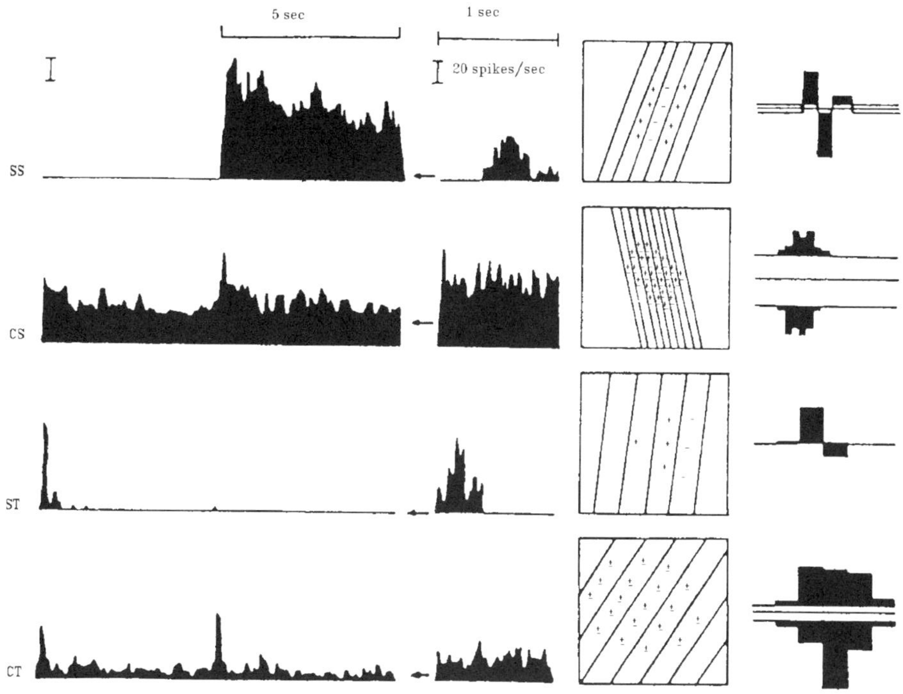


**Figure 1.** Classification of cortical neurons into four types. Column 1: Post-stimulus histograms obtained over 16 presentations from "simple-sustained" (SS), "complex-sustained" (CS), "simple-transient" (ST) and "complex-transient" (CT) cortical cells. The stimulus was an optimal bar reversing in contrast at 0.1 Hz: the overall contrast was 0.4. The horizontal arrows show the (spontaneous) firing rate to the mean luminance background. Second column: post-stimulus histograms obtained over 16 temporal cycles of an optimal grating (contrast 0.4) drifting across the receptive field at 1 Hz. Third column: receptive field plots with optimal bar. Fourth column: receptive field profiles obtained from these receptive field plots. The ordinate is the firing rate in spikes per second. Responses to a bright bar are shown above the abscissa and responses to a dark bar are shown below the abscissa. Spontaneous firing rate is indicated by the horizontal lines cutting the histogram bars. Redrawn from Ikeda and Wright (1975).

of a spot, had larger receptive fields, and they often had a "periphery effect": responses could be evoked from beyond the classical receptive field. On the other hand, "sustained" cells were X-cells according to the criteria of Enroth-Cugell and Robson (1966): that is, they had linear spatial summation. They also gave sustained responses to stationary stimuli, tended to have smaller receptive fields and no periphery effect. Ikeda and Wright (1975) argued that cortical "sustained" cells received predominantly X-cell input, and cortical "transient" cells received predominantly Y-cell input. There were complex cells with sensitivity to high spatial and low temporal frequency (i.e. they were "sustained" or X-like) and that there were on the other hand, some simple cells which had low spatial frequency tuning and bandpass temporal frequency tuning (i.e. they were Y-like). Ikeda and Wright (1975) proposed that the "sustained"/"transient" classification cut across the "simple/complex" classification. Subsequent research has tended to confirm this view, though with some modifications. Mustari, Bullier and Henry (1981) classified cortical cells according to the conduction velocity of their afferents, and this technique established four classes of cells: SX, SY, CX and CY, in line with the classification proposed by Ikeda and Wright on the basis of functional properties. It would appear that these different cell classes differ in their laminar distribution.

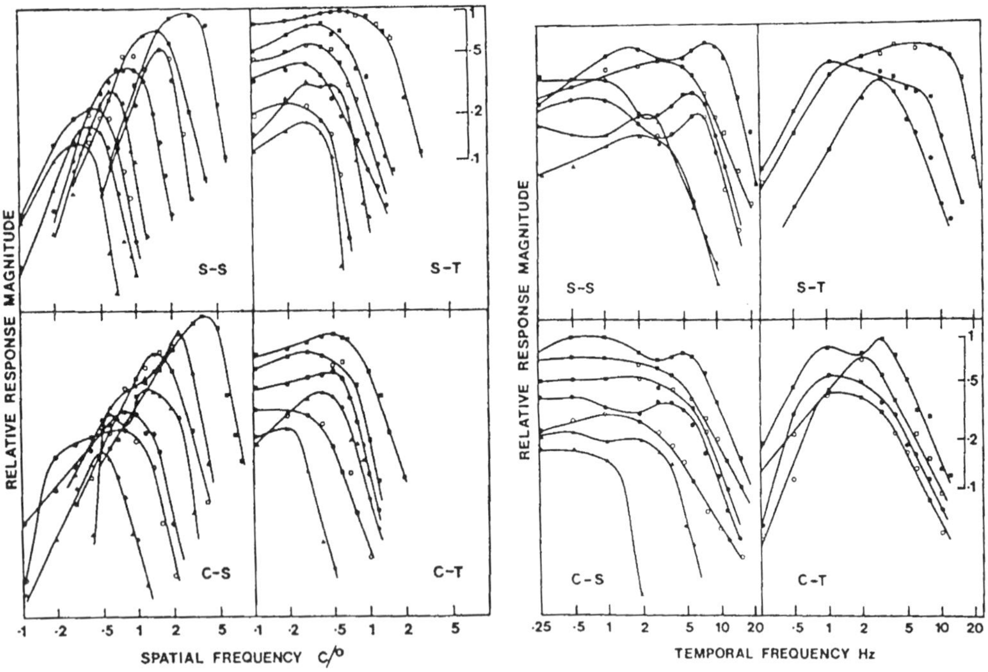


**Figure 2.** (a). Spatial frequency tuning curves of SS, ST, CS and CT cells. The maximum response of each cell is scaled to 1 and each curve is displaced downwards for clarity. Both axes are on log scales. Tuning curves were measured at the optimum orientation using gratings of contrast 0.4 drifting at 1 Hz. (b). Temporal frequency tuning curves of SS, ST, CS and CT cells, same conventions as (a). The grating was at the optimal orientation and spatial frequency for each cell, and its contrast was 0.4. The drift rate in Hz represents the number of grating cycles traversing a given point in 1 sec. Redrawn from Ikeda and Wright (1975).

The organisation of parallel processing streams in visual cortex is too extensive to review here, and there are some differences in the organisation of the visual system in cats and monkeys. However, some relevant conclusions may be drawn from recent literature surveys (Merigan & Maunsell, 1993; Zeki, 1993), of which the scheme in Figure 3 is a simplified summary.

1. There are parallel, morphologically distinct processing streams. The M and P streams in primates may be homologous with the Y and X streams in cats; or else the M pathway in primates may correspond to both Y and X pathways in cats, with the primate P pathway being an additional specialisation.
2. Despite these uncertainties about comparisons across species, the effect of lesions in the M pathway in primates are consistent with this pathway containing cells with spatial and temporal tunings like Ikeda and Wright (1975) "transient" cells, and the effects of lesions in the P pathway in primates are consistent with this pathway containing cells with tunings like Ikeda and Wright's "sustained" cells.

Merigan, Byrne and Maunsell (1991) showed that M pathway lesions due to microinjections of ibotenic acid in the lateral geniculate nucleus cause a large decrease in luminance contrast sensitivity for stimuli of higher temporal frequency and lower spatial frequency. There was no loss of temporal frequency sensitivity at high spatial frequency, nor of spatial frequency sensitivity at low temporal frequency. These results would occur if the M pathway

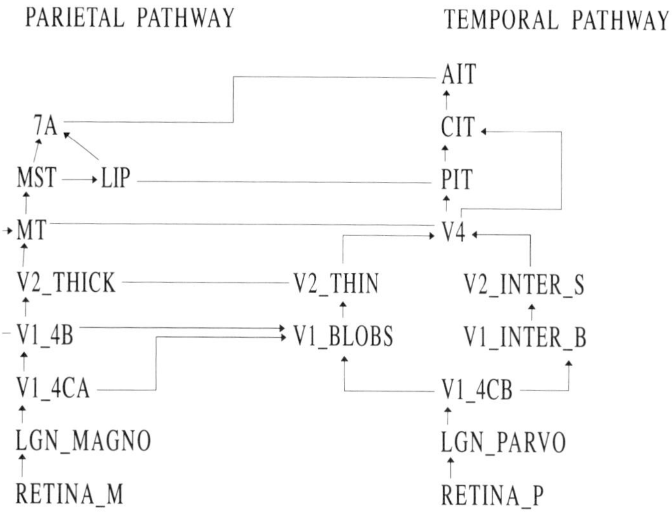


**Figure 3.** Summary diagram of parallel pathways in the visual system of macaque monkeys, (Merigan & Maunsell, 1993).

contains neurons with low pass tuning to spatial frequency and bandpass tuning to temporal frequency, like Ikeda and Wright's transient group.

Merigan and Eskin (1986) showed that P pathway lesions (using acrylamide) reduce contrast sensitivity for stimuli of high spatial and low temporal frequency. P pathway lesions cause an almost complete loss of colour vision in addition to their effects on contrast sensitivity (Merigan & Eskin, 1986). Histological studies confirmed that there was little or no damage other than in the retinogeniculate pathway. The results are expected if the P pathway contains neurons with bandpass tuning curves to spatial frequency and low pass tuning to temporal frequency, like Ikeda and Wright's (1975) sustained group. Lesion studies thus confirm that there are parallel geniculostriate pathways with different spatiotemporal tunings.

However the lesions in the LGN do not produce a global deficit in motion processing: for this to occur, the lesion must be made at a later stage in the M pathway, i.e. area V5 (MT). Lesions of V5 produce severe deficits in motion perception (Newsome and Pare 1988) but not in contrast sensitivity: if MST is included the effects are more severe and permanent. Studies of human patients confirm a specific role for V5 in motion perception. Zihl, von Cramon and Mai (1983) described a patient with a lesion including V5. This patient could see objects, but could not see their motion. She could not pour tea or coffee because "the fluid appeared to be frozen like a glacier", and she could not see the fluid rising in the cup. In a room with several people moving she felt insecure because "people were suddenly here or there, and I had not seen them moving". Crossing a road in traffic was a problem because "when I'm looking at the car it seems first far away. But then I want to cross and suddenly it's very near". She learned that she had to rely on hearing to cross the road. Specific tests of visual acuity, colour vision and object recognition showed no significant deficits, however specific tests of motion perception such as velocity and direction discrimination, anticipation of position and motion after-effects showed profound deficits. These were specific to the visual modality: there was no problem in tracking moving auditory or tactile targets.

# THE APERTURE PROBLEM AND THE COMPUTATION OF VELOCITY

It is known that V5 receives inputs from V1, both direct and indirect, and that V1 cells are tuned to spatiotemporal frequency and orientation. Without V5 subjects cannot judge the speed and direction of moving objects: apparently the information available from V1 cells is insufficient to solve this task, and it is a fair assumption that the computation of speed and direction is carried out by V5 cells on the basis of its inputs from V1.

Why do V1 cells not give an adequate velocity code? There are two problems for V1 cells. The first of these is the so-called aperture problem. Recall that the receptive fields are orientation selective. Providing an edge or bar is aligned with the preferred orientation in the receptive field, the cell will respond. But the edge or bar may be part of an object that is moving in any one of a range of possible directions. All the cell can detect is the component of motion perpendicular to the edge, not the true direction of motion of the object of which that edge is a part. To extract the true direction of motion with oriented receptive fields, at least two samples of the object are required at different orientations, and these samples need to be combined in some way.

Another problem is that V1 cells do not respond to the speed of the objects, but to its spatiotemporal frequency. Recall that

Velocity = Temporal Frequency/Spatial Frequency.

It follows from this equation that if two objects are moving at the same speed, but one is composed predominantly of high spatial frequencies, the other of low spatial frequencies, then different sub-populations of V1 cells will respond to each: there is no explicit code for velocity in V1. The problem of obtaining velocity from the output of V1 cells is non-trivial. Natural objects, unlike sinusoidal gratings, contain a range of spatial frequencies, but for an object moving coherently, these spatial frequencies will produce different temporal frequencies of stimulation. So V1 cells would not give a consistent output, in terms of velocity, for different spatial frequency components of the same moving object.

It might be argued that it is unnecessary for brain cells to encode velocity in an explicit way. Certainly if humans were unable to make consistent judgments of velocity, independent of spatial and temporal frequency, the search for a precise velocity code in the brain would be superfluous. However, Johnson and Wright (1986) showed that human observers can match speeds accurately. There are limits to this ability: stimuli in the periphery of the visual field appear to move more slowly than stimuli in the central visual field (Johnson & Wright, 1985), stimuli of low contrast appear to move more slowly (Thompson, 1982) and judgments of faster moving stimuli are influenced by spatial frequency (Smith & Edgar, 1991). But, within certain limits, humans can judge speeds accurately, independent of spatiotemporal frequency, and it is known that V1 cell responses do not show this independence.

The problem of how velocity is computed has been addressed by us using two techniques. The first is a psychophysical technique in which human observers make judgments about simple stimuli. The second is a neural network technique, in which a model of the neural mechanism is constructed within a computer program.

A sinusoidal grating is a very simple object, defined by a small number of parameters, including orientation, spatial frequency and temporal frequency. For a limited aperture sample, such as a receptive field, a grating is essentially a one-dimensional stimulus: its length, outside the sampling area, is arbitrary. However natural objects are two- or three-dimensional. The simplest possible two-dimensional object (in Fourier terms) is a plaid constructed by adding together two gratings at different orientations. When the component gratings of a plaid move, what we see, provided that the spatial frequency and contrast of the component gratings is similar, is that the combined plaid pattern moves in a single direction. We do not see the motion of the individual components, but of the plaid as a whole.

There is a simple geometric construction, called the Intersection of Perpendicular Constraints, or IOC, which predicts the direction in which observers see the plaid move (Adelson & Movshon, 1982). The perceived perpendicular motion of the grating is compatible with a number of motions of the grating as a whole: providing that those motions all have the same perpendicular component of velocity. If the perpendicular component is represented as a vector, i.e. and arrow pointing in the perpendicular direction, whose length is proportional to the perceived velocity of the component, then all the real motions of the grating which are compatible with the perceived motion are represented by those vectors whose tips lie on a straight line parallel to the grating orientation. This is known as a constraint line. If the same construction is made for the two component gratings of a moving plaid, than two constraint lines will result, and these will intersect at a single point. This is the intersection of constraints (IOC), and the vector that points to it is the unique velocity compatible with both component motions, and it predicts the physical motion of the plaid.

Adelson and Movshon (1982) proposed that this is how velocity is computed by the brain. It is a two-stage process. In the first stage, the individual motion of spatial frequency components is detected. Then these signals are combined, in a second stage, by intersecting the constraints to give the motion of the whole pattern. The model works well for predicting the perceived directions of moving plaids, and for coherently-moving patterns containing more than two spatial frequencies. For a coherent moving object or texture the constraint lines always intersect at a single point (this is physics, not physiology) so there is a unique solution. However if some of the components were moving at the wrong speeds you would get more than one intersection. Under these circumstances, incoherent motion is usually seen: some of the components appear to slide over others. The IOC construction is shown in Figure 4.

There are two unsolved problems with IOC as a model for the computation of motion in the brain. First, is it an accurate model, in the sense of predicting in a quantitative way the thresholds for seeing motion in a complex pattern? Second, how might the brain compute the intersection of constraint lines? There cannot be a little man with a ruler in the head working out where the lines intersect!

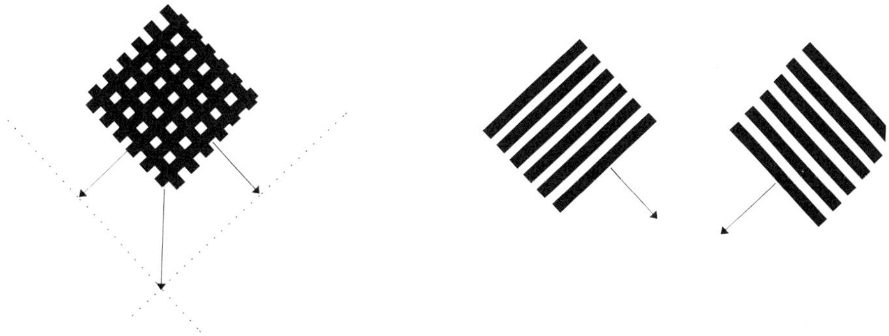

**Figure 4.** Intersection of constraints construction. Drifting gratings (left) are seen as moving in a direction perpendicular to the bars. When the gratings are combined as a plaid, the plaid appears to move as a single pattern. The dashed lines are constraint lines indicating all possible motions consistent with the perpendicular velocity of the component gratings. The point where the constraint lines for the component gratings intersect gives the direction of motion for the plaid pattern.

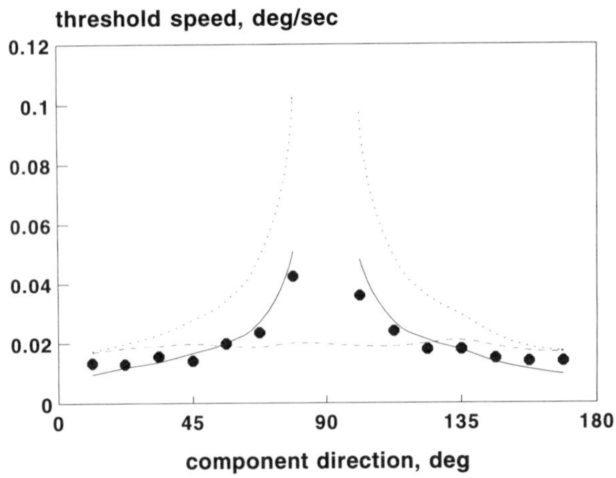


**Figure 5.** Motion discrimination thresholds for moving plaids (circles) are shown as a function of the direction of a component of a symmetrical plaid with respect to the resultant plaid direction. If the threshold depended solely on the speed of the plaid pattern, then it would follow the dotted line. If the threshold depended solely on the speed of the plaid's components, then it would follow the dashed line. A two-stage IOC model (Figure 6) with noise at both the component stage and the pattern stage (solid line) gives the correct prediction. Redrawn from Wright and Gurney (1992).

## PSYCHOPHYSICS OF MOTION ANALYSIS

To address the first question, Wright and Gurney (1992) measured the lower threshold for seeing motion in grating patterns. Moving gratings were presented in different orientations and their speed was adjusted using a two-interval, forced-choice staircase procedure, until the subject could only just discern in which of the two opposing perpendicular directions the grating was moving. This threshold was almost constant, and independent of the grating orientation. Plaids were then constructed by combining two identical gratings at different orientations. IOC predicts that if the two components are moving in a similar direction, then the vector to the intersection point will be similar in length to the perpendicular component vectors. However, keeping the component speeds similar, but make the directions very different, the constraint lines will intersect much further from the origin. The plaid will appear to move much faster, even though the components speeds are the same. This can be expressed formally as follows:

If the plaid pattern velocity has magnitude $V_p$ and direction $O_p$ then, for a component with perpendicular velocity magnitude $V_c$ and direction O

$$V_c = V_p \cos(O - O_p) \tag{1}$$

This gives us two alternative predictions. If the threshold for detecting the plaid direction depends on the speed of the plaid, then it will be proportional to $\cos(O - O_p)$ and so vary with component orientation. If the detection of the plaid motion is limited by the component stage, then the threshold will be constant irrespective of the direction of motion of the components.

The results, shown in Figure 5, are, in a sense, intermediate between the two predictions. A straightforward modification to IOC gives the correct result. Assuming that stage 1 consists of oriented, spatiotemporally tuned channels, and stage 2 consists of cells

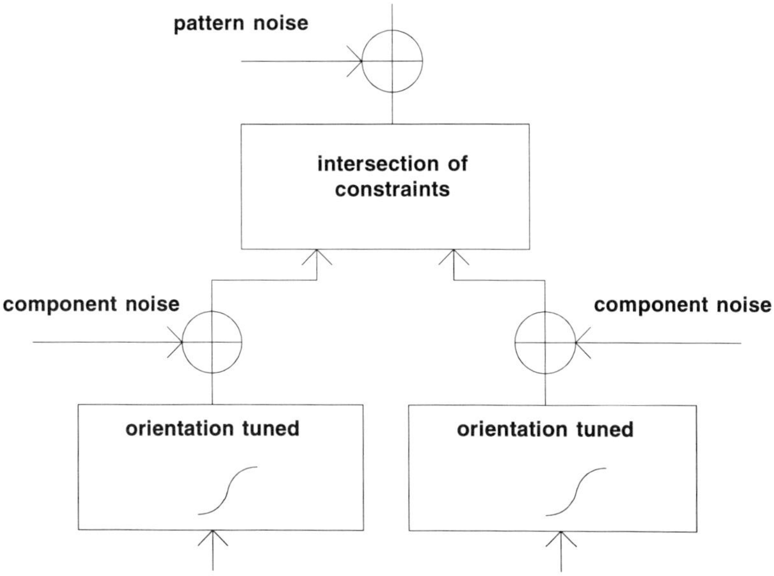


**Figure 6.** Intersection of constraints requires a minimum of two stages. The first stage contains mechanisms sensitive to orientation and spatiotemporal frequency, and across a population of cells, about the speed of motion of an oriented component of a moving pattern. We assume it is subject to intrinsic noise. Signals representing the speed of more than one oriented component (two are shown) are combined in a second stage where an intersection of constraints is implemented. The output represents the speed of the pattern as a whole and is also subject to intrinsic noise.

like those found in cortical area V5 (MT), with the V5 cells receiving their input from the V1 cells, this gives a good quantitative prediction of the result, making one very simple assumption. That assumption is that both the V1 and the V5 stages are noisy, and therefore that both contribute to setting the threshold. The two-stage IOC model of Wright and Gurney (1992) is shown in Figure 6.

## A SELF-ORGANISING NEURAL NET MODEL OF VELOCITY ENCODING

How might the brain be wired up to compute IOC? Of course, we could try out a number of possible neural wiring diagrams to see if they give the right result, but this implies that the wiring diagram must have some unique specification. Gurney and Wright (1992) chose to assume, instead, that the connections in the brain develop through experience. Perhaps, simple exposure to moving stimuli allows the developing brain to make the right connections. The object was to show how, given a very general but physiologically plausible filtering model at the first stage, an artificial neural network based model of the second stage may acquire the necessary velocity mapping by self-organising. The model then has biological plausibility in that the connectivity develops simply as a result of "experience": exposure to inputs derived from rigid translation of textures.

The input cells of the network were idealised versions of V1 cells like those recorded by Ikeda and Wright (1975); that is, they were tuned to spatial frequency, temporal frequency and orientation. These input cells were fully connected to a second layer of cells, representing

V5. Initially, the (synaptic) strengths of all the connections were set at random. The object was to determine whether, through exposure to moving stimuli and the operation of a simple learning rule, the connection strengths could be adjusted so that the second layer combined the inputs from the first layer with the right strengths to compute the intersection of constraints.

Training the network consists of presenting the input layer with simulated moving textures. A texture contains many spatial frequencies, and so will stimulate many spatial frequency channels. Also, since the whole texture is moving at a constant velocity, the different spatial frequencies will be stimulated at different temporal frequencies. The perpendicular velocity of a grating $V_c$ is related to its spatial and temporal frequencies, $f_s$ and $f_t$, according to

$$V_c = f_t/f_s \tag{2}$$

Also, the components will be distributed across different orientations. From equation (1), for a component with perpendicular velocity magnitude $V_c$ and direction O,

$$V_c = V_p\cos(O\text{-}O_p),$$

where $V_p$ and $O_p$ are the velocity and direction of the pattern as a whole. Therefore, combining (1) and (2) we have:

$$f_t = f_s V_p \cos(O_p\text{-}O) \tag{3}$$

where

$$O_p\text{-}n/2 < O < O_p\text{+}n/2$$

and $f_t = 0$ outside this range.

The resulting Fourier Surface is shown in figure 7. The spectrum of the image of any moving object will consist of points sampled from this surface.

It is now possible to compute, for a set of moving objects or textures, the activation of cells in the input layer. Each input cell is defined by a characteristic tuning to spatial frequency, temporal frequency and orientation, and since these are the dimensions of the Fourier Surface, each cell's response to the moving object can be readily computed. In the input layer of the network, four spatial tunings (centred on 1, 2, 4, 8 cycles/degree) and either two (1,6 Hz) or four (1, 2, 4, 8 Hz) temporal tunings were used, at each of 16 equally-spaced orientations. The shapes of the tuning curves and activation functions were made compatible with neurophysiological data.

When the network is stimulated by a texture pattern moving with a given speed and direction, activation of the output layer by the input cells will initially be random, but there will always be one output cell which is the most active. The maximally responsive output cell, together with surrounding cells in an excitatory neighbourhood, are then presumed to learn. They do this by incrementing the strength of the synaptic connections from those input cells which are active and decreasing the synaptic strength from those which are inactive. The details of the learning algorithm are given by Kohonen (1984). With repeated presentation of a set of different patterns, each output cell learns to respond to a different cluster of input patterns. As training proceeds, the size of the excitatory neighbourhood in the output layer is gradually decreased. This technique is known to result in the setting up of topographic feature maps like those found in visual cortex using microelectrode techniques (Kohonen, 1982).

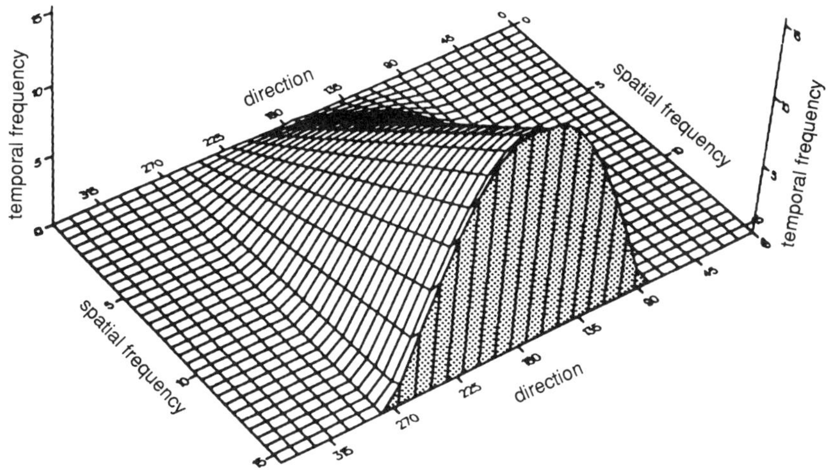


**Figure 7.** Fourier image surface representing the set of all possible spatial components of a rigid object moving in the fronto-parallel plane. From Gurney and Wright (1992).

To recapitulate: in a texture moving at a given velocity, certain spatial frequencies, temporal frequencies and orientations always occur together. The network learns these associations, and in doing so, simplifies the representation of the input from a large number of channels encoding three dimensions (spatial frequency, temporal frequency and orientation) to a smaller number of channels encoding two dimensions (direction and speed).

The results (Figure 8) show that the representation developed in the output layer is topographic: neighbouring cells have similar preferred directions and speeds. The topographic maps exhibit characteristic "swirls" and "fractures" found both in monkey V1 (Blasdel & Salama, 1986) and cat V2 (Bonhoffer & Grinvald, 1991).

The individual nodes in the trained network are tuned to direction and speed, irrespective of spatial and temporal frequency (Figure 9). In this respect they resemble "pattern cells" which have been recorded from V5 by Movshon, et al (1985).

The network as a whole solves the aperture problem. The method is shown schematically in Figure 10. Shown on Cartesian coordinates using direction and speed as axes, the calculated response of the network to a moving grating forms a U-shaped distribution of activity. For two different moving gratings, two U-shaped distributions are formed, and these intersect at a certain point. The point of overlap contains the strongest-firing cells in the output layer, and corresponds to the Intersection of Constraints. We can check that the actual network simulation solves the aperture problem by comparing its responses to two completely different patterns moving at the same speed. The network should give the same response irrespective of the nature of the pattern. The example shown in Figure 11 gives the response to a texture (containing many oriented spatial frequency components) with that to a plaid (containing only two spatial frequency components). The distribution of activity in the net's output layer is almost identical, so the net has encoded true velocity.

To conclude: in mammalian brains, the required connectivity for solving the aperture problem could develop between V1 and V5 simply as a result of exposing the eyes to moving patterns. This assumes that neurons can modify the strength of synaptic connections on the basis of the joint activation of the pre- and post-synaptic cell (Kohonen, 1984). A "systems" version of the neural net model also predicts psychophysical thresholds for direction

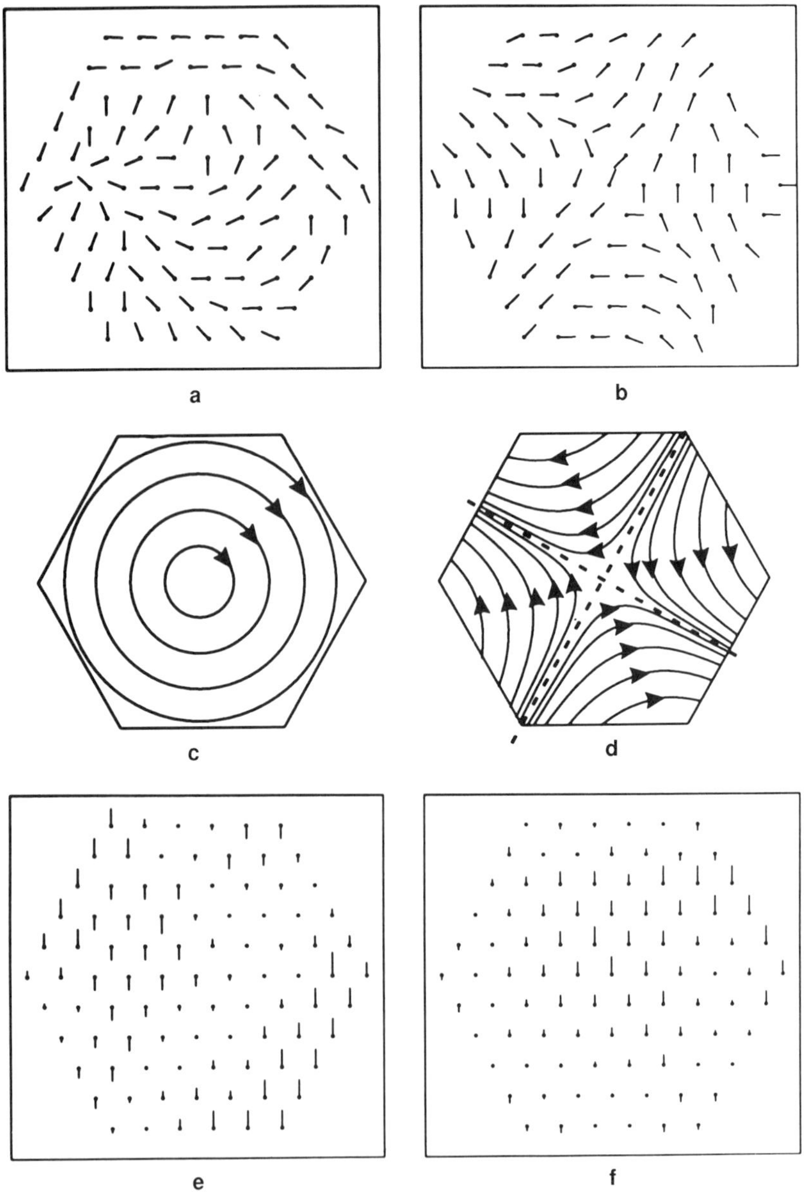


**Figure 8.** Topographic maps of directional tuning in trained networks with four (left column) or two (right column) temporal input channels. Top row: topographic maps of directional tuning. Middle row: idealised versions of the geometry of the above maps. Bottom row: topographic maps of speed tuning (height of bar above (+) or below (-) dot is proportional to log speed). From Gurney and Wright (1992).

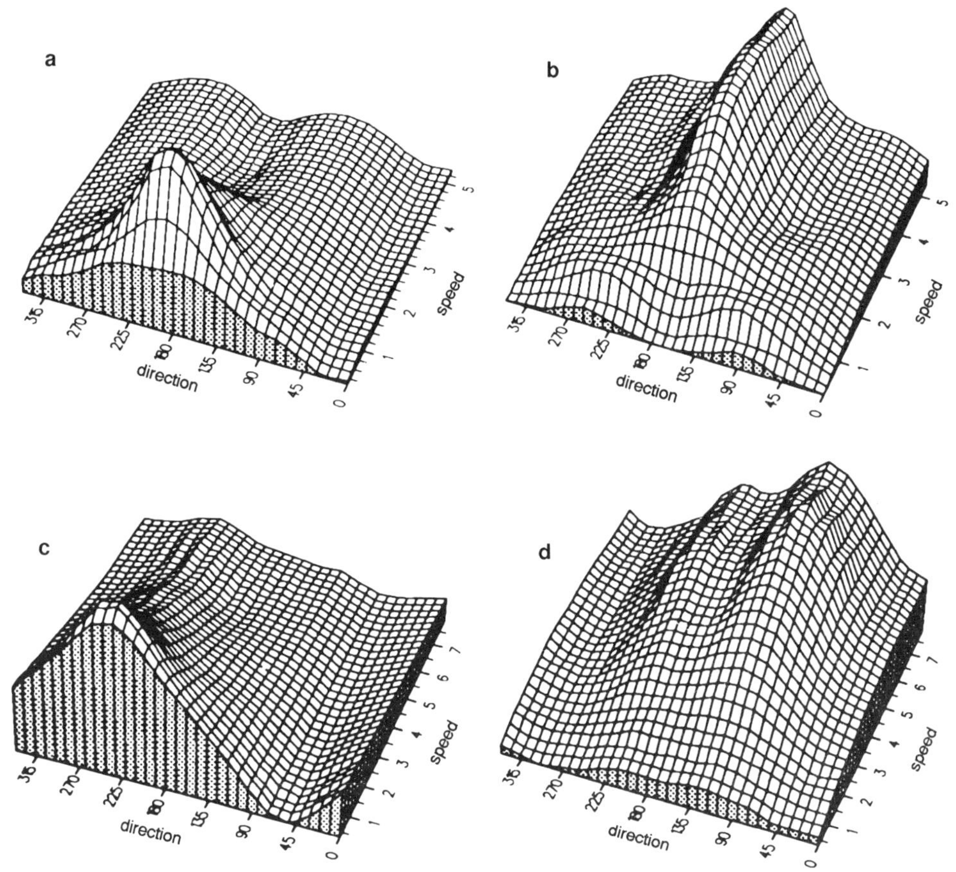


**Figure 9.** Response profiles for trained output nodes in the network. The X and Y axes represent speed and direction and the Z axis represents strength of response. From Gurney and Wright (1992).

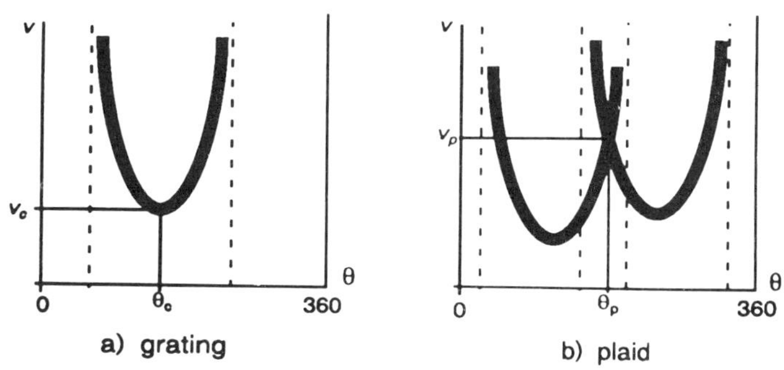


**Figure 10.** The response of a population of velocity-tuned neurons, predicted from IOC, is shown schematically on Cartesian coordinates: ordinate = preferred speed, abscissa = preferred direction. The U-shaped curve (left) shows the response to a grating and the right hand graph shows two U-shaped responses to the components of a plaid which intersect at the speed and direction of the pattern as a whole. From Gurney and Wright (1992).

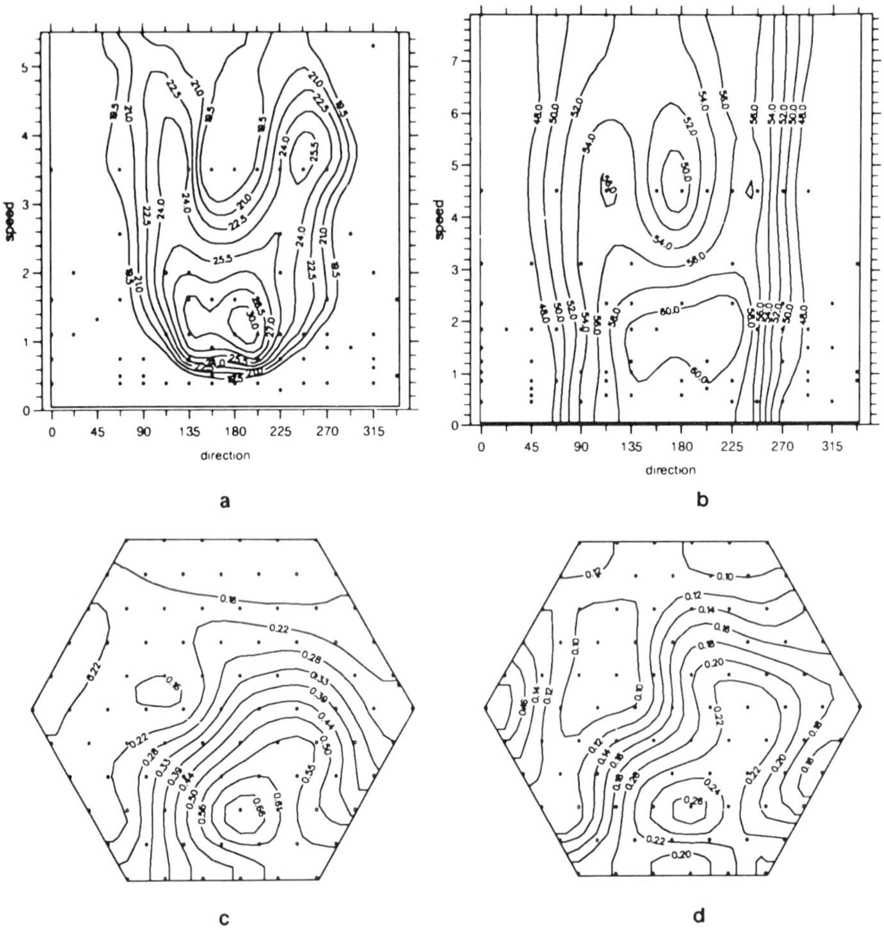


**Figure 11.** Top row: contour maps of activations of trained nodes responding to a single grating component, replotted on Cartesian coordinates. Note the approximately U-shaped distribution. Bottom row, left: contour map of network activation to a texture. Bottom row, right: contour map of network activation to a plaid. The texture and plaid were moving at the same velocity. From Gurney and Wright (1992).

discrimination in two-dimensional patterns. The validity of the approach depends on the extent to which the model accounts for neurophysiological and psychophysical data, and is thus crucially dependent on the data of quantitative neurophysiology, in which Hisako Ikeda was a pioneering worker.

## REFERENCES

Adelson, E.H. and Movshon, J.A. (1982). Phenomenal coherence of moving visual patterns, *Nature* 300:523-525.

Blasdel, G.G. and Salama, G. (1986) Voltage-sensitive dyes reveal a modular organisation in monkey striate cortex, *Nature* 321:579-585.

Bonhoffer, T. and Grinvald, A. (1991) Iso-orientation domains in cat visual cortex are arranged in pinwheel-like patterns, *Nature* 353:429-431.

Enroth-Cugell, C. and Robson, J.G. (1966) The contrast sensitivity of retinal ganglion cells in the cat, *J. Physiol.* 187:517-552.

Gurney, K.N. and Wright, M.J. (1992). A self-organising neural network model of image velocity encoding, *Biological Cybernetics* 68:173-181.

Hubel, D. H. and Wiesel, T.N. (1962) Receptive fields, functional architecture and binocular interaction in the cat's visual cortex, *J. Physiol.* 160:106-154.

Hubel, D.H. and Wiesel, T.N. (1977). The Ferrier Lecture: Functional architecture of macaque monkey visual cortex, *Proc. Roy. Soc. Lond. B.* 198:1-59.

Ikeda, H. and Wright, M.J. (1972) Receptive field organisation of "sustained" and "transient" retinal ganglion cells which subserve different functional roles, *J. Physiol* 227:769-800.

Ikeda, H. and Wright, M.J. (1975) Spatial and temporal properties of "sustained" and "transient" neurons in area 17 of the cat's visual cortex, *Exp. Brain Res.* 22:362-382

Ikeda, H. and Wright, M.J. (1975) Retinotopic distribution, visual latency and orientation tuning of "sustained" and "transient" neurons in area 17 of the cat, *Exp. Brain Res.* 22:385-398.

Ikeda, H. and Wright, M.J. (1976) Properties of LGN cells in kittens reared with convergent squint: a neurophysiological demonstration of amblyopia, *Exp. Brain Res.* 25:63-77.

Johnston, A and Wright, M.J. (1986) Matching velocity in central and peripheral vision, *Vision Res.* 26:1099-1109

Kohonen, T. (1982) Self-organised formation of topologically correct feature maps., *Biological Cybernetics* 43:59-69.

Kohonen, T. (1984) Self-organisation and associative memory. Springer, Berlin, Heidelberg, New York.

Merigan, W.H. and Eskin, T.A. (1986) Spatio-temporal vision of macaques with severe loss of Pb retinal ganglion cells, *Vision Res.*26:1751-1761.

Merigan, W.H. and Maunsell, J.H.R. (1993). How parallel are the primate visual pathways?, *Ann Rev. Neurosci.*, 16, 369-402.

Merigan, W.H. Byrne, C.E. and Maunsell, J.H.R. (1991) Does primate motion perception depend on the magnocellular pathway?, *Neurosci.*11:3422-3429.

Movshon, J.A., Adelson, E.H., Gizzi, M.S. and Newsome, W.T. (1985). The analysis of moving visual patterns. In Chagas, C. and Gross, C.G. (Eds.) Pattern recognition mechanisms (pp 117-151) Vatican Press.

Mustari, M.J., Bullier, J. and Henry, G.H. (1981). Comparison of the response properties of three types of monosynaptic S cell in cat striate cortex, *J. Neurophysiol.* 47:15-741.

Newsome, W.T. and Pare E.B. (1988) A selective impairment of motion perception following lesions of the middle temporal area (MT), *J. Neurosci.* 8:2201-2211.

Henry, G.H. (1986) Streaming in the striate cortex. In Visual Neuroscience, eds. Pettigrew, J.D., Sanderson, K.J. and Levick, W.R. pp 260-279. Cambridge, CUP.

Smith, A.T. and Edgar, G.K. (1991). The separability of temporal frequency and velocity, *Vision Res.* 31, 321-326.

Thompson, P (1982). The perceived rate of movement depends on contrast, *Vision Res.* 22:377-380.

Wright, M.J. and Gurney, K.N. (1992) Lower threshold of motion for one and two dimensional patterns in central and peripheral vision, *Vision Res.* 32:121-134.

Zihl, J. Cramon, D. von and Mai, N. (1983) Selective disturbance of movement vision after bilateral brain damage, *Brain* 106:313-340.

Zeki, S. (1993) A vision of the brain. Blackwell Scientific Publications, Oxford.

# 14

# BINOCULAR INTEGRATION IN THE VISUAL CORTEX

Peter Hammond

Department of Communication and Neuroscience,
Keele University
Keele, Staffordshire ST5 5BG, United Kingdom

## INTRODUCTION

Study of the monocular and binocular properties of neurones in the lightly anaesthetized feline striate cortex reveals a wealth of functional interactions which potentially underpin eye convergence and interocular alignment, and the encoding of visual perspective of three-dimensional objects. Monocularly, neurones' relative preferences for opposing directions of motion, and selectivity for range of directions, are reliant upon the spatial-frequency content of perceived scenes. Each neurone's directionality, spatial-frequency selectivity, eye preference and length selectivity, differs according to whether input is from contralateral or ipsilateral eyes, or binocular.

The motivation for our fundamental research on the visual system of the cat, over the past twenty-five years, has been to gain more detailed insight into normal neuronal function, in order to establish what goes wrong, and at what location, during abnormal function. In the longer term, such understanding can facilitate clinical treatment and the diagnosis of visual impairment.

Traditionally, cortical and subcortical visual neurones have been characterized according to their responsiveness to optimized, simple stimuli (e.g. Hubel & Wiesel, 1962) - a vogue which, inevitably, led to notions of neurones as feature detectors, involved in elementary pattern recognition processes. Our strategy, developed and refined over many years, differs radically from this approach. Firstly, we have exploited repertoires of unconventional, complex (visually textured) stimuli, which highlight the importance of context in the visual environment. Secondly, we have systematically studied neuronal sensitivity to a range of sub-optimal as well as to optimal stimulus configurations.

Through the 1970's, the first approach - the application of visual textures - both challenged and augmented classical understanding of mammalian visual function at that time (Hammond & MacKay, 1975, 1977). It revealed a wealth of neural attributes and intracortical connectivity patterns which could not have been anticipated from conventional studies. The second approach - the systematic application of complex patterns, which revealed profound interactions between visual foregrounds and backgrounds, unexpected refinements in sen-

*Basic and Clinical Perspectives in Vision Research*, Edited by Jon Robbins et al.
Plenum Press, New York, 1995

sitivity to relative motion, and a range of neuronal correlates of adaptation and of motion after-effects - led on, by natural progression, to our current studies of binocular integration in visual cortex.

Pertinent to our work, numerous previous studies establish that the simple and complex neurones of the striate cortex are spatial-frequency selective (e.g. Campbell, Cooper & Enroth-Cugell, 1969; Maffei & Fiorentini, 1973; Ikeda & Wright, 1975; Movshon, Thompson & Tolhurst, 1978; Kulikowski & Bishop, 1981; Tolhurst & Thompson, 1981): neurones respond optimally to gratings of one spatial frequency, less well to coarser or finer gratings. Both simple and complex neurones span the entire spatial-frequency range, with individual members of either group tuned to each of the spatial-frequency channels in the visual system. In general, neurones selective for high spatial frequencies have sharp orientation selectivity, narrow spatial-frequency bandwidth (Tolhurst & Thompson, 1981), and low preferred velocities (Baker, 1990; Hammond & Fothergill, 1994). According to Maffei & Fiorentini (1973), Ikeda & Wright (1975) and ourselves (Hammond & Fothergill, 1994), simple cells are substantially more sharply spatial-frequency tuned than complex cells; whereas Movshon, Thompson & Tolhurst (1978), and Tolhurst & Thompson (1981), claim that these differences are slight.Despite the fact that, in a functioning cortex responsive to real-life sequences of visual images, the substantial majority of neurones must inevitably signal below strength simply because the configurations viewed are not optimum for them, virtually all previous studies have concentrated upon optimized inputs to cortical neurones - best spatial frequency, preferred orientation, direction, velocity or temporal drift frequency of motion - via the dominant eye. Our initial studies systematically investigated responses to non-optimal as well as to optimal stimulus configurations (Hammond, 1979; Hammond & Munden, 1990), direction and velocity (Hammond & Reck, 1980), length (Hammond & Mouat, 1986), spatial-frequency, bandwidth and directionality (Hammond & Pomfrett, 1990b, 1991), for the input from the dominant eye.

On the premise that most cortical neurones are binocular, we have since extended this approach, comparing the monocular inputs to cortex from either eye. Numerous previous studies have demonstrated cortical sensitivity to binocular spatial, orientational and directional disparities, for signalling depth and direction in visual space (Barlow, Blakemore & Pettigrew, 1967; Pettigrew, Nikara & Bishop, 1968; Blakemore, 1970; Bishop, Henry & Smith, 1971; Nelson, Kato & Bishop, 1977; von der Heydt, Adorjani, Hänny & Baumgartner, 1978; Fischer & Krüger, 1979; Maske, Yamane & Bishop, 1984; LeVay & Voigt, 1988); the precision of matching, or otherwise, of inputs between the eyes (Skottun & Freeman, 1984; Ohzawa & Freeman, 1986a, 1986b; Hammond, 1991b); and cortical sensitivity to tilt or to motion in depth (Cynader & Regan, 1982).

We have therefore explored an additional possibility, not previously elaborated explicitly, that neurones might also code for depth, direction, tilt or dynamic rotation in space, on the basis of interocular differences in directional preference, spatial frequency or bandwidth, receptive field dimensions, and length summation or length selectivity.

## METHODS

Methodology is described extensively elsewhere. Only a brief resumé is now given.

*Implantation:* Under surgical anaesthesia - 72.5%:27.5% $N_2O:O_2$, plus 2-5% halothane - adult cats were implanted with a stainless steel peg for atraumatic, stereotaxic restraint, and a recording chamber positioned on the midline, over the striate cortex of either hemisphere. Indwelling leads for monitoring the surface-cortical EEG were incorporated into the wall of the chamber.

*Recording:* After brief periods of surgical anaesthesia, subsequent short recording sessions were performed under light anaesthesia - 0.2-0.5% halothane in 72.5%:27.5% $N_2O:O_2$ - at intervals of 3-4 days. Animals were artificially ventilated via a cuffed endotracheal tube. End-tidal $CO_2$ was maintained between 3.8-4% by adjustment of stroke volume; body temperature at 38.5°C with an electric blanket, controlled by a rectal thermistor probe. Eye position was stabilized by the muscle relaxant, gallamine triethiodide, given intravenously as an initial 40 mg loading dose, with continuous infusion at 7-10mg.kg$^{-1}$.hr$^{-1}$.

Anaesthetic depth was titrated according to the EEG, ECG, heart rate, rectal temperature, neuronal firing rates, and end-tidal $CO_2$. Aseptic precautions were observed throughout. Intramuscular procaine penicillin was given prophylactically before and after each recording session.

Pupils were dilated with 1% atropine sulphate, and the eyelids and nictitating membranes retracted with 10% phenylephrine hydrochloride. The corneas were protected with unpowered contact lenses; 5mm diameter artificial pupils plus corrective lenses were applied for a viewing distance of 57cm.

A small craniotomy was performed at the base of the recording chamber. Under visual guidance, the dura was punctured with a paediatric syringe needle, mounted directly on the micromanipulator. 4M-NaCl micropipette electrodes, inserted through the puncture, were used for extracellular recording. Vertical penetrations were made on successive occasions from the central 10° representation of the lower contralateral visual field in Area 17. Recording stability was enhanced by sealing the craniotomy with 2% sterile agar in 0.9% NaCl.

Visual stimuli were presented on a matched pair of large-screen, X-Y displays with white (P4) phosphor, and were generated electronically by modified "Picasso" Image Synthesizers (Innisfree, Rev. 6). Average luminance was 1.1 log cd.m$^{-2}$; contrast

$$[L(max)-L(min)]/[L(max)+L(min)]$$

was 0.4.

Stimulus configurations - light or dark bars or edges, sine-wave grating patterns, or fields of random visual noise - data logging and preliminary analyses were under computer control. The displays could be viewed monocularly, binocularly or dichoptically; and the eye not under test was either occluded or viewed a uniform field.

## Neuronal Characterization

*Classification.* Receptive fields were mapped as minimum response fields, using sine-wave grating patches of optimal orientation, spatial frequency, direction and speed of motion (Hammond & Ahmed, 1985). Cells were characterized as simple or complex by discreteness of, and spatial summation within, receptive field zones (Hubel & Wiesel, 1962); discreteness (simple) or coextensiveness (complex) of light- and dark-edge discharge centres for motion (Bishop, Coombs & Henry, 1971; Henry, 1977); and lack (simple) or presence (complex) of sensitivity to motion of visual noise (Hammond, 1991a).

*Location.* All neurones sampled were unequivocally from Area 17. Use of recovery techniques precluded reconstruction of electrode penetrations, except following the terminal recording session on each animal. Lamina-of-origin of each neurone was inferred from repeatedly encountered sequences of neurones, especially in relation to the characteristic, monocular simple-cell activity recorded in layer IV.

*Length Summation.* Complex cells were divided into special, intermediate and standard types by length summation, which also defined end-stopping. Special complex cells responded optimally to stimuli much shorter than the receptive field; standard complex neurones responded best to stimuli long in comparison with the receptive field; intermediate complex cells had properties between the two (Gilbert, 1977; Hammond & Ahmed, 1985). Length summation was quantified for sine-wave grating patches of optimal spatial frequency, direction and velocity. Patch length was varied symmetrically about the centre of the minimum response field, between 0.25-10° in pseudo-random sequence; 4 trials were averaged for each length. For binocular neurones, length summation and end-stopping were compared for either eye.

*Directional Tuning.* Orientational/directional tuning was quantified with sine-wave gratings of optimal spatial frequency and velocity, drifting (1sec in either direction) across the receptive field (Hammond & Pomfrett, 1989, 1990a, 1991). Direction was varied in 10° steps, clockwise from upward motion (0°); 4 responses were averaged for each direction. Preferred orientation/direction (orientation orthogonal to direction) was derived from linear regression fits to the flanks of each peak in tuning curves. Resting discharge was determined for uniform grey fields of the same average luminance as gratings. Directional tuning width was derived from the intercept of regression fits with resting discharge.

*Directionality.* Neurones were classified as direction-selective or direction-biased (Hammond & Pomfrett, 1989, 1990a), according to whether the ratio of responses in preferred:opposite directions was respectively greater or less than 20:1.

*Ocular Dominance.* Eye preference was initially determined according to the 7-group scale of Hubel & Wiesel (1962): monocular groups 1 and 7, contralateral and ipsilateral, respectively; groups 2 and 3 strongly and weakly dominated, respectively, by the contralateral eye; groups 6 and 5 similarly dominated by the ipsilateral eye; group 4 driven equally by either. For all binocular neurones, ocular dominance was later quantified by ratio of eye drive at each spatial frequency:

| | |
|---|---|
| weaker/stronger response | for contralateral dominance |
| 2 minus weaker/stronger response | for ipsilateral dominance |

after subtraction of resting discharge (Hammond & Fothergill, 1990).

*Binocular Positional Specificity.* Under dichoptic viewing, each eye viewed an identical sinewave grating, of optimal orientation and direction of motion, and of optimal spatial frequency and speed for the dominant eye. The two gratings were moved in tandem, 1sec in each direction; blocks of 20 trials were averaged. Between blocks, the positional (phase) relationship between the gratings was altered in randomized steps of 1/8 of one grating period. Binocular responses at each spatial phase were derived and compared with responses: to either eye alone; to the algebraic sum of the two monocular responses; and to level of resting discharge.

*Spatial Frequency Selectivity.* Two methods of assessing spatial-frequency tuning were used, for between five and nine spatial frequencies (range 0.25-3.0c/°). In the first, spatial-frequency tuning was derived from the envelope of best responses, across directional/orientational tuning curves obtained for different spatial frequencies, tested in pseudo-random sequence. For binocular neurones, spatial-frequency tuning was derived similarly for either eye.Because preferred orientation is an invariant characteristic of cortical neurones (Hammond, Andrews & James, 1975), unchanging with spatial frequency (Hammond & Pomfrett, 1990b), in the second method preferred direction was first determined from the directional tuning curve at optimal spatial frequency, as above. Then, at that preferred direction and velocity, 20 responses were averaged for each eye in turn and for each spatial frequency, tested in randomized sequence (Hammond & Fothergill, 1990; Hammond & Pomfrett, 1990b).

In either method, velocity was optimal for each neurone; temporal drift frequency was allowed to vary. Grating movement (1sec in each direction, at each orientation) was followed by a 1sec pause. The eye not under test was occluded, or viewed a uniform grey field. Resting discharge was quantified for uniform fields of the same luminance as gratings: average of 16 1sec-periods in the first method; average of two sets of 20 trials, made respectively before and after tests at each spatial frequency, in the second method.

*Repeatability.* In a subset of binocular neurones, entire spatial frequency redeterminations were made later in time, as checks for repeatability.

*Spatial-Frequency Assessment at Constant Velocity.* This strategy was adopted because striate neurones are velocity-selective and velocity low-pass (Movshon, 1975; Orban, Kennedy & Maes, 1981), and their spatial-frequency selectivity is inversely related to preferred velocity (Baker, 1990; Hammond & Fothergill, 1990). In a subset of neurones, for validation, spatial-frequency tuning for each eye was compared at fixed, optimal velocity (allowing temporal frequency to vary), and at constant temporal frequency (allowing velocity to vary) - (Fothergill & Hammond, 1992).

Spatial frequency tuning was closely similar by either method, in all but a few instances where the constant temporal frequency curve was slightly skewed with respect to that at constant velocity, with more pronounced low-frequency attenuation and a resultant shift of optimal spatial frequency to marginally higher values. These trends were slight, and are to be anticipated in velocity low-pass neurones. The envelopes of spatial-frequency tuning at constant velocity and at constant temporal frequency, derived from this subset of neurones overall, were also closely comparable. Thus, on average, spatial-frequency tuning was comparable by either method.

The preliminary observations of Bisti, Carmignoto, Galli & Maffei (1985) had similarly suggested that spatial-frequency tuning of striate neurones is little affected by stimulus velocity. In 13 binocular neurones, spatial-frequency tuning was therefore compared at three different velocities (2, 4 and 8°/sec) straddling the optimal velocity. The curves at each velocity were essentially shape-invariant, except for a tendency for slight narrowing at high velocity, on the high spatial-frequency side of the peak.

*Physiological Recovery.* At the close of each recording session, muscle relaxant was withheld. The specific antidote, neostigmine methylsulphate (0.5ml, at $0.5 mg.ml^{-1}$), preceded by atropine sulphate (0.25ml, at $100 \mu g.ml^{-1}$), was given intravenously. The craniotomy was treated prophylactically with Cicatrin antibiotic powder (Wellcome) in 0.9% saline, and capped. Contact lenses were removed and 10% sulphacetamide eyedrops (Minims: Smith & Nephew) were applied prophylactically, followed by wetting solution to prevent corneal dehydration. Animals were weaned back onto unassisted respiration before discontinuing monitoring.

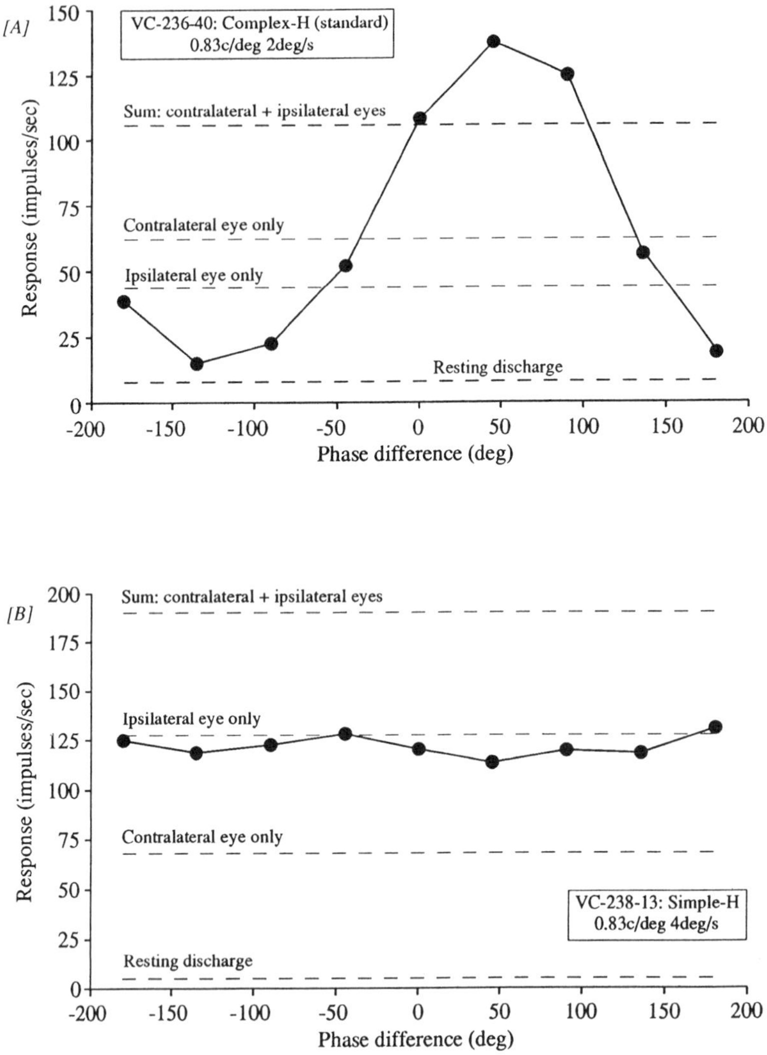


**Figure 1.** Binocular positional specificity: examples of specificity of binocular cortical neurones for interocular positional match/mismatch, in response to dichoptic stimulation by identical sinewave gratings of optimal spatial frequency, velocity and drift direction, moved in tandem, but with variable phase relationships. Responses are illustrated only for the preferred direction of motion. [A] Phase sensitive; [B] phase insensitive.

# RESULTS

## Binocular Positional Specificity

Examples of the binocular positional specificity tests are shown in Figure 1. Some binocular neurones (Figure 1A) showed strong dependence of binocular response on interocular alignment. Their response ranged from binocular facilitation for alignment (binocular response greater than the sum of monocular responses), to binocular occlusion for misalignment (binocular response less than the response to one or both eyes separately). In other

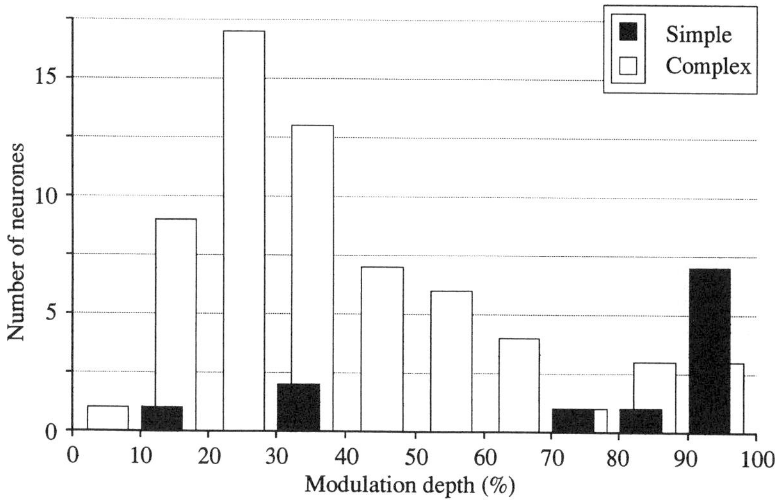


**Figure 2.** Modulation depth vs. neuronal class: comparison of specificity for interocular positional match/mismatch in simple and complex binocular cortical neurones. Percentage modulation depth was derived for dichoptic viewing, from the relative response strengths for sinewave grating stimuli showing worst- and best-case positional match.

neurones (Figure 1B), there was little variation of binocular responses with change in interocular alignment.

In the latter example, the binocular response was no greater than that to the right eye alone, even though a good response could be elicited from either eye. In other phase-insensitive neurones, the binocular response exceeded monocular responses, even though there was little or no dependence on interocular alignment.

Every class of neurone - simple cells; standard, intermediate and special complex cells - was represented in the position-specific and position non-specific categories. Binocular dependency on interocular alignment was quantified by the ratio of best- to worst-phase binocular responses for each neurone (100% modulation corresponds to total binocular occlusion). The distributions for simple and complex neurones (Figure 2) were quite different. Most complex neurones showed relatively low modulation depths; most simple cells showed the anticipated high modulation depths. Special complex neurones, with positionally non-specific monocular responses to short contours, mostly also responded well binocularly, regardless of alignment. However, a minority of complex neurones (all standard, length-summating neurones) showed strong dependency on binocular phase; and a minority of simple neurones were relatively insensitive to misalignment, despite the position-specific segregation into discrete "on" and "off" zones which characterizes their monocular receptive fields (Hammond, 1991b).

## Directionality vs. Spatial Frequency

Directional tuning was compared over a broad spectrum of gratings of different spatial frequency (Hammond & Pomfrett, 1990b, 1991). In two classes of neurones - direction-selective and bidirectional - spatial frequency affected only the magnitude of the response, and was without effect on directionality. In two other classes (Figure 3), spatial frequency profoundly altered the bias for opposing directions. The majority class (Figure 3A) comprised direction-biased neurones, whose bias was labile: the example illustrated was direction-biased at optimal spatial frequency, increasingly direction-se-

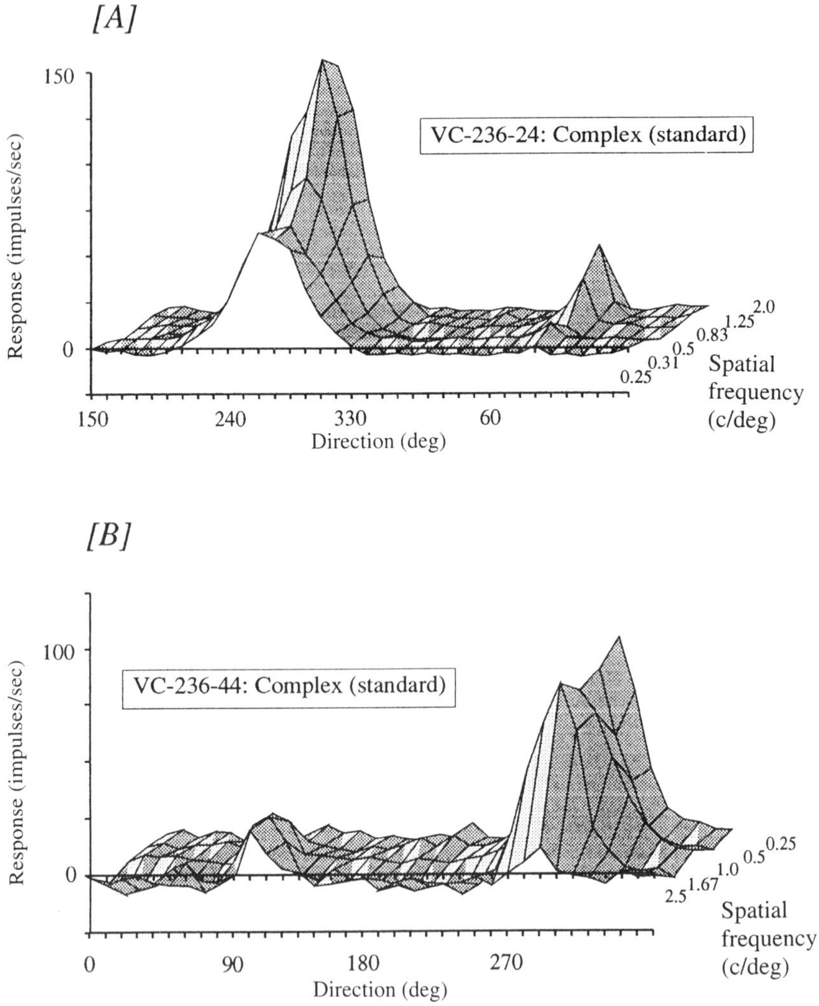


**Figure 3.** Directionality vs. spatial frequency for the dominant eye input to two binocular cortical neurones. [A] direction-biased neurone at optimal spatial frequency, direction-selective at extreme spatial frequencies. [B] direction-biased neurone tuned to different bands of spatial frequencies for opposing directions of motion, with resultant reversal in directional preference at high spatial frequency.

lective at high and at low spatial frequencies. In a rarer class of neurones (Figure 3B), directionality reversed with spatial frequency: the example illustrated was tuned to lower spatial frequencies in the preferred than in the non-preferred direction. Thus, at high spatial frequency the response in the previously non-preferred direction exceeded that in the previously preferred direction.

We therefore made systematic interocular comparisons of spatial-frequency selectivity and directionality in binocular cortical neurones (Hammond & Pomfrett, 1991). Directional bias was often different through the two eyes. The example illustrated in Figure 4 was direction-biased for the dominant, contralateral eye (Figure 4A), direction-selective for the other eye (Figure 4B).

Other binocular neurones were tuned to different bands of spatial frequencies through either eye (Figure 5). Figure 5 (A and B) show montages of directional tuning curves vs.

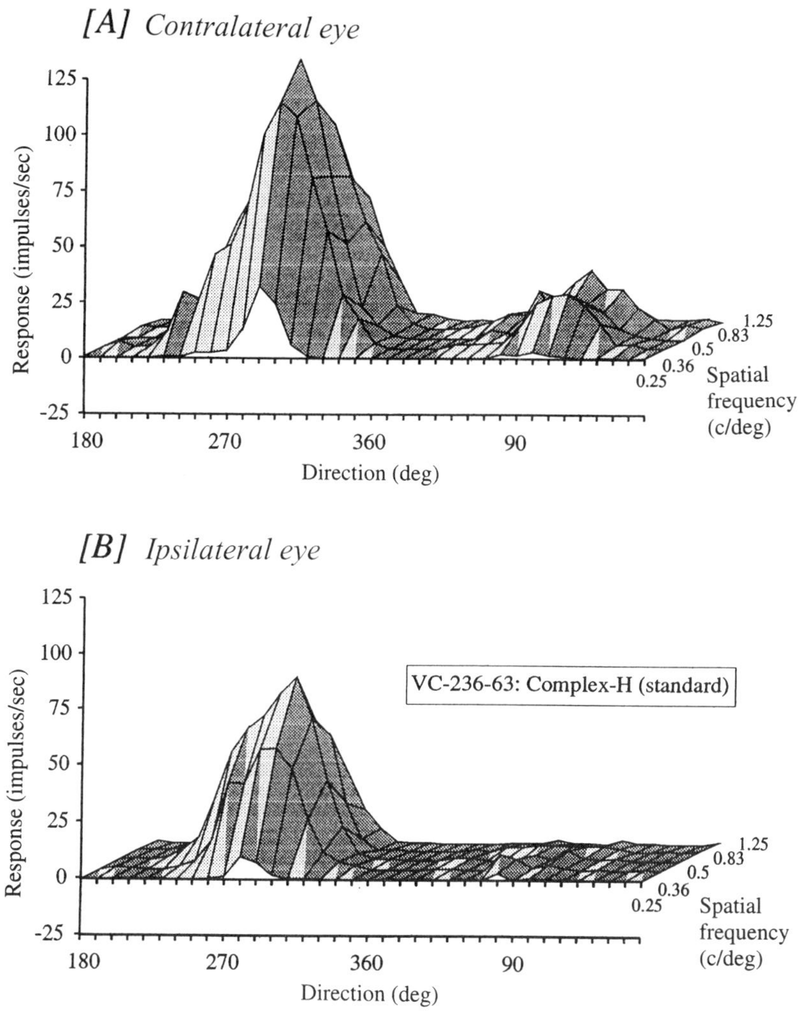


**Figure 4.** Interocular directionality differences: directional tuning vs. spatial frequency, for contralateral and ipsilateral inputs to one binocular cortical neurone. This example was direction-biased through one eye [A], direction-selective through the other [B].

spatial frequency for each eye: (C) shows the spatial-frequency tuning curves for each eye, derived from the envelope of responses to different spatial frequencies in the optimal direction for each set of directional tuning curves, illustrating that the contralateral eye was tuned optimally to higher spatial frequencies than the ipsilateral eye.

## Interocular Spatial Frequency Comparisons

We therefore assessed the range of interocular differences in spatial-frequency tuning of binocular cortical neurones (Hammond & Fothergill, 1990). Four patterns emerged, two of which are illustrated in Figure 6. Spatial-frequency was matched interocularly in only a small minority of neurones (13%). In the majority (87%) there were marked differences between the inputs from the two eyes. These included: spatial-frequency mismatch (Figure

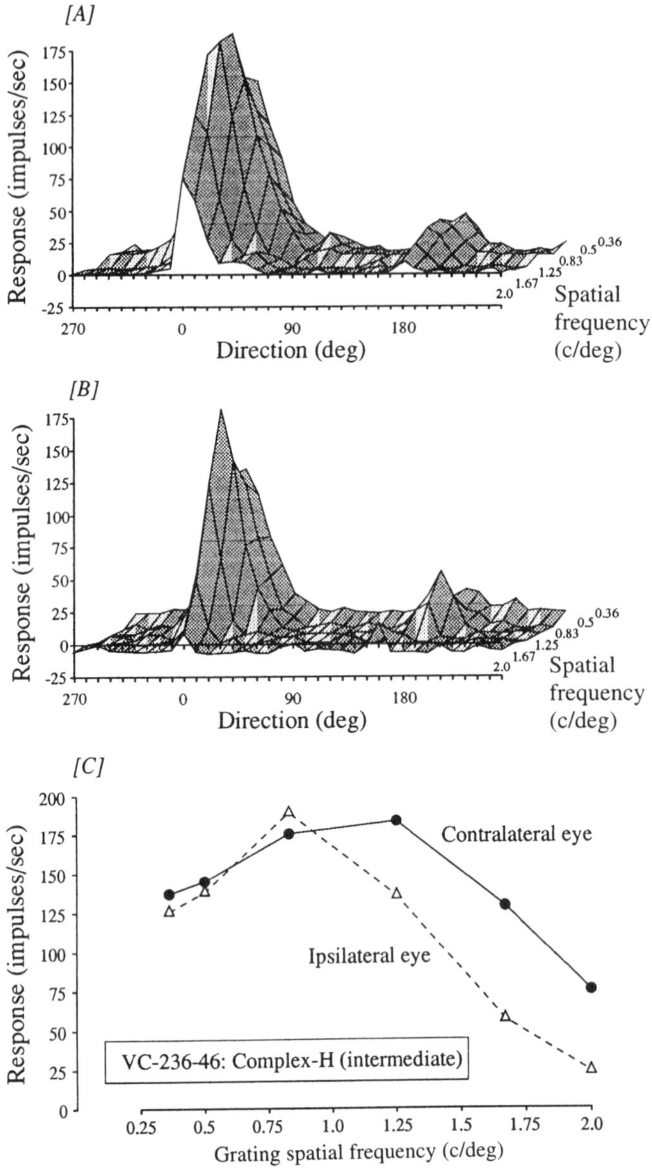


**Figure 5.** Interocular spatial frequency mismatch: directional tuning vs. spatial frequency for contralateral [A] and ipsilateral inputs [B] to a binocular cortical neurone, tuned to dissimilar spatial frequencies through each eye. Spatial frequency tuning for each eye [C] was derived from the envelope of responses in the preferred direction for that eye, over the full range of spatial frequencies tested.

6A), bandwidth mismatch (Figure 6B), and combinations of both spatial-frequency and bandwidth mismatch.

Upper halves of Figure 6 compare spatial-frequency tuning characteristics of two neurones for each eye. Lower halves show changes in ocular dominance consequent upon interocular differences in spatial-frequency tuning. The left-hand scale quantifies comparisons, on the basis of dominance ratio, defined as in methodology; the right-hand scale shows qualitative dominance groups (Hubel & Wiesel, 1962).

Neurones with matched spatial frequency showed no resultant change in eye preference with change in spatial frequency. In neurones with purely spatial-frequency mismatch (Figure 6A), the fact that one eye was tuned to higher spatial frequencies than the other resulted in progressive shifts in eye preference with increasing spatial frequency. Overall,

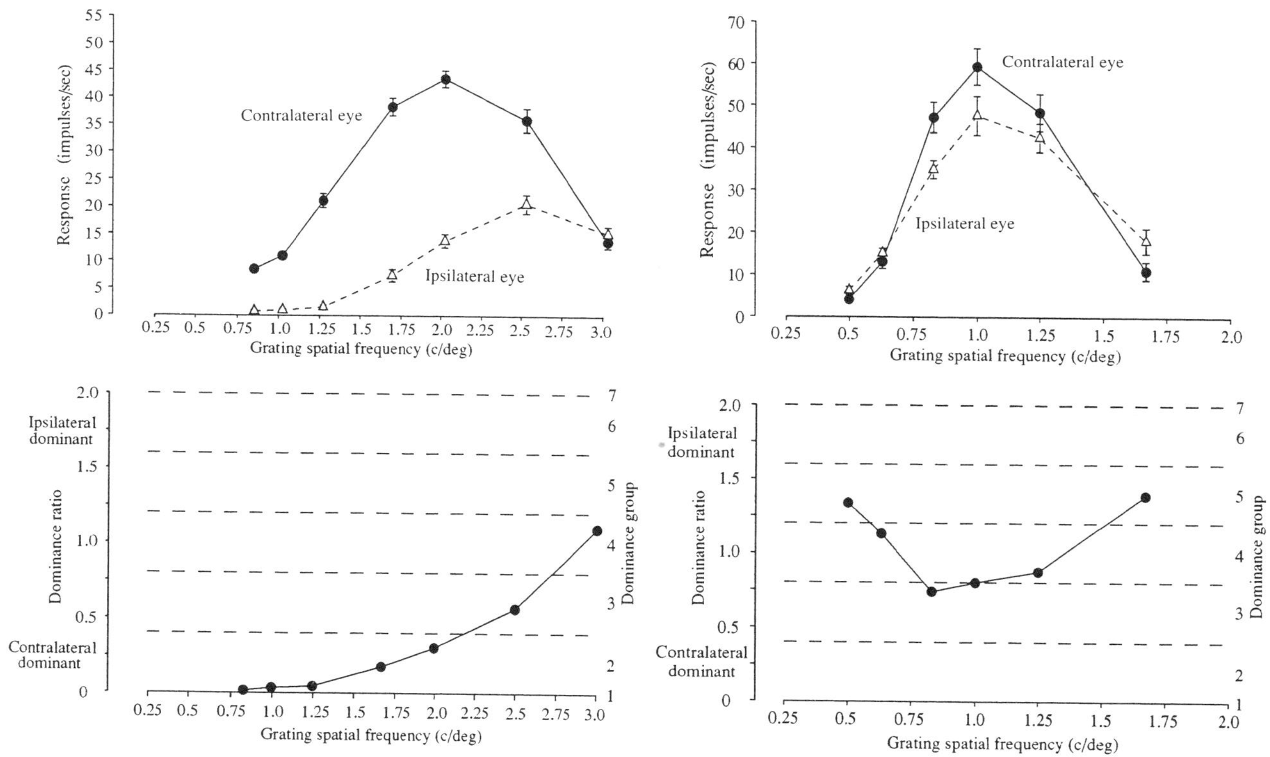


**Figure 6.** Interocular spatial-frequency differences: interocular comparisons of spatial-frequency tuning (upper plots) and resultant ocular dominance (lower plots) in two cortical neurones. Ratio scales to the left of lower plots were derived from weak:strong response for contralateral dominance, 2 minus weak:strong response for ipsilateral dominance. For comparison, horizontal bandings and axes to the right delineate the seven subjective dominance groups (Hubel & Wiesel, 1962). Monocular drive is thus defined by ratios of 0 (contralateral) and 2 (ipsilateral) - dominance groups 1 and 7 respectively. Remaining dominance groups - contralateral (groups 2 and 3), ipsilateral (groups 5 and 6), or equi-dominance (group 4), occupy equal steps of 0.4 on the ratio scale. Error bars are ±1s.e.m.; where not visible, these lie wholly within symbols. For clarity, responses and ocular dominance ratios are shown only for the preferred direction of motion. Detail: [A] spatial-frequency mismatch, neurone VC-237-17, simple, 1°/ sec; [B] bandwidth mismatch, neurone VC-239-15, standard complex, 2°/sec.

as in this example, the contralateral eye was more commonly tuned to higher spatial frequencies than the ipsilateral eye. In neurones with bandpass mismatch, the eye with broader spatial-frequency tuning was relatively more influential at extreme spatial frequencies, the more sharply tuned-eye was relatively more influential at intermediate spatial frequencies (Figure 6B). Neurones exhibiting both spatial frequency and bandwidth mismatch showed changes in eye preference rather similar to those of neurones with dissimilar bandpass tuning.

## Interocular Comparisons of Spatial Frequency and Bandwidth

In order to compare interocular spatial-frequency tuning overall, optimal spatial frequency, and log spatial-frequency bandwidth at half-height, were interpolated by curve fitting to the monocular spatial-frequency tuning curves of each neurone sampled, for each eye. Figure 7A shows the distributions of interocular differences in optimal spatial-frequency; Figure 7B shows bandwidth differences; comparing ipsilateral with contralateral eyes.

Applying an arbitrary criterion difference in spatial frequency or bandwidth of 5%, 28% of neurones were similarly tuned through either eye; the contralateral eye was tuned to higher spatial frequencies in 39%; the converse was true for 33%. This suggested, at most,

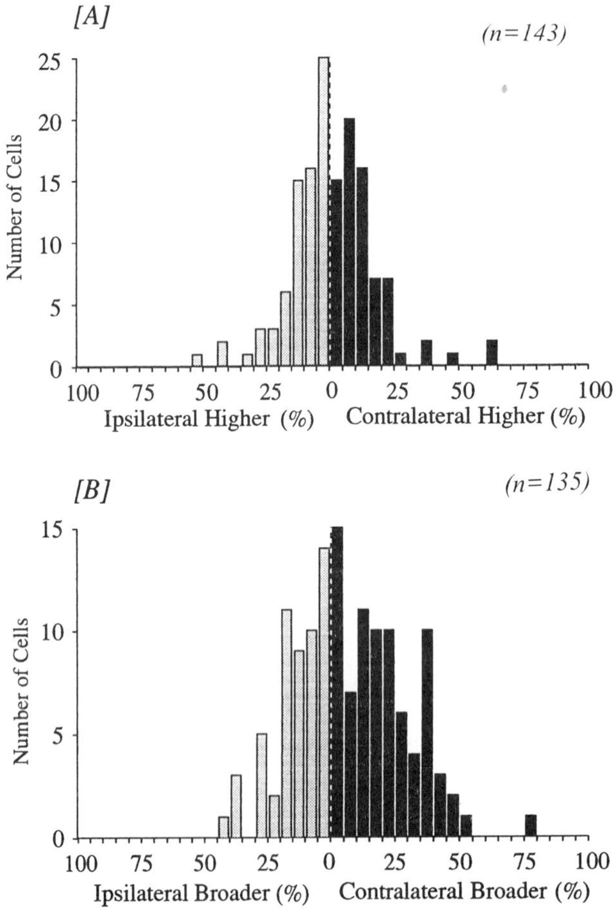


**Figure 7.** Interocular spatial-frequency differences in optimal spatial frequency (n = 143) [A] and bandwidth (n = 135) [B], comparing ipsilateral and contralateral inputs to binocular cortical neurones.

a weak link between preferred spatial frequency and laterality of input, with spatial-frequency differences distributed symmetrically between the eyes.

By contrast, the interocular differences in spatial-frequency bandwidth were substantial, with a clear skew in the distribution. Bandwidth was similar for either eye in only a minority of neurones (21%); in 49% the contralateral eye was more broadly tuned, the opposite being true in only 30%. Thus there was a marked tendency for the contralateral input to be more broadly spatial-frequency tuned.

## Interocular Comparisons of Length Specificity

Using standard tests to assess incremental responses for symmetrical increase in stimulus length, length specificity for each monocular input to binocular cortical neurones was compared. Optimal length ($L_{opt}$), optimal respnse ($R_{opt}$), response to stimuli of maximal length ($R_{max}$), and percentage end-stopping defined as

$$[((R_{opt} - R_{max})/R_{opt}) \times 100],$$

were derived from regression fits to the ascending and plateau or descending portions of each curve (Figure 8).

Each length summation profile was approximated by a pair of linear regression fits: to the ascending phase of the curve; and to the plateau (or descending phase, in cases of end-stopping for stimuli of supraoptimal length). Examples, and derivation of summation height (the optimal length for each neurone), optimal response, and limiting response to stimuli of maximal length, from these regression fits are shown in Figure 8. Percentage end-stopping was calculated from the latter two measures.

Figure 9A illustrates relative summation heights, Figure 9B relative end-stopping, comparing ipsilateral with contralateral inputs to binocular cortical neurones. Overall, there was a tendency for ipsilateral receptive fields of binocular cortical neurones to be larger but less strongly end-stopped than contralateral receptive fields.

# DISCUSSION

Comparisons of interocular positional specificity in binocular cortical neurones suggest that some neurones are exquisitely sensitive to incongruities between the two retinal images. Previous evidence (see Introduction) indicates that such neurones may be concerned with processing of depth or direction in visual space, or with stereopsis. Our work identifies numerous binocular neurones in the cortex, predominantly complex though occasionally simple, which show relatively broad tolerance for interocular misalignment in both position and orientation - neurones which respond well binocularly, regardless of the interocular phase relationship. The difference between our analysis (Hammond, 1991b) and that of Ohzawa & Freeman (1986a,b) is that the latter authors identified all binocular simple cells as position specific; with approximately equal numbers of position-specific and position-independent binocular complex cells. We saw a minority of complex cells requiring precise interocular alignment - and a minority of simple cells which lacked such specificity, despite the positional requirements which characterize their monocular receptive fields. Based on this evidence, we propose a two-stage model, in which binocular neurones with low positional specificity (primarily complex neurones) signal approximate interocular correspondence between the two retinal images, leaving position-specific (predominantly simple) cells to carry out the fine tuning.

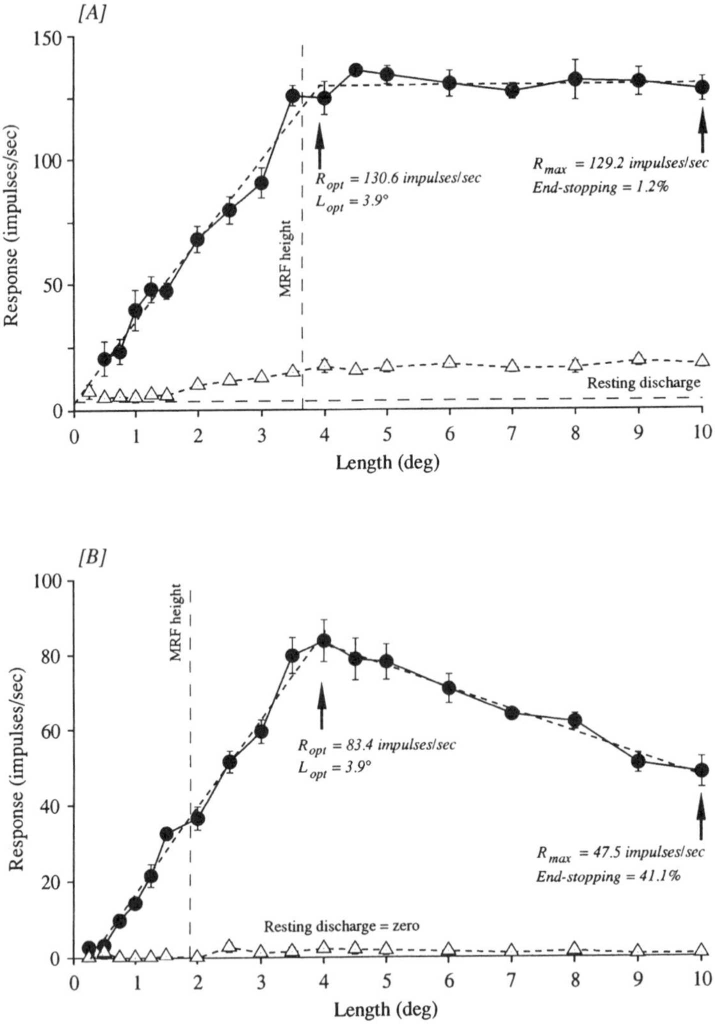


**Figure 8.** Examples of length specificity in cortical neurones. [A]: length summation in a non-endstopped neurone; [B]: length summation, and response decrement for stimuli of supraoptimal length, in an end-stopped neurone. Optimal response ($R_{opt}$), optimal length ($L_{opt}$), response to stimuli of maximal length ($R_{max}$) and percentage end-stopping. Preferred and opposite directions: circles and triangles, respectively. MRF: minimum response field.

An additional feature of our data, not hinted at previously, is the diversity of interocular differences in directional bias, optimal spatial frequency and spatial-frequency bandpass seen in a high proportion of binocular neurones. Whilst the preferred direction of motion of cortical neurones is invariant (Hammond, Andrews & James, 1975), their directional and orientational tuning, and their bias for opposing directions, are not fixed entities but change systematically with spatial frequency (Hammond & Pomfrett, 1990b). Typically, such neurones also become more sharply orientation-and direction-tuned at high spatial frequency (Vidyasagar & Sigüenza, 1985; Hammond & Pomfrett, 1990b).

Directionality in two minority classes - direction-selective neurones and bidirectional neurones- was not influenced by spatial frequency. However, in two other classes of neurones

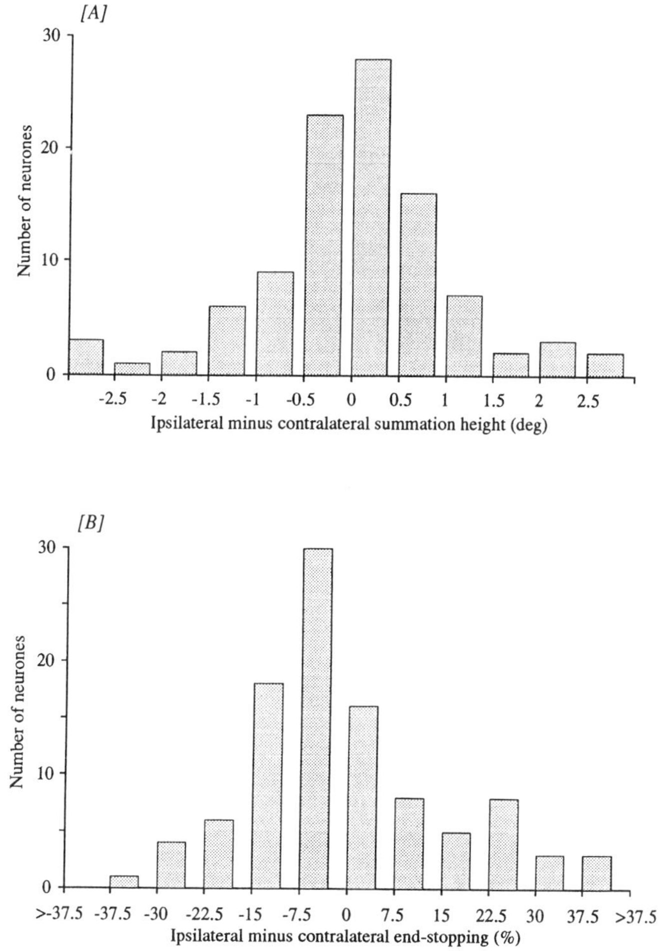


**Figure 9.** Interocular length specificity: interocular differences in summation height [A] and percentage end-stopping [B], comparing ipsilateral with contralateral inputs to binocular cortical neurones.

- both direction-biased - which represented the majority, bias varied, or even reversed, with spatial frequency. The pessimistic view is that such differences are no more than capricious - manifestations of quirks in wiring - imperfections of the kind prevalent in all biological systems. However, whilst there can be no absolute guarantee that such differences underlie perceptual abilities, sufficient evidence is beginning to accumulate that they might play significant roles in some of the more complex attributes of visual processing. However, the commonness of occurrence of such events and their potential function make it unlikely that they could simply reflect purely capricious, random shortcomings in connectivity.

Separate evidence relates to the interocular match or mismatch in spatial-frequency tuning, directionality, receptive field structure and dimensions. Each of these factors could, potentially, be associated with the neural encoding of three-dimensional perspective - sensitivity to objects tilted out of the fixation plane, or rotating dynamically in visual space. Many cortical neurones indeed show mismatch in directionality for each eye. Moreover, the majority are mismatched interocularly for spatial frequency, bandwidth, or both spatial frequency and bandwidth - differences which result in often dramatic variation in eye preference with change in spatial frequency.

Optimal spatial frequencies for ipsilateral and contralateral inputs to binocular cortical neurones are distributed symmetrically: thus, each eye is equally likely to be tuned to higher spatial frequencies than the other. However, bandpass is skewed, so that contralateral inputs are commonly more broadly tuned than ipsilateral inputs - perhaps reflecting the predominance of contralaterally-dominated neurones in the cortex, associated with greater pooling of inputs from lower order neurones.

That a majority of binocular cortical neurones are differentially tuned for spatial frequency through the two eyes might also simply reflect imperfect matching of inputs: a perfect match would require formidable precision in monocular convergence patterns. Again, on balance, we consider such explanations unlikely, for several reasons. First, spatial-frequency tuning for either eye was smooth, precise and repeatable over time in most neurones. Second, it would be difficult to conceive of an imperfect system that could, by chance, achieve such perfect matching in even a minority of neurones. Third, spatial-frequency and bandwidth mismatches represent potentially useful perceptual cues for evaluating perspective and surface texture of solid objects, as fundamental as the positional and orientational mismatch widely thought to be associated with coding of depth and direction in space.

Although changes in eye preference are purely secondary to the measured interocular spatial-frequency differences, they underscore the importance of assessing ocular dominance quantitatively rather than subjectively. Eye preference specified for optimal input inevitably carries the rider that eye allegiance in many neurones alters radically with changing conditions, whilst those neurones are still contributing to the co-operative activity of neuronal ensembles. One obvious caution concerns radical fluctuations occurring at extreme spatial frequencies, simply because the drive from one or both eyes is weak. Against this is the empirical observation that ocular dominance shifts consequent upon spatial-frequency or bandwidth mismatch were typically smooth and progressive over the entire spatial-frequency range, rather than abrupt.

As indicated above, interocular differences in spatial-frequency selectivity potentially provide a basis for coordinating the different perspective views presented by three-dimensional objects to the two eyes: for objects viewed either off-axis or tilted, statically or dynamically, out of the fixation plane.

The association with off-axis viewing is weak. Receptive fields reported lay within 10° of the fixation point. Over this range, differences in angular subtense of binocularly viewed objects at each eye are trivial - 1% or less, which is scarcely measurable experimentally - whereas the actual differences in optimal spatial frequency and bandpass were typically much greater than this, as were differences in receptive field size and relative end-stopping (Hammond & Fothergill, 1992).

By contrast, when objects are tilted statically, or rotate dynamically about axes having a vertical component of orientation within the plane of fixation (Figure 10A), the mismatch in spatial frequency and differences in temporal sequence of events at the two eyes can be substantial. The spatial frequency content of a rotating surface will tend towards infinity, its angular subtense to zero, as that face becomes edge-on to the viewing eye (Figure 10B). This can never occur at both eyes simultaneously, so that the temporal sequence of events will always be out of phase (Figure 10C): in particular, relative phase will be opposite for clockwise or counterclockwise rotation. Each eye will thus receive an input whose average spatial frequency ramp-increases, then decreases, over time, according to the cosine of rotation angle. Additionally, over part of the total 360° rotation, direction of spatial-frequency change at either eye will be opposite in sign, as objects flip from front to rear view and back again.

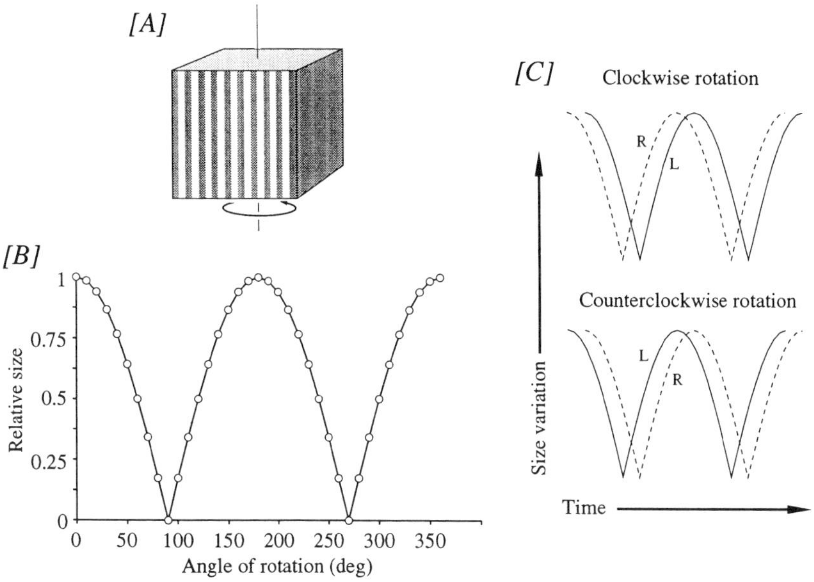


**Figure 10.** Size variation vs. angular rotation: [A]: rotation of three-dimensional objects about a vertical axis lying within the plane of fixation induces significant spatial-frequency mismatch between the eyes. [B]: change in angular subtense accompanying object rotation. [C]: schematic representation of the temporal sequence of size changes (reciprocal of spatial- frequency content) of the paired retinal images (L and R), as a binocularly-viewed object rotates clockwise or counterclockwise about a vertical axis, alternately out of and into the fixation plane.

This illustrative example is for rotation about a vertical axis in the fixation plane. Interocular differences would, of course, depend critically upon the orientation of the axis of rotation, diminishing towards zero as that axis approached horizontal. The convolutions involved are highly complex, and it would be premature to speculate in detail upon the neural mechanisms necessary for encoding them. The general principle, however, is an important one. Equally, it would be naïve to consider contributions only from isolated neurones. Co-operative activity in neuronal ensembles is much more likely to yield useful perceptual correlates - a factor underscoring our motivation for study of neuronal properties and the pooling of inputs under non-optimal as well as optimal conditions, on the premise that individual neurones make meaningful contributions in either situation.

Although such coding could conceivably be performed by purely monocular neurones, binocular mechanisms are altogether more elegant, more efficient and less ambiguous. Rotation and speed of rotation could be encoded in relative phase and frequency information at each eye. Dynamic tilt additionally involves velocity mismatch between the eyes. However, static tilt might derive simply from differences in angular subtense or orientation: cues which could be signalled efficiently by neurones with different optimal length and end-stopping requirements through each eye, of the kind identified experimentally.

Spatial frequency differences alone are unlikely to contribute to coding of motion in depth, which generates opposing motion at the two eyes. Their involvement in coding of static depth, however, remains a distinct possibility. Neurones with interocular spatial-frequency or bandwidth mismatch would respond differentially through the two eyes to different depths in space. Inevitably such schemes could work well only for (complex) neurones with poor binocular positional specificity (Hammond 1991b); for position-specific neurones (Ohzawa & Freeman, 1986a, b; Hammond, 1991b) the conflict between binocular positional and spatial-frequency matches would be unresolvable.

That approximately equal numbers of neurones are tuned to higher patial frequencies through either eye is not surprising. A system capable of signalling perspective would actually require matched sets of neurones, to encode all possible perspective views of objects tilted out of the fixation plane, about whatever axis.We have no way of knowing whether such cues are actually used for perceptual judgments. What we do know is that the neural substrate is there. The next stage of our investigation is to make binocular simulations of spatial-frequency and size changes at each eye during object tilt or rotation, to establish whether the experimental results match predictions from the monocular data.

## ACKNOWLEDGMENTS

Research supported by grants G8920072N and G8924156N from the Medical Research Council. Dr. Christopher J.D. Pomfrett, Lindsey K. Fothergill and Jong-Nam Kim participated in some experiments. David G. Glover provided invaluable technical assistance. Pharmaceutical preparations, and supplies of sterile disposable materials, were generously donated by: Abbott Laboratories Ltd.; CNS-Respiratory Division of Astra Pharmaceuticals Ltd; Imperial Chemical Industries Ltd.; Monoject Division of Sherwood Medical; Roche Products Ltd.; Smith and Nephew Medical Ltd.; Smith and Nephew Pharmaceuticals Ltd.; Smith and Nephew Textiles Ltd.; and The Wellcome Foundation Ltd.

## REFERENCES

Baker, C.L. (1990). Spatial- and temporal-frequency selectivity as a basis for velocity preference in cat striate cortex neurones. *Visual Neurosci.* 4:101-114.

Barlow, H.B., Blakemore, C., and Pettigrew, J.D. (1967). The neural mechanism of binocular depth discrimination, *J. Physiol.* 193:327-342.

Bishop, P.O., Coombs, J.S. and Henry, G.H. (1971). Responses to visual contours: spatio-temporal aspects of excitation in the receptive fields of simple striate neurones, *J. Physiol.* 219:625-657.

Bishop, P.O., Henry, G.H. and Smith, C.J. (1971). Binocular interaction fields of single units in cat striate cortex, *J. Physiol.* 216:39-68.

Bisti, S., Carmignoto, G., Galli, L. and Maffei, L. (1985). Spatial frequency characteristics of neurones of area 18 in the cat: dependence on the velocity of the visual stimulus, *J. Physiol.* 359:259-268.

Blakemore, C. (1970). The representation of three-dimensional visual space in the cat's striate cortex, *J. Physiol.* 209:155-178.

Campbell, F.W., Cooper, G.F. and Enroth-Cugell, C. (1969). The spatial selectivity of the visual cells of the cat, *J. Physiol.* 203:223-235.

Cynader, M. and Regan, D. (1982). Neurones in cat visual cortex tuned to the direction of motion in depth: effect of positional disparity, *Vision Res.* 22:967-982.

Fischer, B. and Krüger, J. (1979). Disparity tuning and binocularity of single neurones in cat visual cortex, *Exp. Brain Res.* 35:1-8.

Fothergill, L.K. and Hammond, P. (1992). Spatial frequency selectivity of striate cortical neurones in the anaesthetized cat: velocity vs. temporal frequency, *J. Physiol.*452:289P.

Gilbert, C.D. (1977). Laminar differences in receptive field properties of cells in cat primary visual cortex, *J. Physiol.* 268: 391-421.

Hammond, P. (1979). Stimulus-dependence of ocular dominance and directional tuning of complex cells in area 17 of the feline visual cortex., *Exp. Brain Res.* 35:583-589.

Hammond, P. (1991a). Letter to the Editors. On the response of simple and complex cells to random dot patterns: a reply to Skottun, Grosof and DeValois, *Vision Res.* 31:47-50.

Hammond, P. (1991b). Binocular phase specificity of striate cortical neurones, *Exp. Brain Res.* 87:615-623.

Hammond, P. and Ahmed, B. (1985). Length summation of complex cells in cat striate cortex: a reappraisal of the "special"/"standard" classification, *Neurosci.* 15, 639-649.

Hammond, P., Andrews, D.P. and James, C.R. (1975). Invariance of orientational and directional tuning in visual cortical cells of the adult cat, *Brain Res.* 96: 56-59.
Hammond, P. and Fothergill, L.K. (1990). Ocular dominance of striate cortical neurones in the anaesthetized cat: variation with spatial frequency, *J. Physiol.* 430:106P.
Hammond, P. and Fothergill, L.K. (1994). Cat striate cortex: monocular and interocular comparisons of spatial-frequency selectivity. *Anais da Academia Brasilaira de Ciencias* 66:45-113.
Hammond, P. and MacKay, D.M. (1975). Differential responses of cat visual cortical cells to textured stimuli, *Exp. Brain Res.*22:427-430.
Hammond, P. and MacKay, D.M. (1977). Differential responsiveness of simple and complex cells in cat striate cortex to visual texture, *Exp. Brain Res.*30:275-296.
Hammond, P. and Mouat, G.S.V. (1986). Influence of stimulus length on directional bias of complex cells in cat striate cortex, *Neurosci.*18:25-30.
Hammond, P. and Munden, I.M.E. (1990). Areal influences on complex cells in cat striate cortex: stimulus-specificity of width and length summation, *Exp. Brain Res.* 80:135-147.
Hammond, P. and Pomfrett, C.J.D. (1989). Directional and orientational tuning of feline striate cortical neurones: correlation with neuronal class, *Vision Res.* 29:653-662.
Hammond, P. and Pomfrett, C.J.D. (1990a). Directionality of cat striate cortical neurones: contribution of suppression, *Exp. Brain Res.* 81:417-425.
Hammond, P. and Pomfrett, C.J.D. (1990b). Influence of spatial frequency on tuning and bias for orientation and direction in the cat's striate cortex, *Vision Res.* 30:359-369.
Hammond, P. and Pomfrett, C.J.D. (1991). Interocular mismatch in spatial frequency and directionality characteristics of striate cortical neurones, *Exp. Brain Res.* 85:631-640.
Hammond, P. and Reck, J. (1980). Influence of velocity on directional tuning of complex cells in cat striate cortex for texture motion, *Neurosci. Lett.* 19:309-314.
Henry, G.H. (1977). Receptive field classes of cells in the striate cortex of the cat, *Brain Res.* 133:1-28.
Hubel, D.H. and Wiesel, T.N. (1962). Receptive fields, binocular interaction and functional architecture in the cat's visual cortex, *J. Physiol.* 160:106-154.
Ikeda, H. and Wright, M.J. (1975). Spatial and temporal properties of "sustained" and "transient" neurones in area 17 of the cat's visual cortex, *Exp. Brain Res.* 22:363-384.
Kulikowski, J.J. and Bishop, P.O. (1981). Linear analysis of the responses of simple cells in the cat visual cortex, *Exp. Brain Res.* 44:386-400.
LeVay, S. and Voigt, T. (1988). Ocular dominance and disparity coding in cat visual cortex, *Visual Neurosci.* 1:395-414.
Maffei, L. and Fiorentini, A. (1973). The visual cortex as a spatial frequency analyser, *Vision Res.* 13:1255-1267.
Maske, R., Yamane, S. and Bishop, P.O. (1984). Binocular simple cells for local stereopsis: comparison of receptive field organization for the two eyes, *Vision Res.* 24:1921-1929.
Movshon, J.A. (1975). The velocity tuning of single units in cat striate cortex, *J. Physiol.* 249:445-468.
Movshon, J.A., Thompson, I.D. and Tolhurst, D.J. (1978). Spatial and temporal contrast sensitivity of neurones in areas 17 and 18 of the cat's visual cortex, *J. Physiol.* 283:101-120.
Nelson, J.I., Kato, H. and Bishop, P.O. (1977). Discrimination of orientation and position disparities by binocularly activated neurones in cat striate cortex, *J. Neurophysiol.* 40:260-283.
Ohzawa, I. and Freeman, R.D. (1986a). The binocular organization of simple cells in the cat's visual cortex, *J. Neurophysiol.* 56:221-242.
Ohzawa, I. and Freeman, R.D. (1986b). The binocular organization of complex cells in the cat's visual cortex, *J. Neurophysiol.* 56:243-259.
Orban, G.A., Kennedy, H. and Maes, H. (1981). Response to movement of neurones in areas 17 and 18 of the cat: velocity sensitivity, *J. Neurophysiol.* 45:1043-1058.
Pettigrew, J.D., Nikara, T. and Bishop, P.O. (1968). Binocular interaction on single units in cat striate cortex: simultaneous stimulation by single moving slit with receptive fields in correspondence, *Exp. Brain Res.* 6:391-410.
Skottun, B.C. and Freeman, R.D. (1984). Stimulus specificity of binocular cells in the cat's visual cortex: ocular dominance and the matching of left and right eyes, *Exp. Brain Res.* 56:206-216.
Tolhurst, D.J. and Thompson, I.D. (1981). On the variety of spatial frequency selectivities shown by neurones in area 17 of the cat, *Proc. Royal Soc. B* 213:183-200.
Vidyasagar, T.R. and Sigüenza, J.A. (1985). Relationship between orientation tuning and spatial frequency in neurones of cat area 17, *Exp. Brain Res.* 57:628-631.
von der Heydt, R., Adorjani, C., Hänny, P. and Baumgartner, G. (1978). Disparity sensitivity and receptive field incongruity of units in the cat striate cortex, *Exp. Brain Res.* 31:523-546.

15

# VISUAL CORTICAL PLASTICITY AND NEUROTROPHIC FACTORS

Michela Fagiolini,[1] Tommaso Pizzorusso,[1] and Lamberto Maffei[1,2]

[1] Scuola Normale Superiore
P. zza Cavalieri 56127, Pisa, Italy
[2] Istituto Neurofisiologia CNR
via S. Zeno, 51 56127, Pisa, Italy

## INTRODUCTION

In a series of landmark experiments, Hubel and Wiesel (1963) demonstrated that the development of primary visual cortex could be disrupted by a variety of sensory perturbation. If a mammal is monocularly deprived (MD) of vision for several days during the early period of postnatal development, the animal permanently becomes ambliopic in that eye (Baker et al., 1974; Boothe et al., 1985; Domenici et al., 1991). The visual acuity of the deprivated eye is decreased and its contrast sensitivity depressed. Most visual cortical neurones become unresponsive to visual stimulation of the deprived eye and the ocular dominance distribution of cells shifts in favour of the eye receiving normal visual input (Giffin and Mitchell, 1978; Harwerth et al., 1989; Domenici et al., 1991a, c). Anatomically, MD performed during the critical period determined the reduction of the territories occupied in the primary visual cortex by the afferents from the deprived laminae of the LGN and the expansion of the territories occupied by the terminals from the non deprived laminae. There is also a shrinkage of the soma size of LGN projection cells in the binocular portion of the deprived laminae (Shatz and Stryker, 1978; LeVay et al., 1980; Guillery and Stelzner, 1970; Sherman et al., 1974).

The effects of monocular deprivation are thought to be the outcome of an unbalanced competition between the visual inputs from the two eyes on the cortical binocular cells. The hypothesis has been advanced that two visual inputs compete for the acquisition of a neurotrophic factor, whose production or release from cortical neurones is dependent upon the electrical activity of the input. The electrical activity in the LGN fibers driven by the deprived eye would be insufficient or most probably inappropriate for triggering off the production and/or release of a neurotrophic factor.

It has been recently shown that one of the neurotrophic factors, involved in this mechanism, are NGF and other neurotrophic factors of the NGF family (Domenici et al., 1991; Maffei et al., 1992; Berardi et al., 1992). These neurotrophic factors, together with their mRNA and their receptors are present in the visual cortex during the period of major plasticity of the visual system (critical period, P14-P40) and in adult life. Their production

*Basic and Clinical Perspectives in Vision Research*, Edited by Jon Robbins et al.
Plenum Press, New York, 1995

is dependent upon visually driven electrical activity (Large et al, 1986; Bozzi et al., 1993; Thoenen, 1991; Castren et al., 1992). It has been reported that a continuos infusion of NGF prevents the effects of MD:

a. the shift in ocular dominance distribution of visual cortical neurones in monocularly deprived rats or cats does not take place;
b. visual acuity and contrast sensitivity, evaluated by visual evoked potentials or by behavioural methods are unaffected by MD;
c. shrinkage of lateral geniculate cells in MD rats is absent.

It has also been shown that administration of neurotrophins of NGF family are active for the survival of degenerating neuronal cells and interestingly they have beneficial effects in animal models of Alzheimer disease (Phelps et al., 1989; Knusel et al., 1992; Hefti, 1986), Parkinson disease (Knusel et al.,1992; Lapchak et al., 1993) and ageing related deficits in learning and memory (Fischer et al., 1991).

Neurotrophins, however, do not cross the blood-brain barrier and therefore their application for therapeutical treatment is unlikely. Their exogenous infusion in man is practically impossible over an extended period of time. An attempt to circumvent this difficult problem would be the implantation into the brain of biological mini-pumps producing the neurotrophins directly.

This possibility has prompted the experimentation of biological producers of neurotrophic factors transplanted into the brain.

Schwann cells (Sc) and genetically modified fibroblasts seem to give promising results. Their transplant in the CNS have been shown to prevent the loss of axotomized cholinergic septal neurones and to promote their regeneration (Kromer & Cornbrooks, 1985; Kawaja et al., 1992). Sc transplanted into the eye have been shown to partly prevent or delay the loss of retinal ganglion cells after optic nerve section (Maffei et al., 1990).

As yet the effects of the transplants of producers of neurotrophic factors have been mainly tested by neurochemical or anatomical techniques in order to asses for instance whether process of neuronal degeneration were arrested or whether regeneration took place. All these controls are obviously essential. It seems to us, however, that particularly in view of possible clinical applications of transplants, functional tests are also needed. The analysis of the physiological properties of a given class of neurones can be a very sensitive method to see whether these neurones maintained or restored their normal function under the influence of transplants.

Since it is well known that Sc are able to produce many substance, among which neurotrophic factors such as NGF and BDNF (Bandtlow et al., 1987; Acheson et al., 1991), we tested the effects of Sc transplant in the lateral ventricles of MD rats. In particular, our experiments aimed at establishing whether the Sc transplant can prevent the effects of MD without altering the physiology of the visual system.

We found that transplant of Sc 1) do not cause any pathological effects in the animals in which they have been transplanted and in particular they do not cause any alteration in the physiology of their visual cortex. 2) They prevent completely the anatomical and functional effects of monocular deprivation in the rat (Pizzorusso et al, 1994; Ferrari et al., in press).

## MATERIALS AND METHODS

### Primary Neonatal Sc Cultures and Transplant Procedure

Sc were prepared from neonatal rat sciatic nerves according to the method of Brockes et al. and maintained in culture as described (Ferrari et al., 1991). In one preparation

Fluoro-Gold (Fluorochrome Inc., Englewood CO, USA) 0.45 mg/ml was added to culture medium to label Sc (Daniloff, 1991). 45-60' before transplantation, cells were counted and washed by three cycles of centrifugation and suspension in Hank's solution. The pellet was re-suspended in few microliters of Hank's solution. The volume was then measured by a Gilson P20 and the concentration of cells determined. In order to inject the same volume (4.5-6 μl), albeit with different cells concentrations, in all transplanted rats an adequate volume of Hank's solution was added to the suspension. Four different amounts of Sc ($10^5$, 3.3 $10^5$, 6.3 $10^5$, $10^6$ Sc) were injected bilaterally at P14-15 in the lateral ventricles of monocularly deprived animals. Rats were anaesthetised with ether. The injection was performed with a 25ul Hamilton syringe connected with a 21 gauge needle acutely inserted 1 mm lateral to bregma (Maffei et al., 1992). In two rats transplanted with $10^6$ Sc we transplanted 2 $10^6$ aD11 hybridoma the functionality of visual cortical cells (Berardi et al., 1993; Berardi et al., 1992).

## Primary Adult Sc Cultures and Transplant Procedures

In six rats we transplanted Sc taken from the sciatic nerves of adult rats. Adult rat Sc were prepared from crushed sciatic nerves 10 days after injury as described by Ferrari et al. (in press). Briefly, two-cm. lenghts of crushed sciatic nerve were pooled into phosphate-buffered saline (PBS), and after removal of the epineurium nerves were teased into bundles, cut, centrifuged for 5 min at 100x g, transferred to Dulbecco's Modified Eagle's Medium (DMEM) containing 15% fetal calf serum (FCS). 25 mM HEPES, 1.25 U/ml dispase (Boehringer, D), 80 U/ml collagenase (Type I, Worthington Biochemicals Corp.), 1 mg/ml hyaluronidase (Sigma), 0.5 U/ml penicillin G, and 0.5 mg/ml streptomycin (GIBCO) and incubated overnight at 37 °C . The next day the tissue was centrifuged for 5 min at 100x g, the resulting pellet was washed 3 times with DMEM containing 15% FCS and then resuspended in DMEM/15% FCS. After 48 hr the medium was replaced with DMEM containing 3% FCS and cultures were reefed 3 times per week. Under these conditions many bipolar, spindle-shaped cells were present after 30 days in vitro. These cells were immunopositive for S-100 and 192-IgG, two well-known markers for Sc. Transplantation was achieved as for neonatal Sc.

## Animals Treatment

Experiments were performed on 51 Long-Evans hooded rats. 7 were left undeprived while the other 44 were monocularly deprived by suturing the eyelids of the right eye at P14-15. Eyelids suture was performed under ether anaesthesia in the same session of the Sc transplant. Among MD rats, 5 were left untreated, 34 were transplanted with Sc, 2 were transplanted with Sc together with αD11 hybridoma cells and 3 were transplanted only with αD11 as control. After 20-30 days of monocular deprivation electrophysiological and anatomical analyses were carried out. One of the 11 rats treated with $10^6$ neonatal Sc and one of the six rats treated with $10^6$ adult Sc revealed complete absence of the effects of Sc and were excluded from our sample. Transplant failure could be due to inability of Sc to enter the brain because of cerebrospinal fluid and/or blood outflow during injection or to an enhanced immune reaction due to a major tissue damage or a sensitised status of the animal (Sloan et al., 1991).

## Lateral Geniculate Nucleus (LGN) Neurones Shrinkage

MD is known to induce shrinkage of deprived LGN neurones. The effect is restricted to neurones projecting to the binocular segment of the primary visual cortex (Domenici et al., 1993; Guillery & Stelzner, 1970). Neurones innervated by the ipsilateral eye totally project to the binocular part of the primary visual cortex and are therefore affected by MD. In the rat segregation between the ipsilateral and contralateral subfields of LGN cannot be revealed by usual histological stainings (Reese & Jeffery, 1983). However, injecting HRP in one eye allows an easy identification of the ipsi and contra subfields (Reese, 1988). Relay neurones located in the ipsilateral subfield of the LGN form a population that is homogeneously affected by MD. We measured the effects of MD in this part of the LGN in untreated and Sc transplanted rats. Adopting the protocol of Domenici et al. with little variations we anterogradely labelled retinal fibers with horseradish peroxidase (HRP Grade I, Boehringer D) in Sc-treated and control rats, and we compared soma size in the portion innervated by the ipsilateral eye of the right (deprived) and the left (non-deprived) LGN. Briefly, 24 hours after the injection of 10-14 ml of HRP (30% in saline plus 2% dimethylsulfoxide) in the non-deprived, left eye, rats were perfused with saline followed firstly by 500 cc. of a 1.25% glutaraldehyde/1% paraformaldehyde solution for 30 minutes and secondly by 500 cc. of a 25% sucrose/0.1 M phosphate buffer solution. Brains were then cut by a $CO_2$ freezing-microtome in 40 microns slices and HRP revealed using paraphenilendiamine/pyrocathecol as chromogens to achieve a stable reaction product able to tolerate the following cresylviolet counterstaining (Hanker et al., 1977). Cells in the ipsilaterally receiving subfields showing a clear visible nucleus and nucleolus were drawn by means of a camera lucida. Drawings were fed into a computer and soma size measured using an image analysis program.

## Fluoro-Gold Visualisation

Rats were perfused with 4% paraformaldehyde/0.1M phosphate buffer. Their brains were removed, post-fixed overnight in the same fixative and put for 12 hours in a 25% sucrose/0.1 M phosphate buffer solution. Brains were then cut by a $CO_2$ freezing-microtome in 40 microns slices. Slices were mounted and observed by a fluorescence microscope.

## Electrophysiology

Recordings were performed as described (Maffei et al., 1992; Domenici et al., 1991). We recorded single cell activity from the primary visual cortex of normal, MD and MD plus Sc rats. In the same session Visual Evoked Potentials (VEP) from 5 normal, 4 MD, and 5 MD plus Sc rats were recorded in response to alternating gratings of moderate luminance (12 cd/$m^2$) and temporal frequency varying in contrast and spatial frequency. Recordings were performed under urethane anaesthesia (6 cc/kg , 20% in saline) by means of a glass micropipette, filled with NaCl 3M, and inserted in the binocular portion of the primary visual cortex (Area Oc1B) contralateral to the deprived eye. Both eyes were fixed with metal rings. At the end of the experiment pupils were dilated and the optic disks were backprojected onto the screen where all the receptive fields had been plotted. For single cell recordings, location of the receptive field in the visual space, its size and organisation, optimal stimulus orientation and direction of movement, ocular dominance class and response type were determined for each cell according to standard criteria (Hubel & Wiesel, 1962; Parnavelas et al., 1981). Only cells whose receptive field fell within the binocular visual field (farther than 30 deg nasal from the optic disk and in the upper visual field) were included in our sample. The mean receptive field position with respect to the vertical meridian was 12±9 deg in normal, 10±8 deg in untreated MD, 11±10 deg in Sc transplanted. To avoid sampling biases resulting from the organisation of area Oc1B as far as ocular dominance is concerned (Thurlow & Cooper, 1988), electrode penetrations were angled and at least two well spaced penetrations were performed for each animal. Neurones in ocular dominance class 1 were driven only by the controlateral eye; neurones in classes 2 and 3 were binocular and preferentially driven by the contralateral eye; class 4 neurones were equally driven by two eyes, neurones in classes 5 and 6 were binocular and preferentially driven by the ipsilateral eye, class 7 neurones were driven exclusively by the ipsilateral eye. , non deprived eye; , deprived eye. To summarise the effect of MD an ipsilaterality index was attributed to each rat. This index was defined as: ipsilaterality index = (cells in classes 5-7) / (total number of responsive cells). VEP recordings were performed as described (Domenici et al., 1991). The binocular index was defined as = (cells in classes 2-6)/(total number of responsive cells). The orientation index was defined as = (orientation selective + orientation biased cells)/(total number of responsive cells).

## RESULTS

### The Transplant of Sc Does not Cause Pathological Effects

Our initial experiments were aimed first at controlling the survival of Sc in the brain and second at investigating whether the transplants induced pathological effects.

The survival of Sc was assessed in 4 rats transplanted with Fluoro-Gold labelled Sc. Both neonatal and adult-derived Fluoro-Gold labelled cells were found in the vicinity of the lateral ventricles at the end of the deprivation period, i.e. 30 days from transplantation.

The following observations suggest that Sc transplants do not cause pathological alterations:

1. The behaviour of treated animals ($10^6$ Sc) and their weight were normal, the gross anatomy of the brain and the size of ventricles were indistinguishable from controls.
2. The spontaneous activity of visual cortical neurones was in the normal range both in terms of frequency and pattern of discharge; cortical neuronal responsiveness to visual stimulation was indistinguishable from control animals. No unresponsive cells were found.

   The overall receptive field organisation was normal. The receptive field sizes of visual cortical neurones of transplanted rats was not significantly different from that of normal rats (Table 1).

   Orientation selectivity of cortical neurones was in the range of normality (Table 1).
3. The visual acuity of transplanted rats, evaluated by means of visual evoked potentials (VEP) in response to sinusoidal gratings of various spatial frequencies, was in the order of 1 cycle/deg which is in the normal range (Table 1).

### The Transplant of Sc Prevents the Effects of Monocular Deprivation

The duration of MD covered the entire rat critical period (from the day of eye opening until P35-45). MD is known to induce a shift in the ocular dominance distribution of cortical cells towards the non deprived eye (Maffei et al., 1992, Berardi et al., 1993), an impairment of visual acuity in the deprived eye (Domenici et al., 1991) and shrinkage of deprived LGN neurones (Domenici et al., 1993). All these quantities were tested and found to be normal at the end of the deprivation Period in Sc rats

### Ocular Dominance Distribution of Cortical Cells

The ocular dominance distribution of visual cortical neurones of 7 normal and 5 MD untreated rats is shown in Fig. 1 (top). The shift in ocular dominance ic completely prevented

**Table 1.** Mean Binocular index, mean ipsilateral index, mean orientation index, mean receptive field (RF) size, mean visual acuity are reported for each experimental group

| | MD+ neonatal-derived SC | MD+ adult-derived SC | MD untreated | Normal |
|---|---|---|---|---|
| Mean binocular index | 0.92 | 0.89 | 0.35 | 0.83 |
| SD | 0.056 | 0.05 | 0.153 | 0.04 |
| N | 5 | 4 | 4 | 5 |
| Mean ipsilateral index | 0.067 | 0.125 | 0.8 | 0.11 |
| SD | 0.02 | 0.06 | 0.154 | 0.04 |
| N | 5 | 4 | 4 | 4 |
| Mean orientation index | 0.621 | 0.69 | 0.605 | 0.66 |
| SD | 0.13 | 0.08 | 0.05 | 0.06 |
| N | 5 | 4 | 4 | 4 |
| Mean receptive field size (deg) | 6.35 | 6.40 | | 6.7 |
| SD | 2.16 | 2.5 | | 2.5 |
| cells | 15 | 15 | | 32 |
| Visual acuity (c/deg) | 0.85 | 1.05 | 0.4 | 0.94 |
| SD | | | 0.055 | 0.06 |
| N | 2 | 1 | 4 | 4 |
| % shrinkage of LGN neurones | 1.575 | 2.3 | 20.5 | |
| SD | 1.2 | | 0.73 | |
| N | 4 | 1 | 3 | |

N = number of animals, SD = standard deviation. Percentage of shrinkage = [(mean diameter of the neurones of non-deprived lamina - mean diameter of the neurones of the deprived lamina)/ mean diameter of the neurones of the non-deprived lamina] x 100. Ipsilateral index is drastically increased (t-test, $p<0.05$) and binocular index is significantly decreased (t-test, $p<0.05$) by MD. These alterations are not present in rats transplanted with $10^6$ neonatal-derived Sc and $10^6$ adult-derived Sc. Their indexes are not different from those of normal animals (t-test, $p>0.05$). None of the treatments decreased the selectivity for orientation of the stimulus (t-test, $p>0.05$ between each group). RF size and visual acuity are not affected by Sc transplants (t-test: $10^6$ neonatal-derived Sc transplanted rats vs normal animals, $p>0.05$ and $p>0.05$; $10^6$ adult-derived Sc transplanted rats vs normal animals, $p>0.05$).In these rats ($10^6$ Sc) the mean soma size of the neurones of the deprived and non-deprived LGN laminae were not significantly different (t-test between the diameter distribution of deprived and undeprived LGN neurones, $p<0.05$).

in neonatal-derived $10^6$ Sc and adult-derived $10^6$ Sc transplanted rats (Fig. 1, bottom). the effect of the transplant of adult-dreved Sc was quite similar to that obtained with neonatal-derived Sc.

## Effects of Sc on MD as a Function of the Number of Transplanted Sc

We found that transplants of $10^5$ or 3.3 $10^5$ Sc gave no or little effects. transplants of 6.3 $10^5$ Sc gave variable effects. $10^6$ Sc were sufficient to prevent the effect of MD. The results are shown in Fig.2. where the ipsilaterality index is reported as a function of the number of transplanted Sc. The ipsilterality index (see methods for definition) summarises the ocular dominance distribution. This index is low (Table 1) in normal rats on account of the natural bias in favour of the contralateral eye (Maffei et al., 1992) and it increases in the cortex contralateral to the deprived eye of untreated rats (Table 1). It can be noted that transplantation of increasing amounts of Sc are progressively more effective in preventing the shift of the ipsilaterality index towards the high values typical of MD animals.

## Visual Performance of Sc Transplanted Rats

As in other mammals (Giffin & Mitchell, 1978) MD induces in the rat a strong reduction of visual acuity in the deprived eye. Visual acuity can be estimated by measuring

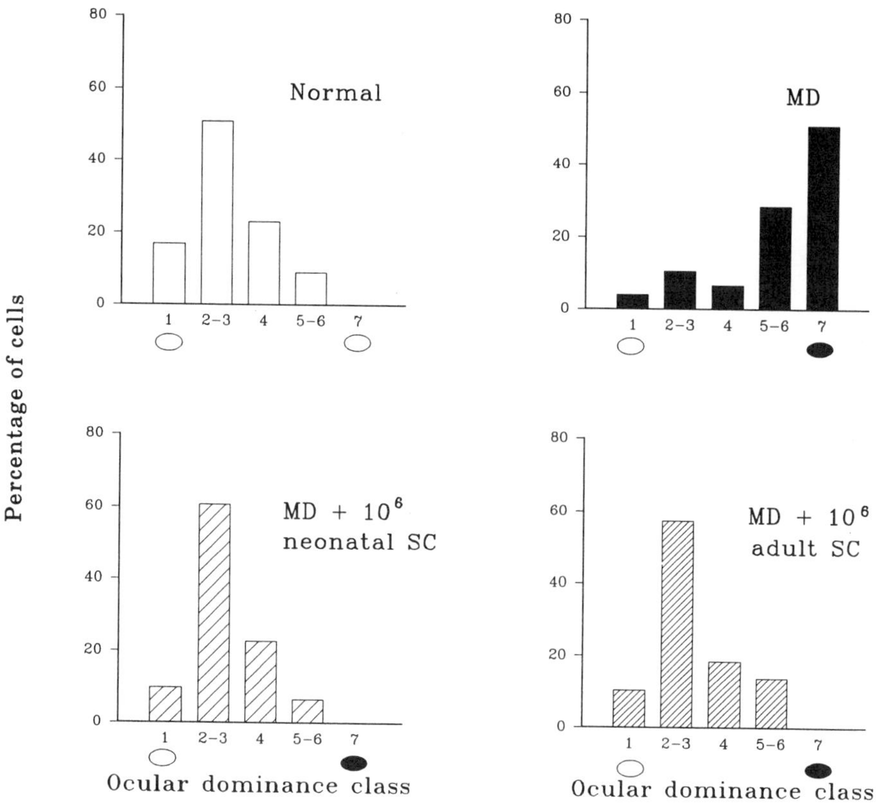


**Figure 1.** Ocular dominance distribution of visual cortical neurones in area 17. Top: Data from 6 normal rats (164 cells, left) and from 5 untreated MD rats (108 cells, right). Bottom: Data from 6 MD rats transplanted with $10^6$ neonatal Sc (100 cells, left) and from 5 MD rats transplanted with $10^6$ adult-derived Sc (87 cells, right).MD shifts the ocular dominance towards the non deprived, ipsilateral eye (classes 5-7). Both MD + $10^6$ neonatal-derived Sc and MD + $10^6$ adult-derived Sc rat ocular dominance distributions of visual cortical neurones are not significantly different from that of normal ($X^2$ test, $p>0.05$, $p>0.1$ respectively) and they differ from MD untreated rats ($X^2$ test, $p>0.05$ for both). Adult-derived Sc are as effective as neonatal-derived Sc in preventing the shift of ocular dominance. The ocular dominance distributions of MD rats transplanted with neonatal-derived or adult-derived are not significantly different ($X^2$ test, $p>0.05$).

visual evoked potentials (VEP) in response to sinusoidal gratings of various spatial frequencies. Visual acuity is taken as the highest spatial frequency of the stimulus still evoking a VEP signal above noise level. Fig. 3 shows the mean visual acuity of normal rats, of the deprived eye of MD untreated rats, and of the deprived eye of MD rats transplanted with $10^6$ Sc. In untreated rats, MD affects the visual performance of the deprived eye reducing its visual acuity by nearly a factor-of-three. The visual acuity of the deprived eye of Sc transplanted rats does not differ from that of normal animals (t-test, $p>0.05$).

## Sc Prevent the Anatomical Effects of MD in the LGN

The soma size of neurones in the ipsilaterally receiving lamina of the LGN was measured in untreated MD rats and in MD rats transplanted with different quantities of

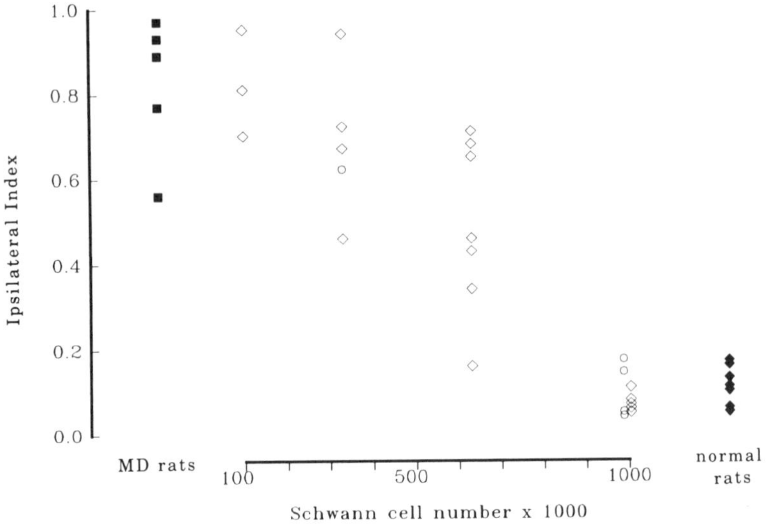


**Figure 2.** The effects of Sc on MD rats depends on the number of transplanted Sc. Filled square: MD untreated rats. Hollow diamonds: $10^6$ neonatal-derived Sc transplanted rats. Hollow circles: $10^6$ adult-derived Sc transplanted rats. Filled diamonds: normal rats. On the ordinate is reported the ipsilateral index. This index summarises the ocular dominance distribution. Each point reports the results obtained in one rat. On the abscissa is reported the number of transplanted Sc. The increase in ipsilateral index induced by MD is gradually prevented by increasing amounts of Sc. No statistical difference is present between the mean ipsilateral index of MD untreated rats and MD + $10^5$ Sc rats (t-test, $p>0.9$). A slight, but not significant (t-test, $p>0.2$) decrease in this index is present in 3.3 105 Sc rats. The mean ipsilateral index of 6.3 $10^5$ Sc rats is intermediate between MD and normal rats being significantly different from both (t-test, $p>0.01$). $10^6$ Sc rats have a mean ipsilateral index not significantly different from those of normals (t-test, $p>0.15$).

neonatal-derived and adult-derived Sc. Table 1 summarises the results: MD causes a mean shrinkage of 20.6 %. A comparable amount of shrinkage is observed in rats treated with an insufficient number of Sc. No shrinkage was observed in rats treated with 106 neonatal-derived Sc. In these rats (106 Sc) the mean soma size of the neurones of the deprived and non-deprived LGN laminae were not significantly different (t-test between the diameter

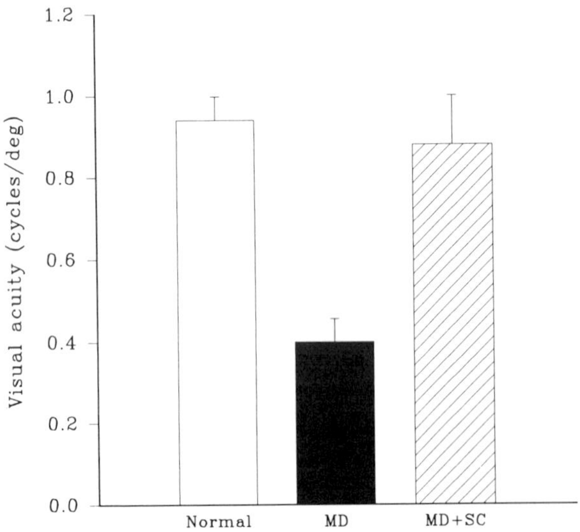


**Figure 3.** $10^6$ Sc prevent the reduction of visual acuity induced by MD. Visual acuity estimated by VEP for normal (N=5), monocularly deprived rats (N=4) and monocularly deprived plus $10^6$ Sc rats (N=3). Recordings were performed from the area 17 contralateral to the stimulated eye. In monocularly deprived rats the stimulated eye was the deprived eye. MD significantly reduces visual acuity of the deprived eye compared to normal eyes (t-test, $p>0.05$). $10^6$ Sc prevent this reduction (t-test, $p<<0.001$) and the visual acuity of the deprived eye dose not differ from that of a normal eye (T-test, $p>0.05$).

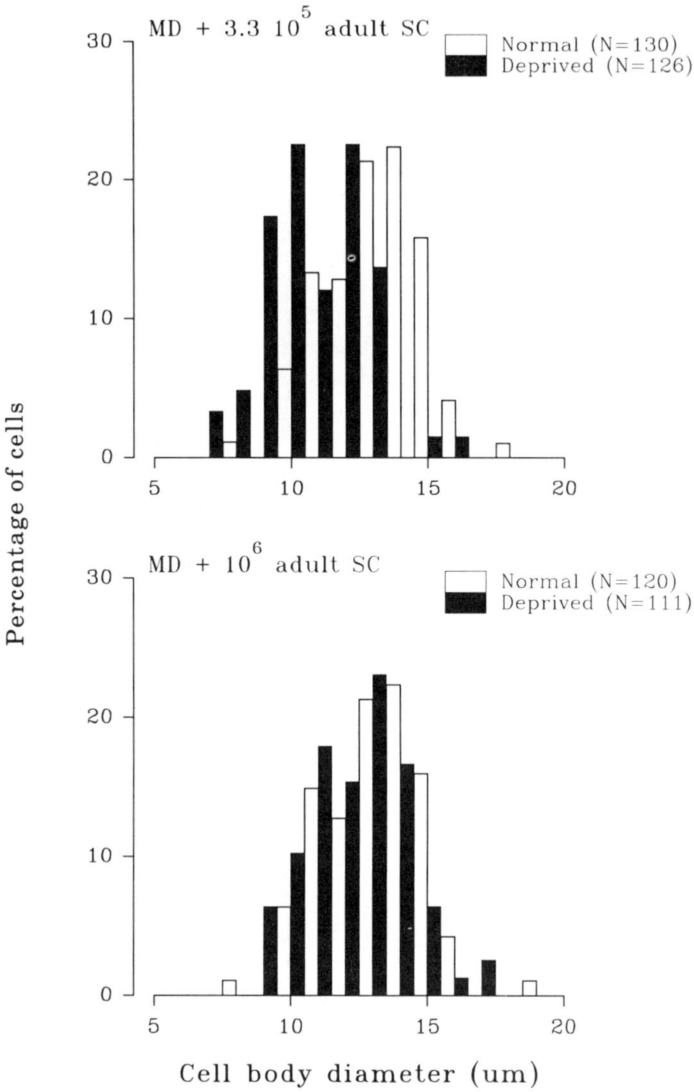


**Figure 4.** Cell soma size distribution in the lateral geniculate nucleus (LGN) of 3.3 $10^5$ adult-derived Sc (top) and 106 adult-derived Sc (bottom) MD transplanted rats. The percentage (%) of neurones has been plotted as a function of cell diameter (μm) for the LGN-deprived laminae (solid columns) and the LGN-non-deprived laminae (open columns). The mean soma size distribution for the ipsilateral deprived LGN lamina is significantly different from that of the ipsilateral non-deprived LGN lamina in 3.3 $10^5$ adult-derived Sc transplant MD rats (t-test, $p<0.01$), while the soma size distributions for the deprived and non-deprived LGN lamina in $10^6$ adult-derived Sc overlap extensively and the difference between the averaged medians is not significant (t-test, $p>0.05$).

distribution of deprived and non-deprived LGN neurones, $p<0.05$). Fig.4 shows the soma size distribution of LGN neurones of 3.3 $10^5$ adult-derived Sc (top) and 106 adult-derived Sc (bottom) MD transplanted rats. The effect of MD is completely prevented by the transplant of the highest number of adult-derived Sc, while transplant with 3.3 $10^5$ adult-derived Sc is ineffective.

## Effects of the Co-transplant of Sc and Anti-NGF Producing Hybridoma Cells

Sc are known to produce several trophic factors including NGF (Bandtlow et al., 1987; Acheson et al., 1991; Friedman et al., 1992). We tried to investigate whether the ability of Sc transplants in preventing the effect of MD is mediated by NGF. This problem was faced by using the hybridoma cell line aD11. aD11 produces an antibody that is able to block NGF binding to its receptor (Cattaneo et al., 1988). This antibody does not inhibit the biological activity of BDNF and NT-3 in vitro (Y.A. Barde, personal communication). Hybridoma cell transplant in the CNS as a means for counteracting the effects of NGF has been recently developed in our laboratory (Berardi et al., 1993; Berardi et al., 1992). The quantity of hybridoma cells necessary to counteract NGF in vivo was found to be of the order of 2 106 aD11. The effects of the transplants of hybridoma cells in the CNS are specific since transplants of parental myeloma cells or control hybridoma are ineffective (Schnell & Schwab, 1990; Berardi et al., 1993). Transplants of aD11 hybridoma cells per se are not able to prevent the effects of MD (3 rats, ipsilaterality indexes 0.7, 0.76, 0.81).

We co-transplanted 2 rats with $10^6$ Sc, which are sufficient to prevent the effects of MD, and 2 106 aD11 which are sufficient to block NGF biological activity. We found that the transplantation of hybridoma cells counteracts at least in part the action of Sc. Results are shown in Fig. 5. It can be noted that distribution is drastically shifted towards the non-deprived eye.

# DISCUSSION

The aim of the present experiments was to investigate the possible effects of the transplant of biological mini-pumps producing neurotrophic factors in vivo. The final goal of our research is to attempt to apply these biological mini-pumps in the therapy of CNS diseases, where the lack of neurotrophic factors is crucial. In this context our results seems to be promising for the following reasons:

1. In our experiments transplants of Schwann cells do not induce phatological effects: general health, gross behaviour and weight of treated animals are in the range of controls. The responsiveness, size of the receptive field, orientation selectivity of visual cortical neurones as well as visual acuity of the deprived eye estimated by VEP show normal values.

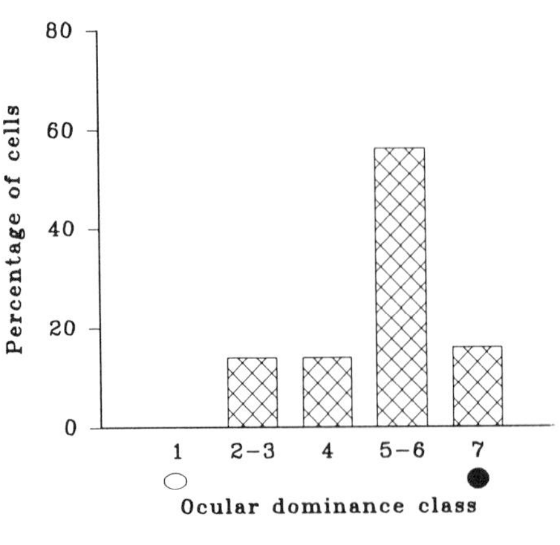


**Figure 5.** Transplants of hybridoma cells producing antibody against NGF counteract the effects of Sc transplants on the ocular dominance distribution of MD rats. The ocular dominance distribution of cortical neurones of two rats co-transplanted with $10^6$ Sc and 2 $10^6$ hybridoma cells is reported. The distribution is significantly different from the ocular dominance distribution of normal rats and untreated rats ($X^2$ test, $p<0.01$). The distribution is clearly shifted towards the ipsilateral, non deprived, eye.

2. Transplants of Sc hinder the effects of MD. 106 Sc prevent the shift in ocular dominance, impairment of visual acuity and shrinkage of LGN neurones soma size. The repeatability of the effects of the Sc transplants was indeed very high since only two out of the seventeen treated animals were not successful.
3. The mechanism of action of Sc is at least in part known, probably related to the production of NGF or neurotrophic factors of NGF family. This hypothesis, suggested by our results, is supported by the following facts a) continuous infusion of NGF in the ventricles prevents the effects of MD (Maffei et al., 1992; Berardi et al., 1993; Domenici et al., 1993); b) it is known that Sc produce NGF, BDNF as well as other neurotrophic factors in vitro (Acheson et al., 1991; Carmignoto et al., 1989; Matsuoka et al., 1991) and in vivo ( Bandtlow et al., 1987). This property is probably maintained when Sc are transplanted into the CNS as suggested by the recent observation that a considerable amount of NGF was found to be present in pieces of sciatic nerve transplanted in the CNS (Messersmith et al., 1991). It is possible that the NGF found in the transplanted nerve effectively originates from sciatic Sc; c) Sc transplants, NGF or BDNF exogenous administration prevent the reduction of axotomized basal forebrain and retinal ganglion cells with comparable efficacy (Knusel et al., 1992; Kromer & Cornbrooks, 1985; Kawaja et al., 1992, Maffei et al., 1990; Carmignoto et al., 1989); d) transplants of hybridoma cells producing antibody against NGF counteract, at least in part, the effect of Sc.
4. Transplants of Sc are still alive and active 1 month after implantation.

A possible obstacle in the use of transplants is the reaction of the immune system. The effects of Sc on MD and the persistence of Fluoro-Gold labelled cells one month after the transplant do not exclude reactions of the immune system at later times. Moreover, clinical application is generally performed in adults where the probability of rejection is greater. One way of overcoming this difficulty is suggested however by results presented here showing that adult-derived Sc are as effective as neonatal-derived Sc. At the moment we are actively investigating autologous transplants of Sc in adult animals.

## ACKNOWLEDGMENTS

We wish to thank Prof. Antonino Cattaneo for the kind gift of the alpha-D11 cell line, Prof. J. Nicholls for helpful discussions, Prof. A. Fiorentini and Mr. P. Martini for critical reading of the manuscript and Dr. M.C. Cenni for excellent technical help. We are indebted with international Institute for Paraplegia and HFSPO grant 93/93 for financial support.

## REFERENCES

Acheson A., Baker P.A., Alderson R.F., Miller F.D. and Murphy R.A. (1991) Detection of Brain-Derived Neurotrophic Factor-like activity in fibroblasts and Schwann cells: inhibition by antibodies to NGF, *Neuron* 7:265-275.

Baker F.H., Grigg P. and von Noorden G.K. (1974) Effects of visual deprivation and strabismus of the response of neurons in the visual cortex of the monkey, including studies on the striate and prestriate cortex in normal animals, *Brain Res.* 6:185-208.

Bandtlow C., Heumann R., Schwab M.E. and Thoenen H. (1987) Cellular localization of NGF synthesis by in situ hybridization, *EMBO J.* 6:891-899.

Berardi N., Domenici L., Parisi V., Pizzorusso T., Cellerino A. and Maffei L. (1993) Monocular deprivation effects in the rat visual cortex and lateral geniculate nucleus are prevented by Nerve Growth Factor (NGF). I. Visual cortex, *Proc. R. Soc. Lond.* 251:17-23.

Berardi N., Cellerino A., Domenici L., Fagiolini M., Pizzorusso T., Cattaneo A. and Maffei L. (1993) Monoclonal antibodies to NGF affect the postnatal development of the visual system, *Proc. Natl. Acad. Sci. USA* (in press).

Berardi N., Cattaneo A., Cellerino A., Domenici L., Fagiolini M., Maffei L. and Pizzorusso T. (1992) Monoclonal antibodies to NGF affects the postnatal development of the rat geniculocortical system, *J. Physiol.* 452:293P.

Boothe, R.G., Dobson, M.V. and Teller, D.Y. (1985) Postnatal development of vision in human and non human primates, *Ann. Rev. Neurosci.* 8:495-545.

Bozzi, Y., Pizzorusso, T., Cremisi, F., Comelli, M.C., Berardi, N. and Maffei, L. (1993) Monocular deprivation decreases the expression of BDNF mRNA in rat visual cortex, *Soc. Neurosci. Abstr.* 19:6.

Brokes, J.P., Fields, K.L. and Raff, M.C. (1979) Studies on cultured rat Schwann cells. I: Establishment of purified population from cultures of peripheral nerve, *Brain Res.* 165:105-118.

Campbell, F.W. and Maffei, L. (1970) Electrophysiological evidence for the existance of orentation in size detectors in the human visual system, *J. Physiol. (Lond.)* 207:635-652.

Carmignoto, G., Canella, R., Candeo, P., Comelli, M.C. and Maffei, L. (1993) Effects of NGF on neuronal placicity of the kitten visual cortex, *J. Physiol. (Lond.)* 464:343-360.

Carmignoto G., Maffei L., Candeo P., Canella R. and Comelli M.C. (1989) Effect of NGF on the survival of retinal ganglion cells after section of the optic nerve J. Neurosci. 9: 1263-1272.

Castren, E., Zafra, F., Thoenen, H. and Lindholm, D. (1992) Light regulates expression of brain-derived neurotrophic factor mRNA in rat visual cortex, *Proc. Natl. Acad. Sci. USA* 89:9444-9448.

Cattaneo A., Rapposelli B. and Calissano P. (1988)Three distinct types of monoclonal antibodies after long-term immunization of rats with mouse NGF, *J. Neurochem.* 50:1003-1010.

Daniloff, J.K. (1991) A novel assay fpor the, *in vivo* study of Schwann cells, *Exp. Neurol.* 114:140-143.

Domenici, L., Berardi, N., Carmigntot, G., Vantini, G. and Maffei, L. (1991) Nerve growth factor prevents the amplyopic effects of monocular deprivation, *Proc. Natl. Acad. Sci.* USA 88:. 8811-8815.

Domenici, L., Cellerino, A. and Maffei, L. (1993) Monocular deprivation effects in the rat visual cortex and lateral geniculate nucleus are prevented by NGF. II. Lateral geniculate nucleus, *Proc. R. Soc. Lond B* 251:25-31.

Ferrari G., Fabris M., Polato P., Skaper S.D., Fiori M.G. and Yan Q. (1991) Rat NGF receptor is recognized by Tumor-associated antigen monoclonal antibody 217c, *Exp. Neurol.* 112: 183-194.

Fischer W., Bjorklund A. Chen K. and Gage F.H. (1991)NGF improves spatial memory in aged rodents as a function of age, *J. Neurosci.* 11:1889-1906.

Friden P.M., Walus L.R., Watson P., Doctrow S.R., Kozarich J.W., Backman C., Bergman H., Hoffer B., Bloom F. and Granholm A.C. (1993) Blood-brian barrier penetration and in vivo activity of an NGF conjugate,*Science* 259:373-377.

Friedman B., Scherer S.S., Rudge J.S., Helgren M., Morrisey D., McClain J., Wangf D., Wiegand S.J., Furth M.E., Lindsay R.M. and Ip N.Y. (1992) Regulation of Ciliary Neurotrophic Factor expressio in myelin-related Schwann cells in vivo, *Neuron* 9: 295-305.

Giffin F. and Mitchell D.E. (1978) The rate of recovery of vision after early monocular deprivation in kittens, *J. Physiol. (Lond.)* 274:511-537.

Guillery R.W. and Stelzner D.J. (1970) The differential effects of unilateral lid closure upon the monocular and binocular segments of the dorsal genicualte nucleus in the cat, *J. Comp. Neurol.* 139:413-422.

Hanker J.S., Yates P.E., Metz C.B. and Rustioni A. (1977) A new specific, sensitive and non-carcinogenic reagent for the demonstration of horseradish peroxidase, *Histochem. J.* 9:789-792.

Harwerth R.S., Smith E.L. III, Crawford M.L.J. and von Noorden G.K. (1989) the effects of reverse monocular deprivation in monkeys. I:Psychophysical experiments, *Exp. Brain Res.* 74:327-337.

Hefti F. (1986) NGF promotes survival of septal colinergic neurons after fimbrial transections, *J. Neurosci.* 6:2155-2162.

Hubel D. H. and Wiesel T. N. (1963) Receptive fields of cells in the striate cortex of very young, visually inexperienced kittens, *J. Neurophysiol.* 26:994-1002.

Hubel D. H. and Wiesel T. N. (1962) Receptive fields, binocualr interactionand functional architecturein the cat's visual cortex, *J Physiol (Lond)* 160: 106-154.

Kawaja M.D., Rosenberg M.B., Yoshida K. and Gage F.H. (1992) Somatic gene transfer of NGF promotes the survival of axomized septal neurons and the regeneration of their axons in adult rats, *J. Neurosci.* 12:2849-2864.

Knusel B., Beck K.D., Winslow J.W., Rosenthal A., Burton L.E., Widmer H.R., Nikolics K. and Hefti F. (1992) BDNF administration protects basal forebrain cholinergic but not nigral dopaminergic neurons from degenerative changes after axotomy in the adult rat brain, *J. Neurosci.* 12:4391-4402.

Kromer L.F. and Cornbrooks C.J. (1985) Transplants of Schwann cell cultures promote axonal regeneration in the adult mammalian brain, *Proc. Natl. Acad. Sci. USA* 82:6330-6334.

Lapchak P.A., Beck K.D., Araujo D.M., Irwin I., Langston J.W. and Hefti F. (1993) Chronic intranigral administration of BDNF producers striatal dopaminergic hypofunctioin in unlesioned adult rats and fails to attenuate the decline of striatal dopaminergic function following medial forebrain bundle transection, *Neurosci.* 53:639-650.

Large T.H., Bodary S.C., Clegg D.O., Weskamp G., Otten U. and Reichardt L. F. (1986) Nerve growth factor gene expression in the developing rat brain, *Science* 234: 352-355.

LeVay S., Wiesel T.N. and Hubel D.H. (1980) The development of ocular dominance columns in normal and visually deprived monkeys, *J. Comp. Neurol.* 191: 1-51.

Maffei L., Carmignoto G., Perry V.H., Candeo P. and Ferrari G. (1990) Schwann cells promote the survival of retinal ganglion cells after optic nerve section, *Proc. Natl. Acad. Sci. USA* 87:1855-1859.

Maffei L., Berardi N., Domenici L., Parisi V. and Pizzorusso T. (1992) Nerve Growth Factor (NGF) prevents the shift in ocular dominance distribution of visual cortical neurons in monocularly deprived rats, *J. Neurosci.* 12:4651-4662.

Matsuoka I., Meyer M. and Thoenen H. (1991) Cell-type-specific regulation of NGF synthesis in non-neuronal cells: comparison of Schwann cells with other cell types, *J. Neurosci.* 11:3165-3177.

Messersmith D.J., Fabrazzo M., Mocchetti I. and Kromer L.F. (1991)Effects of sciatic nerve transplants after fimbria-fornixlesion: examination of the role of NGF, *Brain Res.* 557:293-297.

Parnavelas J.G., Burne R.A. and Lin C.S. (1981) Receptive field properties of neurons in the visual cortex of the rat, *Neurosci. Lett.* 27:291-296.

Phelps C.H., Gage F.H., Growdon J.H., Hefti F., Harbaugh R., Johnston M.V., Kachaturian Z.S., Mobley W.C., Price D.L., Raskind M., Simpkins J., Thal L.J. and Woodcock J. (1989) Potential use of NGF to treat Alzheimer's disease, *Neurobiol. Aging* 10:205-207.

Pizzorusso T., Fagiolini M., Fabris M., Ferrari G. and Maffei L. (1994) Schwann cells transplanted in the lateral ventricles prevents the functional and anatomical effects of monocular deprivation in the rat, *Proc. Natl. Acad. Sci. USA* 91:2572-2576.

Reese B.E. (1988) "Hidden lamination" in the dorsal lateral geniculate nucleus: the functional organization of this thalamic region in the rat, *Brain Res Rev.* 13:119-137.

Reese B.E. and Jeffery G. (1983) Crossed and uncrossed visual topography in dorsal lateral genicualte nucleus of pigmented rat, *J. Neurophysiol.* 49:877-885.

Schnell L. and Schwab (1990) Axonal regeneration in the rat spinal cord produced by an antibody against myelin-associated neurite growth inhibitors, *Nature* 343:269-272.

Shatz C.J. and Stryker M.P. (1978) Ocular dominance in layer IV of the cats visual cortex and the effects of visual deprivation, *J. Physiol.* 281:267-283.

Sherman S.M., Guillery R.W., Kaas J.H. and Sanderson K.J. (1974) Behavioral, electrophysiological and morphological studies of binocular competition in the development of the geniculo-cortical pathways of cats, *J. Comp. Neurol.* 158:1-18.

Sloan D.J., Wood M.J. and Charlton H.M. (1991) The immune response to intracerebral neural grafts, *Trends Neurosci.* 14:341-346.

Thoenen H. (1991) The changing scene of neurotrophic factors, *Trends Neurosci.* 14:165-170.

Thurlow G. A. and Cooper R. M. (1988) Metabolic activity in striate and extrastriate cortex of the hooded rat: controlateral and ipsilateral eye input, *J. Comp. Neurol.* 274:595-607.

Wiesel T. N. and Hubel D. H. (1963) Single-cell responses in striate cortex of kittens deprived of in one eye, *J. Neurophysiol.* 26:1003-1017.

Wiesenfeld Z. and Kornel E. (1975) Receptive fields of single cells in the visual cortex of the hooded rat, *Brain Res.* 94:401-412.

16

# RESIDUAL VISUAL FUNCTION IN THE ABSENCE OF THE HUMAN STRIATE CORTEX

Keith H. Ruddock

Departments of Physics (Biophysics) and Biology
Imperial College
London, SW7 2BZ, United Kingdom

## INTRODUCTION

The organisation of visual areas in the macaque cortex suggests that different attributes of the visual image are registered in modular fashion. Thus neurones responsive to a particular attribute such as movement are associated with histologically identifiable regions of the visual cortex. Evidence for such an arrangement in humans has been obtained by functional mapping using positron emission tomography or functional magnetic resonance imaging, although species differences between the locations of areas with equivalent functional specialisation have been noted. Results of studies on human patients with cortical lesions are also consistent with the existence of functional specialisation in different visual areas. In this article, I discuss residual visual function in patients who suffer extensive field losses associated with lesions of the striate cortex. I examine in detail data for a patient, GY, who can detect transient light stimuli presented within the 'blind' region of his visual field. His responses to such stimuli reveal that he can discriminate between transient stimuli which differ in flicker rate or speed of movement, and he can also identify certain colours. He cannot, however, identify spatial patterns presented to his 'blind' field. I examine the neuronal mechanisms responsible for these responses and I discuss them with reference to the organisation of the visual cortex.

The analysis of visual responses in patients with cortical lesions requires a consideration of the anatomy and function of the neural pathways involved, and the brief account given here is based on a more extensive review which contains full details of appropriate references (Ruddock, 1991).

The visual pathways form the major input to the primate brain, and an estimated 50% of the cortical surface is concerned with the processing of visual signals. Early stages of visual processing are performed by the neural retina, the output of which projects via the optic nerve to a number of brain areas. The principal retinal projections are those via the dorsal lateral geniculate nucleus (dLGN) to layer *IV* of the striate cortex and that to the superficial layers of the superior colliculus. The axons which form these projection pathways

*Basic and Clinical Perspectives in Vision Research*, Edited by Jon Robbins et al.
Plenum Press, New York, 1995

arise in several histologically and functionally distinct classes of ganglion cells, including the P- and M- types which project, respectively, to the parvocellular and magnocellular layers of the dLGN, and thence to the striate cortex (visual area VI). Some 10% of the M-type ganglion cells, together with others classed histologically as $P_\gamma$ or $P_\varepsilon$, form the projection to the superior colliculus.

Each visual half field is mapped topographically on to the contralateral dLGN and thence to the striate area VI, with a marked over-representation of the foveal region of the field. In macaque, there is an estimated 1 deg. overlap at the vertical meridian in the projections to the two hemispheres. Further representations of the visual field are found in the prestriate cortical areas corresponding to Brodmann's histological areas 18 and 19, and in addition to their major input via visual area VI, they also receive a sparse direct projection from dLGN, involving some 1% of the afferent geniculate fibres. Each half-field is mapped topographically on to the superficial layers of the contralateral superior colliculus, with an enlarged central representation similar to that of the VI map. The superior colliculus projects retinotopically to the pulvinar, and this latter also receives a direct, ordered retinal projection. The pulvinar projects to both the striate and pre-striate cortical areas, and to the inferotemporal lobe, and these retinotopic projections are reciprocal. There are also retinal projections to several regions in the pretectum, and axons from this area also innervate the pulvinar. Other retinal projections terminate in the ventral lateral geniculate nucleus and the nucleus paralemniscalis of the optic accessory system. The retinotopically organised collicular-pulvinar projection pathway is of particular significance for the interpretation of visual responses which occur in the absence of striate cortical signals.

Neuronal responses of the P- and M- projection pathways differ from each other in several respects. Both have circularly symmetric receptive fields, but those of the P- type ganglion and geniculate cells are smaller than those of the M- type, their responses are wavelength selective and are more sustained than those of the M-type cells. This division of ganglion cell responses into sustained and transient types, a functional division which is maintained at cortical level, is particularly important in the analysis of the visual responses examined in this paper and was first noted for the cat by Ikeda and Wright (1972, 1975).

In the macaque, the separation between the projections of these two classes of ganglion cells is maintained in the dLGN and can be traced onwards through the striate and prestriate cortical areas. Some P-type cells project via the so-called blobs, areas of high cytochrome oxidase uptake in VI, via the thin stripes of similar high oxidase activity in V2, to area V4. This pathway yields colour selective, non-orientation specific responses. The other P-type cells, which project via the interblob regions of VI and the interstripe regions of V2, yield orientation selective, non-spectrally opponent responses. The M-type pathway projects via the interblob region of VI and the thick stripes of V2 to areas V3 and V5. This pathway is characterised by high contrast sensitivity, and is responsive to transient stimuli, including those which move in a specific direction. Some orientation sensitive cells of both M- and P-type pathways exhibit sensitivity to retinal disparity. The division of functions between different areas of the prestriate cortex, with movement sensitivity strongly represented in V3 and V5 and colour sensitivity in V4, has been emphasised by Zeki (1993). Neurones in the superficial layers of the superior colliculus have large receptive fields (up to 225 $deg^2$ ) and are selectively sensitive to transient, particularly fast moving targets presented at high luminance contrast. These neurones lack orientation selectivity, have broad-band spectral sensitivities and only some 10% are sensitive to direction of movement.

The field losses observed in patients with lesions of the striate cortex can be predicted from the mapping of the field on to VI, and involvement of the whole area in one hemisphere leads to loss of the contralateral hemifield, a condition known as hemianopia. Such field losses are generally regarded as permanent, and indeed have formed the basis for the analysis of visual field respresentation in the striate area (Holmes, 1918; Teuber et al., 1960). Recovery of visual function has, however, been recorded on a number of occasions, and although this may be reflected in enlargement of the visual field, functional responses are sometimes observed in the absence of general recovery of fields. Thus Riddoch (1917) reported movement sensitivity within scotomatous areas of the visual field in five soldiers with brain damage and Pöppel et al. (1973) found a weak correlation between direction of gaze and the position of a light stimulus flashed for 100ms in the 'blind' field areas of four patients with incomplete hemianopias. Weiskrantz et al. (1974) reported that a patient, DB, with a nearly complete left homonymous hemianopia, could point accurately at a target flashed in his 'blind' hemifield, and similar observations were reported by Perenin and Jeannerod (1975, 1978). The patient GY, whose visual functions form the principal topic of this paper, can also locate accurately transient stimuli presented within his 'blind', right hemifield (Barbur et al. , 1980; Blythe et al., 1987).

There are two distinct types of response associated with sensitivity to transient light stimuli presented within 'blind' areas of the visual field. One group of patients responds by guesswork and experiences no sensation associated with the light stimulus. This phenomenon is known as 'blindsight', a term first introduced by Weiskrantz et al. (1974) in their report on DB, whose visual functions remain the best-known example of this phenomenon (Weiskrantz, 1987). Other patients, exemplified by GY, report a perceptual experience, usually described as a dark shadow, which is elicited by light stimulation of the 'blind' field. It has recently been reported that, dependent on the stimulus parameters, GY can respond either on the basis of his residual visual perception, or by 'blindsight' (Weiskrantz et al., 1995). The relationship between these two response modes has yet to be resolved and their relative incidence amongst patients with striate lesions is also uncertain. Five of twenty-five patients examined by Blythe et al. (1987) demonstrated residual vision responses within 'blind' areas of the visual field, and the incidence of 'blindsight' in those with striate lesions has been variously estimated as 20% (Marzi et al. ,1986) 25% to 50% (Ptito et al., 1987), 63% (Weiskrantz, 1980) and 100% (Pizzamiglio et al., 1984).

There is considerable variation between individual patients in the visual capacities associated with the 'blind' field. The prevalence of the ability to detect and localise light stimuli presented transiently in the 'blind' field has already been reviewed. Other reported responses include detection of real and apparent movement (Barbur et al., 1980; Blythe et al., 1986), discrimination of spatial orientation and resolution of spatial structure (Weiskrantz et al., 1974; Weiskrantz, 1986), spectral responses involving rod and cone function (Stoerig and Cowey, 1989, 1991) and spectral discrimination (Blythe et al.. 1987; Stoerig, 1987; Brent et al. , 1994).

The retinal projection pathways responsible for these various responses have not been identified unequivocally, but the observed parametric characteristics of the detection and localisation of transient stimuli are consistent with activity involving visually driven neurones of the superior colliculus. Other responses suggest the involvement of the direct projections from dLGN to the pre-striate visual cortex (Benevento and Yoshida, 1981; Fries, 1981). This projection survives ablation of the monkey striate cortex (Yukie and Iwai, 1981), the remaining ganglion cells projecting to the degenerated dLGN, where they contact, via an interneurone, surviving neurones which project to the pre-striate visual cortex (Kisvárday et al., 1991).

In this paper, results of the extensive studies performed on a patient GY, who has well-established residual vision associated with his 'blind' hemfield, are reviewed and interpreted with reference to his cortical lesions and the known functional organisation of the primate visual cortical areas.

# METHODS

## Observers

GY is a 37 year old male, who has right hemianopic field losses, homonymous for the two eyes, associated with traumatic damage to the left striate and pre-striate cortical areas incurred in a traffic accident at age 8 years. Magnetic resonance imaging (MRI) of the brain (Fig.1) reveals an extensive lesion of the left hemisphere, involving the lingual gyrus, the striate and peristriate cortex below the calcarine fissure to within about 1cm of the occipital pole, and the similar cortex above the calcarine fissure, not quite extending to the occipital pole. The fusiform gyrus is spared, but medial aspects of the cuneous, to the sulcus parieto-occipitalis anteriorly, a small strip of the posterior aspect of the pre-cuneous in the parietal lobe and the caudal extremity of the parahippocampal gyrus are also involved. There is also a much more restricted lesion of the right hemisphere involving mainly the inferior parietal, part of the supra-marginal gyrus and juxta cortical white matter of the adjacent part of the superior parietal lobule. Much of the left geniculo-calcarine tract is involved, but that on the right is intact and no visual abnormalities associated with the right hemispheric lesion have been observed. Thus his increment threshold sensitivity and colour vision in his left hemifield are normal. A detailed visual field plot (Barbur et al., 1980) reveals a right hemianopia with sparing of sensitivity extending up to some 3.5 deg. to the right of the

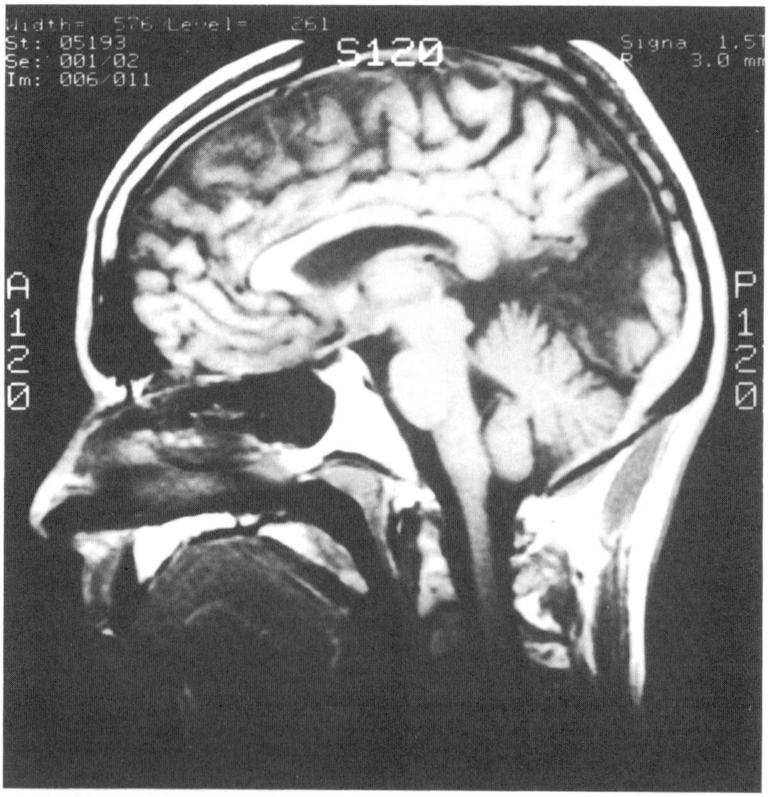


**Figure 1.** a) Parasaggital MRI scan (T2 weighted) through GY's left cerebral hemisphere, showing the sparing around the occipital pole.

fixation point, and the corresponding sparing of the left striate cortex is apparent in the MRI scans of Fig. 1a. Positron emission tomography (PET) reveals that stimulation of the 'blind' hemifield by moving stimuli activates pre-striate areas corresponding to V3 and V5 of the macaque, but not the striate cortex (Barbur et al., 1993).

Comparison data are given for a second hemianope, CB, who lacks residual visual capacity in the 'blind' hemifield, and for a number of normal subjects, approximately age matched to GY, all of whom have normal visual function, including colour vision.

## Equipment

In order to minimise effects of scattered light, all light stimuli were generated with a three-beam Maxwellian-view optical system (Barbur and Ruddock, 1980). The various stimulus configurations were provided by placing high contrast (>95%) photographic slides in the focal planes of the instrumental beams. Target movement was obtained by mechanical displacement of the target slide and the luminance of the target beam could be varied continuously with a neutral density wedge, the setting of which was under computer control.

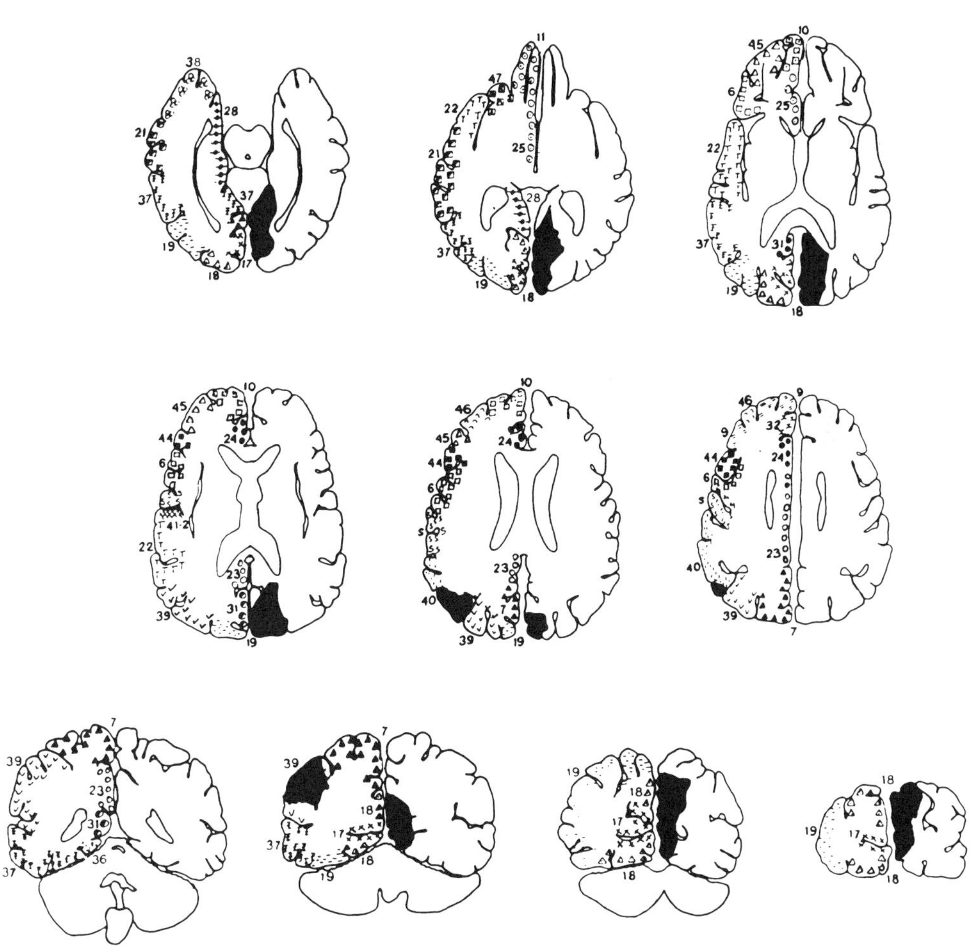

**Figure 1.** b) Axial templates taken from Damasio and Damasio (1989), with the Brodmann areas numbered on the right hemisphere and the locations of GY's lesions shown in black.

The luminances of the other beams were controlled to within 0.1 log unit accuracy by neutral density filters. The spectral bandwidths of the beams were controlled by spectrally selective filters, either narrow bandwidth Balzer B40 interference filters or broad bandwidth Wratten filters. Sinusoidal flicker was obtained by a combination of a fixed and a rotating Polaroid disc. Presentation times were set by computer controlled electronic shutters. The observer's eye was centred relative to the 3mm instrumental exit pupil and position was maintained by the observer's biting on a dental clamp. All stimuli were calibrated with a Macam Photometer/Radiometer.

## Procedures

All experimental sessions commenced with 15 minutes dark adaptation, and measurements were taken in a dark room. GY used his right eye for observations, for which the nasal retina projects to the damaged left hemisphere. Thresholds were obtained by a double staircase method (Cornsweet, 1962) to an accuracy of 0.03 log units. Sets of stimuli used in discrimination or identification tasks were presented in random sequence. All responses were based on residual vision involving conscious preception of the light stimulus.

## RESULTS

Localisation by saccadic eye movements of light flashes presented in random sequence at different locations on the horizontal meridian are illustrated in Figure 2. Data are given for patient GY, for a hemianopic patient CB who lacks residual vision, and for a normal control. The results demonstrate GY's ability to localise light flashes located within his 'blind' hemifield. Detection thresholds made in response to a moving target are illustrated in Figure 3a, in which target luminance required for its detection is plotted as a function of target velocity, v, for GY's blind hemifield and for a normal observer. The increase in GY's sensitivity as the target velocity increases is clearly demonstrated, whereas for normal vision,

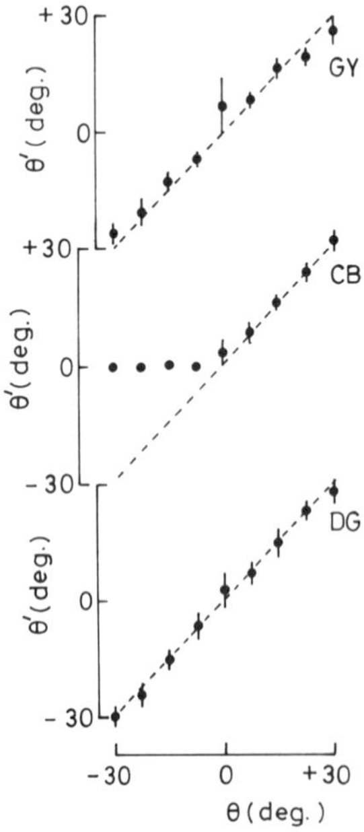


**Figure 2.** Angular eye position, θ′, established by a saccade made in response to a light flash, presented for 100ms at angular position θ, relative to fixation at θ′= 0 Data for two hemianopes GY and CB, and a normal control, DG. Error bars denote ±SE in the mean of 10 readings. *Stimulus* Circular (diameter min 23.): Wavelength; 565nm: luminance; 56.2 cd. $m^{-2}$: duration 100 ms. After Blythe et al., 1987.

threshold sensitivity shows only limited variation over the velocity range investigated. Threshold luminance required for detection of a target flash presented to GY's 'blind' hemifield falls linearly as the area of the target increases (Fig. 3b), but under conditions which isolate cone activity in the 'blind' hemifield (see Fig. 4b), a different response pattern is observed. Thus, with a 612 nm flash presented on a uniform white background, threshold falls linearly only for areas up to about 4 $deg^2$, (corresponding to a diameter of about 2 deg), and for larger areas, threshold is approximately proportional to the square root of the target area (Fig. 3b).

Spatial responses recorded for a monochromatic light flash, presented 30 deg. off axis in GY's blind hemifield, follow the rod spectral response function, and are unaffected by changes in the chromatic background field which in normal vision induce a Purkinje shift (Fig. 4a). In contrast, measurements made with a monochromatic target superimposed on a monochromatic background, located 10° off axis in GY's blind hemifield, reveals activity of both the red-sensitive ($\pi_5$) and green-sensitive ($\pi_4$) spectral response mechanisms (Fig. 4b). Data obtained for GY's blind hemifield under experimental conditions which in normals isolate the blue-sensitive ($\pi_3$) response mechanism, however, corresponded to the rod ($\pi_0$) spectral response.

The ability of patients such as GY to discriminate between and identify light stimuli presented within the 'blind' hemifield is of particular importance for the analysis of cortical visual organisation. He can discriminate normally between the speeds of two targets which move horizontally in the same direction across his 'blind' hemifield (Fig. 5a), but his identification of movement direction is poor (Fig. 5b). He discriminates between flickering lights of different frequencies with near normal performance (Fig. 5c). GY can identify monochromatic lights presented in the form of a semicircle (diameter 20 deg.) to his blind hemifield, his responses being most accurate for long wavelength stimuli (Fig. 6 a,b,c). He

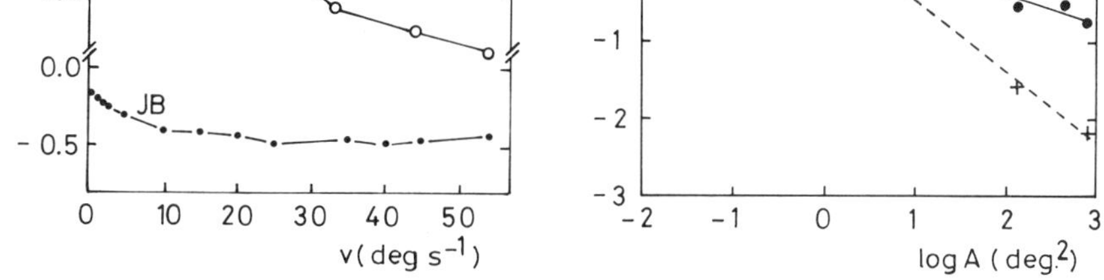


**Figure 3.** Target luminance, $I_t$, required for its detection. a) $I_t$, in log trolands, plotted against the velocity of a circular target, T, moving along the horizontal meridian at v deg. $s^{-1}$. Data for GY ('blind' right hemifield) and a normal control JB. *Stimulus* (See inset). Target T: Circular (5.7 deg. diameter): White light. Background: Circular (diameter 48 deg.): White light: luminance; 1.0 log trolands. Fixation spot, F, diameter 0.6 deg. Movement along the horizontal axis was centred 30 deg. from fixation, and was 12 deg. in amplitude, between the two broken lines. b) $I_t$ (relative units) plotted against target area A. Data for a red (612nm) target, presented to GY's 'blind' hemifield (open circles) or to his normal hemifield (full circles). Crosses refer to a white target. The best lines through the data sets are fitted by least squares method; a slope of -1 indicates full summation of target flux. *Stimulus* Target: Circular, of area A: Duration; 100 ms: Location; 10 deg off-axis on the horizontal meridian. Background: Circular (diameter 20 deg.): White light: luminance; 0.9 log trolands.

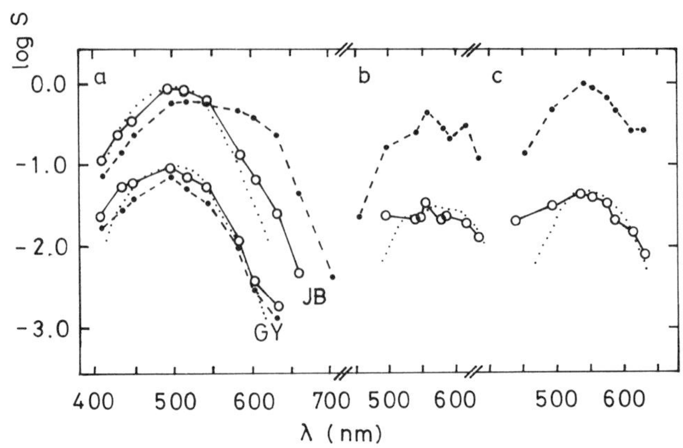


**Figure 4.** Sensitivity, S, for detection of a target, plotted as a function of wavelength. a) Values for a circular, moving target of wavelength λ, presented against a background of wavelength 495 nm (full circles) or 628 nm (open circles). Data for GY's 'blind' hemifield recorded for a general white light background of luminance 1.0 log trolands, which was increased to 3.0 log trolands for the normal control JB. The dotted line denotes the relative spectral sensitivity of the rod ($\pi_0$) response mechanism (Wyszecki and Stiles, 1982). S.E. in means were typically 0.04 (GY) and 0.03 (JB). *Stimulus* Target: Circular (5.7 deg. diameter): Movement through 12 deg., at 40 deg. $s^{-1}$, centred 30 deg. off-axis on the horizontal meridian. Background: Circular (48 deg. diameter). After Barbur et al., 1980. b) S values for detection a flashed, 650nm target presented against a background of wavelength λ. S is expressed in terms of the radiance of a background field required to bring to threshold a target presented 0.4 log units above its threshold in the absence of the background. Data for GY's 'blind' hemifield (open circles) and normal hemifield (full circles). The dotted line denotes the normal $\pi_{5}$- spectral sensitivity function (Wyszecki and Stiles, 1982). *Stimulus* Target: Circular (4.5 deg diameter): Duration 100 ms.: Location; 10 deg. off-axis on the horizontal meridian. Background: Circular (17 deg, diameter): Wavelength λ. An auxiliary background stimulus, wavelength 515 nm, luminance 1.0 log trolands, was added to suppress rod activity. After Brent et al. (1994). c) As b, for but for a 515 mn target, and the dotted line represents the normal $\pi_4$- spectral sensitivity function (Wyszecki and Stiles, 1982).

has limited saturation discrimination for long wavelength stimuli (Fig. 6d), but if the stimulus size is reduced to a hemifield of 10 deg. diameter, he fails to respond correctly even to those stimuli which were accurately identified for the larger field ( Fig. 6f).

GY performs poorly in recognition of spatial patterns presented to his 'blind' hemifield, but he can identify the orientation of an elongated flickering bar (Fig. 7a,b). This latter discrimination capacity disappears, however, if a flickering grating consisting of parallel bars is presented within a circular aperture. It appears that in this latter case, the extensive summation evident in threshold detection of simple targets (Fig. 3b) acts to average the light flux over the whole stimulus, thereby removing information about the orientation of the grating bars. Under certain conditions, GY can also identify the orientation of a single edge, formed either by a simple luminance step, or as a boundary between two semicircular fields which differ in spectral composition or in flicker frequency (Fig. 7,c,d). These orientation discriminations enable him to distinguish between large differences in the relative orientation of 2-D patterns, such as squares (Fig. 7e), but he appears unable utilise other spatial parameters to perform pattern discrimination (Barbur et al., 1980).

## DISCUSSION

The localisation and detection of light stimuli presented transiently within 'blind' areas of the visual field are common features of residual visual or blindsight responses observed in patients with striate lesions (Ruddock, 1991). GY's threshold responses reveal the activity of two response mechanisms (Barbur et al., 1980; 1994), one with very large receptive fields and the other with band-pass spatial and temporal frequency responses

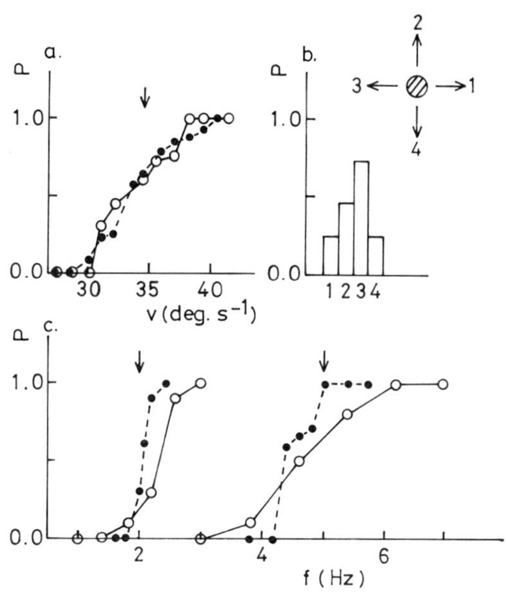


**Figure 5.** Discrimination data for GY's 'blind' hemifield. a) The probability, P, of identifying the speed, v, of a circular target as greater than that of a reference target travelling at 34.5 deg. s$^{-1}$ (denoted by arrow). Data for GY's 'blind' hemifield (open circles) and for JB, a normal control (full circles). *Stimulus* Target: Circular (5.7 deg. diameter): Luminance; 2.8 log trolands: White light: Location; movement through 12 deg., along the horizontal meridian, centred 30 deg. off axis. Background: Circular (diameter 48 deg.): White light: Luminance; 1.0 log troland. After Barbur et al. (1980). b) Probability, P, of identifying correctly the direction of movement of a circular target, relative to that of a similar reference stimulus which was presented immediately prior to target presentation, moving in direction 1. Only the performance in detection of the target moving in direction 3 (see insert) is above chance at 0.01 significance level. *Stimulus* Circular (3 deg. diameter): Luminance; 3 log trolands: Speed 15 deg s$^{-1}$: Location; horizontal movement through 10 deg centred 30 deg. off-axis along the horizontal meridian. After Blythe et al., 1987. c) Probability, P, of identifying a target as flickering at a faster rate than a similar, reference field, which flickered at 2 Hz or 5 Hz. Data for GY's 'blind' hemifield (open circles) and for a normal control JW (full circles). *Stimulus* Target: Rectangular (18deg. vertical, 12 deg. horizontal): Mean luminance; 2.8 log trolands: White light: Location; centred 16 deg. off-axis along the horizontal meridian.

similar to those of the ST2 response mechanism found in normal vision (Holliday and Ruddock, 1983).

The contributions of these two pathways to threshold detection are evident in the variation of detection sensitivity with target area which, dependent on the stimulus conditions, reveals summation extending over the full area of stimulation, or summation restricted to a circular area of diameter about 2 deg. (Fig. 3b). The latter is consistent with spatial integration of light flux over the receptive field centre of a response mechanism with a

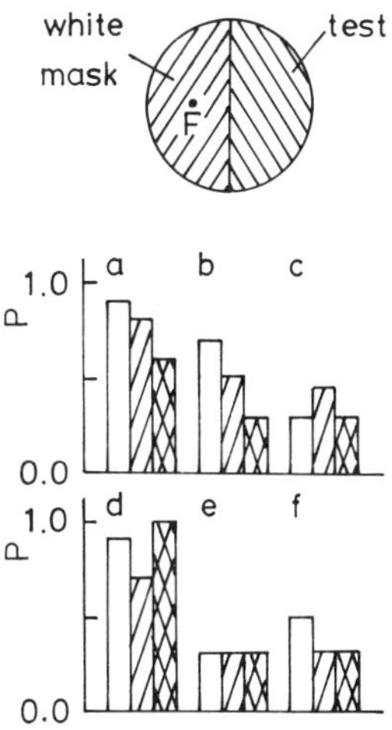


**Figure 6.** The probability of GY's identifying the colour of a semicircle (diameter 40 deg.) presented to his 'blind' hemifield by the same name as that used when it was presented to the normal hemifield. Each data set refers to three stimuli, each presented 10 times in random sequence. The three stimuli are identified in the order illustrated, with P values corresponding to the open, hatched and double hatched bars respectively. a) 665 nm; 535 nm; 440 nm. b) 615 nm; 585 nm; 535 nm c) 535 nm; 490 nm; 455 nm. d) 665 nm; 665 nm mixed with an equal photometric amount of white; white. e) 440 nm; 440 nm mixed with an equal photometric amount of white; white. f) As a), but for a semicircular field of diameter 10 deg. *Stimulus* (see inset). Test field Semicircle (diameter 40 deg): Luminance; 2.1 log troland: Spectral composition; variable: Duration: 500 ms: Location; vertical edge 17 deg from fixation , F. The other hemifield was white: Luminance; 1.5 log troland: continuously visible. After Brent et al., 1994.

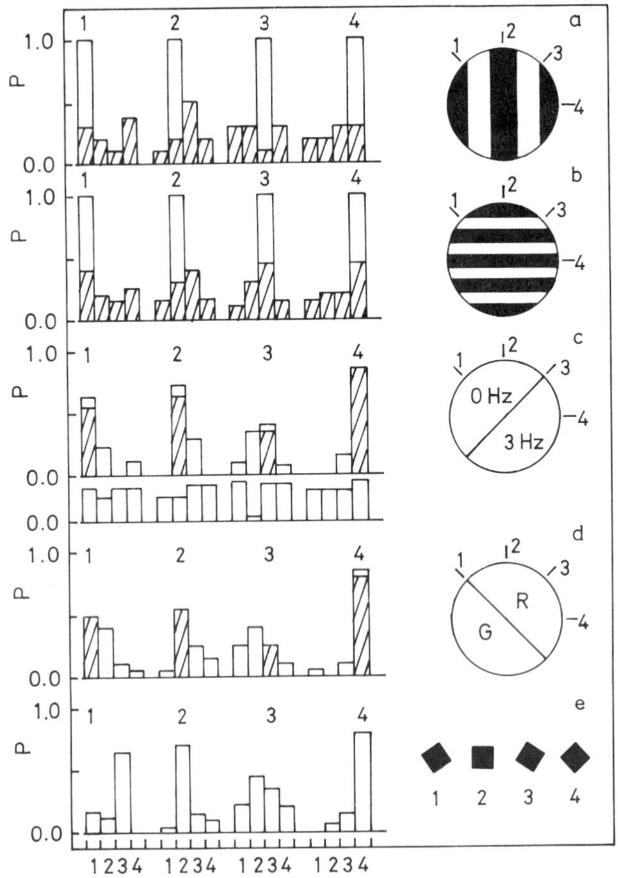


**Figure. 7.** Orientation discrimination data for GY's 'blind' hemifield. Stimuli were presented at four orientations, numbered as indicated to the right of the probability histograms, which specify the probability of identifying correctly the stimulus as having the different orientations (40 presentations per stimulus). The correct orientation is denoted above the histograms. a) Open histogram bars; values for a single flickering bar (3Hz; 15 deg. long; 3.6 deg. wide). Hatched histogram bars; values for a square waveform grating presented in a circular field (diameter 15 deg.), the bars of which had the same dimensions as the single bar. *Stimulus* White light: Luminance; 3.5 log trolands: Location; centred 225 deg. off axis along the horizontal meridian: Duration 3s. b) As a, but for bars of width 1.15 deg. c) Open histogram bars; values for the edge formed between the steady (0Hz) and flickering (3Hz) stimuli. Hatched histogram bars; the probability of identifying correctly the relative positions of the two flickering components. The lower histogram shows the values obtained when both fields were steady (0Hz). *Stimulus* Circular (25 deg. diameter): White light: Luminance; 3.5 log trolands: Location; centred 28 deg. along the horizontal meridian: Duration 3s. d) As c, but for a field divided by colour into a red (Wratten filter) and green (Wratten filter) half. e) Data for a square, flickering at 3 Hz, presented at four different orientations. *Stimulus* Square (side, 13 deg.): White light: Luminance; 3.5 log trolands: Location; centred 22.5 deg. along the horizontal meridian: Duration 3 s. In these experiments the visual field from the fovea to the stimulus itself was illuminated by white light (luminance 3.4 log trolands).

frequency response which peaks at about 1cycle deg$^{-1}$ as observed for GY by Barbur et al. (1994).

Visual discriminations on the basis of differences in temporal characteristics of the light stimuli have been reported in several studies of 'blindsight' or residual vision. GY performs almost normally in discrimination of speed (Fig. 5a) and flicker frequency (Fig. 5c), although his ability to identify the direction of movement of a single spot target is markedly impaired (Fig. 5b). Other experimental data not illustrated in this paper establish that GY can identify correctly the direction of apparent ($\phi$-) motion although his spatial resolution for this response is much greater for motion along the horizontal meridian than for that along the vertical or, particularly, at oblique angles (Blythe et al., 1986; Ruddock, 1991). The retention of discriminations for transient stimuli is consistent with the PET evidence that appropriate stimuli generate activity in prestriate areas equivalent to the motion sensitive areas V3 and V5 of the macaque cortex.

Spectral discrimination requires activity of at least two spectral classes of cones, and the results of two colour increment threshold measurements show that both the red ($\pi_5$-) and green ($\pi_4$-) sensitive mechanisms associated with normal colour vision (Stiles, 1978; Wyszecki and Stiles, 1982) contribute to GY's residual vision. Correspondingly, GY can distinguish between stimuli which differ in spectral composition, his performance being most robust for long wavelength stimuli, for which he exhibits a degree of saturation discrimination (Fig. 6). His colour discrimination capacity is, however, very limited and is revealed only with very large fields. In contrast, the hemianopes studied by Stoerig (1987) could utilise 'blindsight' to make red-green discriminations for relatively small (36' diameter) stimuli. Lesions which selectively disturb colour vision (Meadows, 1974; Zeki, 1990; Kennard et al., 1995) are localised to the lingual and fusiform gyri and it appears that the non-involvement of the fusiform gyrus in GY's cortical damage allows him to perform limited colour discrimination.

Very few patients with striate lesions have demonstrated the capacity to recognise or to discriminate between spatial patterns presented within 'blind' areas of the visual field. Orientation discrimination and restricted grating resolution was observed in the original 'blindsight' patient DB, and limited pattern discrimination was found in two hemidecorticate patients by Perenin (1978). GY was unable to differentiate between a circle, a triangle and a square adjusted in size to give the same integrated flux (Barbur et al., 1980), but he is able to discriminate to varying degrees the orientations of a steady or flickering bar-shaped stimulus, and those of a single edge formed by a contrast step or by change in spectral composition or flicker frequency (Fig. 7a,b,c,d). It should be emphasised that these discrimination measurements refer to extended edges and that GY's discrimination performance is, in some cases, significant only in the detection of horizontal and, to a lesser degree, vertical edges. Thus the response asymmetries evident in his detection of apparent motion (Blythe et al. 1986; Ruddock, 1991) are also present in GY's orientation discrimination. GY's excellent orientation discrimination for a single, high contrast flickering bar disappears completely with parallel bars bounded symmetrically by a circular aperture, such that the flux distribution is symmetric around the centre of the aperture (Fig. 7a,b). This result indicates that the response mechanism which at threshold yields integration over large areas of the visual field is also dominant in the detection of high contrast, supra-threshold gratings and submerges the orientation selective signals associated with the component bars. Summation of light flux over large areas of the 'blind' field appears to be the underlying cause of GY's poor performance in tests of pattern recognition and shape discrimination, although orientation signals associated with squares provide the basis for the limited pattern discrimination evident in the results of Fig. 7e. GY has good orientation discrimination for a single, high contrast edge but for edges defined by colour contrast (Fig. 7d) or flicker contrast (Fig. 7c), orientation discrimination is weak, even though with sequential presentations, he can readily discriminate the differences in colour or flicker frequency which exist between the fields on either side of the edge. The mechanisms which underly GY's ability to discriminate colour or flicker frequency do not, therefore, provide strong orientation selective signals. Orientation selectively is a prominent feature of striate neuronal organisation, and it appears that in the absence of this contribution, colour and flicker sensitive pathways do not signal edge orientation effectively. Our experimental results show that not only is GY incapable of accurate orientation discrimination, but his identification of the topographical relationship between the two components of the compound stimuli which form the edges is also poor (Fig. 7c,d , hatched bars). Thus the striate input to the pre-striate visual areas is also, apparently, essential for establishing the topographical distributions of flicker and colour selective responses.

In summary, detailed experimental studies on GY's residual vision have established that in addition to topographical localisation of isolated, transient visual stimuli, he can discriminate between stimuli on the basis of differences in their spectral or temporal parameters. It has been argued previously by Barbur et al. (1980) that topographical information in the absence of striate cortex is provided by the mapping of the visual field on to the superficial layers of the superior colliculus. Signals from the superior colliculus are relayed to pre-striate cortex both via the pulvinar (Benevento and Fallon, 1975; Benevento and Rezak, 1976) and via the lateral geniculate nucleus (Kisvárday et al. , 1991), thus utilisation of this topographical information may involve only the superior colliculus itself, or additionally its projections to the extra-striate visual areas. The retention by GY of discrimination capacity for transient stimuli is consistent with the preservation of activity in areas analogous to the macaque prestriate areas V3 and V5 (Barbur et al., 1993), and his weak colour discrimination can be attributed to non-involvement of the fusiform gyrus in his cortical lesions. The experimental results also show that GY has poor orientation discrimination and topographical localisation for pairs of stimuli which, when seen sequentially, he can distinguish on the basis of differences in spectral composition or flicker frequency. It is proposed that these functional impairments arise because of the absence of input from the striate cortex, and that these findings imply a critical role for striate signals in providing topographical localisation and orientation sensitivity for pre-striate mechanisms sensitive to colour and flicker. The high level of topographical organisation observed in striate area V1, and its absence in areas V4 and V5 of the macaque prestriate cortex (Zeki, 1993) provide some support for this interpretation of GY's responses.

## ACKNOWLEDGMENTS

The work reported here has been carried out over a period of some twenty years with the willing co-operation of the principal subject, GY, and I am grateful for his continued collaboration and interest in the research. The research has also involved collaboration with a number of colleagues, past and present, including Dr A.B. Morland, J.A. Ogilvie and Julie R. Wright, who have kindly agreed to my inclusion of unpublished results of our current research. I am also pleased to acknowledge the support given for this work by the Wellcome Trust.

## REFERENCES

Barbur, J.L. and Ruddock, K.H. (1980). Spatial charateristics of movement detection mechanisms in human vision. I. Achromatic vision, *Biol. Cyber.*, 37:77-92.

Barbur J.L., Ruddock, K.H. and Waterfield, V.A. (1980). Human visual responses in the absence of the geniculo-calcarine projection, *Brain* 103:905-928.

Barbur, J.L., Watson, J.D.G., Frackowiak, R.S.J. and Zeki, S. (1993). Conscious visual perception without V1, *Brain* 116:1293-1302.

Barbur, J.L., Harlow, A.J. and Weiskrantz, L. (1994). Spatial and temporal response properties of residual vision in a case of hemianopia, *Phil. Trans. Roy. Soc. Lond. B.* 343:157-166.

Benevento, L.A. and Fallon, J.H. (1975). The ascending projections of the superior colliculus in the rhesus monkey (*Macaca mulatta*), *J. Comp. Neurol.* 160:339-362.

Benevento, L.A. and Rezak, M. (1976). The cortical projections of the inferior pulvinar and adjacent lateral pulvinar in the rhesus monkey (*macaca mulatta*). An autoradiographic study, *Brain Res.* 127:197-218.

Benevento, L.A. and Yoshida, K. (1981). The afferent and efferent organisation of the lateral geniculo-prestriate pathways in the macaque monkey, *J. Comp. Neurol.* 203:455-474.

Blythe, I.M., Bromley, J.M., Kennard, C. and Ruddock, K.H., (1986). Visual discrimination of target displacement remains after damage to the striate cortex in humans, *Nature* 320:619-621.

Blythe, I.M., Kennard, C. and Ruddock, K.H., (1987). Residual vision in patients with retrogeniculate lesions of the visual pathways, *Brain* 110:887-905.

Brent, P.J., Kennard, C. and Ruddock, K.H. (1994). Residual colour vision in a human hemianope: spectral responses and colour discrimination, *Proc. Roy. Soc. Lond. B* 256:219-225.

Cornsweet, T.N. (1992). The staircase method in psychophysics, *Am. J. Psychol.* 75:485-491.

Damasio H. and Damasio A.R. (1989). Lesion Analysis in Neurophysiology. Oxford: University Press.

Fries, W. (1981). The projection from the lateral geniculate nucleus to the prestriate cortex of the macaque monkey, *Proc. Roy. Soc, Lond. B* 213:73-80.

Holliday, I.E. and Ruddock, K.H. (1983). Two spatio-temporal filters in human vision. I. Temporal and spatial frequency response characteristics, *Biol. Cybern.* 47:173-190.

Holmes, G. (1918). Distubances of vision by cerebral lesions, *Br. J. Ophthalmol.* 2:353-384.

Ikeda, H. and Wright, M.J. (1972). Receptive field organisation of 'sustained ' and 'transient' retinal ganglion cells which subserve different functional roles, *J. Physiol.* 227:769-800.

Ikeda, H. and Wright. M.J. (1975). Spatial and temporal properties of 'sustained' and 'transient' neurones in area 17 of the cat's visual cortex, *Expl. Brain Res.* 22:363-383.

Kennard, C., Lawden, M., Morland, A.B. and Ruddock, K.H, (1995). Colour identification and colour constancy are impaired in a patient with incomplete achromatopsia associated with prestriate cortical lesions, *Proc. Roy. Soc. Lond. B* 260:169-175.

Kisvárday, Z.F., Cowey A., Stoerig, P. and Somogyi, P. (1991). Direct and indirect retinal input into degenerated dorsal lateral geniculate nucleus after striate cortical removal in monkey: implications for residual vision, *Expl. Brain Res.* 86:271-292.

Marzi, C.A., Tassinari, G., Aglioti, S. and Lutzemberger, L. (1986). Spatial summation across the vertical meridian in hemianopics: a test of blindsight, *Neuropsychologia* 24:749-758.

Meadows, J.C. (1974). Disturbed perception of colours associated with localized cerebral lesions, *Brain* 97:615-632.

Perenin, M.T. (1978). Visual function within the hemianopic field following early cerebral hemidecortication in man. II. Pattern discrimination, *Neuropsychologia* 16:697-708.

Perenin, M.T.and Jeannerod, M. (1975) Residual vision in cortically blind hemifields, *Neuropsychologia* 13:1-7.

Perenin, M.T. and Jeannerod, M. (1978).Visual function within the hemianopic field following early cerebral hemidecortication in man. I. Spatial location, *Neuropsychologia* 16:1-13.

Pizzamiglio, L., Antonucci, G. and Francia. A. (1984). Response of the cortically blind hemifields to a moving visual scene, *Cortex* 20:89-99.

Pöppel, E., Held, R. and Frost, D. (1973). Residual visual function after brain wounds involving the central visual pathways in man, *Nature* 243:295-296.

Ptito, A., Lassonde, M., Leporé, F. and Ptito, M. (1987). Visual discrimination in hemispherectomized patients, *Neuropsychologia* 25:869-879.

Riddoch, G. and Ruddock, K.H. (1991). Spatial vision after cortical lesions. in Regan, D.M. (Ed.) Spatial Vision, Vol. 10 of Vision and Visual Dysfunction, London, Macmillan, 261-289.

Stiles, W.S. (1978). Mechanisms of colour vision. London: Academic Press.

Stoerig, P. (1987). Chromaticity and achromaticity: evidence for a functional differentiation in visual field defects, *Brain* 110:869-886.

Stoerig, P. and Cowey, A. (1989). Wavelength sensitivity in blindsight, *Nature* 342:916-918.

Stoerig, P. and Cowey, A. (1991). Increment - threshold sensitivity in blindsight: evidence for colour opponency, *Brain* 114:1487-1512.

Teuber, H-L., Battersby, W.S. and Bender, M.B. (1960). Visual Field Defects after Penetrating Missile Wounds of the Brain. Cambridge Mass: Harvard University Press.

Weiskrantz, L. (1980). Varieties of residual experience, *Q. J. Exp. Psychol.* 32:365-386.

Weiskrantz, L. (1986). Blindsight: A case study and implications. Oxford: Clarendon Press.

Weiskrantz, L. (1987). Residual vision in a scotoma: a follow-up study of 'form' discrimination, *Brain* 110:77-92.

Weiskrantz, L, Warrington, E.K., Sanders, M.D. and Marshall, J. (1974). Visual capacity in the hemianopic field following a restricted occipital ablation, *Brain* 97:709-728.

Weiskrantz, L., Barbur, J.L. and Sahraie, A. (1995). Parameters affecting conscious versus unconscious visual discrimination with damage to the visual cortex (V1), *Proc. Natl Acad. Sci. U.S.A.* 92:6122-6126.

Wyszecki, G. and Stiles, W.S. (1982). Color science: concepts and methods, quantitative data and formulae, 2nd. edn. New York: Wiley.

Yukie, M. and Iwai, E. (1981). Direct projection from the dorsal lateral geniculate nucleus to the prestriate cortex in macaque monkeys, *J. Comp. Neurol* 201:81-97.

Zeki, S. (1990). A century of cerebral achromatopsia, *Brain* 113:1721-1777.

Zeki, S. (1993). A Vision of the Brain. Oxford: Blackwell.

# CONTRIBUTORS

**John L. Barbur**
Applied Vision Research Centre
The City University
London EC1V 7DD, UK

**Ivan Bodis-Wollner**
Department of Neurology
State University of New York
Health Sciences Center at Brooklyn
450 Clarkson Avenue, Box 1213
Brooklyn, NY 11203

**Artur V. Cideciyan**
Department of Ophthalmology
University of Miami School of Medicine
Bascom Palmer Eye Institute
Miami, Florida

*Permanent address:*
Scheie Eye Institute
University of Pennsylvania
51 North 39th Street
Philadelphia, PA 19104

**Robert J. Clarke**
School of Optometry
Department of Physiological Optics
University of Alabama at Birmingham
924 South 18th Street
Birmingham, Alabama 35294-4390

**Mustfa B. A. Djamgoz**
Department of Biology
Imperial College
Prince Consort Road
London SW7 2BB, UK

**John E. Dowling**
Department of Molecular and Cellular Biology
Harvard University
16 Divinity Avenue
Cambridge, Massachusetts 02138

**Michela Fagiolini**
Scuola Normale Superiore
P. zza Cavalieri 56127
Pisa, Italy

**Gertrude Falk**
Department of Physiology
University College
London WC1E 6BT, UK

**P. D. R. Gamlin**
School of Optometry
Department of Physiological Optics
University of Alabama at Birmingham
924 South 18th Street
Birmingham, Alabama 35294-4390

**K. N. Gurney**
Department of Human Sciences
Brunel University
Uxbridge, Middlesex UB8 3PH, UK

**Peter Hammond**
Department of Communication and Neuroscience
Keele University
Keele, Staffordshire ST5 5BG, UK

**Mark W. Hankins**
Gunnar Svaetichin Laboratory
Department of Biology
Imperial College at Silwood Park
Ascot Berks SL5 7PY, UK

**Sohan S. Hayreh**
Department of Ophthalmology
University of Iowa
Iowa City, Iowa

**Hisako Ikeda**
Vision Research Unit
Rayne Institute
St Thomas' Hospital (UMDS)
London SE1 7EH, UK

**Samuel G. Jacobson**
Department of Ophthalmology
University of Miami School of Medicine
Bascom Palmer Eye Institute
Miami, Florida

*Permanent address:*
Scheie Eye Institute
University of Pennsylvania
51 North 39th Street
Philadelphia, PA 19104

**Colin M. Kemp**
Department of Ophthalmology
University of Miami School of Medicine
Bascom Palmer Eye Institute
Miami, Florida

**Helga Kolb**
University of Utah
Departments of Ophthalmology
John A. Moran Eye Center
University of Utah Health Sciences Center
50 North Medical Drive
Salt Lake City, Utah 84132

**Lamberto Maffei**
Instituto Neurofisiologia CNR
via S. Zeno 51 56127
Pisa, Italy

**Jeremy Nathans**
Departments of Molecular Biology and Genetics and Neurosciences
Howard Hughes Medical Institute
Johns Hopkins University School of Medicine
Baltimore, Maryland

**Michael J. Neal**
Department of Pharmacology
St Thomas' Hospital (UMDS)
Lambeth Place Road
London SE1 7EH, UK

**Renata Petruv**
Department of Biology
Imperial College
Prince Consort Road
London SW7 2BB, UK

**Tommaso Pizzorusso**
Scuola Normale Superiore
P. zza Cavalieri 56127
Pisa, Italy

**Jon Robbins**
Pharmacology Group
Biomedical Sciences Division
King's College
Manresa Road
London SW3 6LX, UK

**Keith Ruddock**
Biophysics Section
Physics Department.
Imperial College
London SW7 2BZ, UK

**Richard A. Shiells**
Department of Physiology
University College
London WC1E 6BT, UK

**Malcolm Slaughter**
Departments of Biophysical Sciences and Ophthalmology
School of Medicine
State University of New York
Buffalo NY 14214

**Ning Tian**
Departments of Biophysical Sciences and Ophthalmology
School of Medicine
State University of New York
Buffalo NY 14214

**Michael J. Wright**
Department of Human Sciences
Brunel University
Uxbridge
Middlesex UB8 3PH, UK

**Jian Zhang**
Departments of Biophysical Sciences and Ophthalmology
School of Medicine
State University of New York
Buffalo NY 14214

Participants in the meeting held in August 1993 to mark the retirement of Professor Hisako Ikeda. Photograph taken outside St. Thomas's Hospital, London, by The Audio Visual Department, St. Thomas's Hospital.

# INDEX